Physics Problems for Aspiring Physical Scientists
With Hints and Full Solutions

An essential part of studying to become a physi⋯⋯⋯⋯⋯⋯ g
how to solve problems. This book contains over ⋯⋯ ⋯⋯⋯⋯⋯ ⋯ ⋯ ⋯ ⋯s
with hints and *full* solutions. The author demonstrates how to break down a problem
into its essential components, and how to chart a course through them to a solution.
With problem-solving skills being essential for any physical scientist or engineer, this
book will be invaluable to potential and current undergraduates seeking a career in
these fields.

The book is divided into three parts: questions, hints, and solutions. The first part
is subdivided into fifteen chapters, each centred on a different area of physics, from
elementary particles, through classical physics, to cosmology. The second part provides brief hints, whilst the third sets out a full and explicit solution to each problem.
Most begin with thoughts students might have after reading a problem, allowing the
reader to understand which questions they should be asking themselves when faced
with unfamiliar situations.

KEN RILEY is a retired lecturer in physics at the University of Cambridge's Cavendish Laboratory.
He is also a fellow of Clare College, where he was both the Senior Tutor and Admissions
Tutor, and taught physics and mathematics for over forty-five years. He has served on many
committees and panels concerned with the teaching and examining of these subjects at all
levels of tertiary and university education. His research was centred on nuclear physics at
Harwell and then elementary particle physics at Brookhaven (NY), the Rutherford Labora-
tory, and Stanford. He is the lead author of *Mathematical Methods for Physics and Engineering*
(Cambridge University Press, 3rd edn, 2006), and a joint author of both *Foundation Mathematics for the Physical Sciences* (Cambridge University Press, 2011) and *Essential Mathematical
Methods for the Physical Sciences* (Cambridge University Press, 2011). He is a co-author of *200
Puzzling Physics Problems* (Cambridge University Press, 2001) and the Consultant Editor for
200 More Puzzling Physics Problems (Cambridge University Press, 2016).

Physics Problems for Aspiring Physical Scientists and Engineers

With Hints and Full Solutions

Ken Riley

Clare College, University of Cambridge

CAMBRIDGE
UNIVERSITY PRESS

University Printing House, Cambridge CB2 8BS, United Kingdom

One Liberty Plaza, 20th Floor, New York, NY 10006, USA

477 Williamstown Road, Port Melbourne, VIC 3207, Australia

314–321, 3rd Floor, Plot 3, Splendor Forum, Jasola District Centre, New Delhi – 110025, India

79 Anson Road, #06-04/06, Singapore 079906

Cambridge University Press is part of the University of Cambridge.

It furthers the University's mission by disseminating knowledge in the pursuit of education, learning, and research at the highest international levels of excellence.

www.cambridge.org
Information on this title: www.cambridge.org/9781108476690
DOI: 10.1017/9781108592123

© Ken Riley 2019

First published 2019

Printed in the United Kingdom by TJ International Ltd. Padstow Cornwall

A catalogue record for this publication is available from the British Library.

ISBN 978-1-108-47669-0 Hardback
ISBN 978-1-108-70130-3 Paperback

Contents

Preface   *page* vii

How To Use This Book   xi

1 Dimensions
   Stick to Your Own Kind!   1

2 Statics
   Hold It There!   4

3 Dynamics
   Keep It Moving!   9

4 Rotation
   Round and Round We Go!   16

5 Gravity
   To Be Taken Seriously!   22

6 SHM
   Here We Go Again!   27

7 Waves
   We're Not Phased!   33

8 Solids and Liquids
   All Together Now!   41

9 Electrical Circuits
   Go With the Flow!   46

10 Electrostatics
   Reach Your Potential!   54

11 Magnetic Fields
   Stay in the Loop!   59

12 Electromagnetism
   A State of Flux!   63

13 Heat Transfer
   Some Like It Hot!   67

14 *Gases*
      Little Things Matter!   71
15 *Particles and Atoms*
      All Bits and Pieces!   75
Hints   80
Solutions   97
Physical Constants   335

Preface

The principal aim of this book of physics problems is to allow those advanced school students who seek a future in one of the physical sciences or engineering to develop the skill of being able to 'chart their own course' through a problem or investigation.

For reasons which I well understand, advanced level examination questions are nowadays broken up into many parts, each with its own (small) mark. This has to be done because, with many examiners for any one examination paper, the need for uniformity of marking, guided by a mark scheme, and the necessity of being able to provide an open and justifiable paper trail when appeals are made, make it almost inevitable.

However, the result of such a procedure is effectively to lead every student by the hand, including both those who, on their own, would not know what to do next, and those with a clear understanding of the science involved, who could find their way to the solution without any guidance whatsoever. This set of physics problems is for the latter group – and for those who have the ability and aspiration to join them; they are the scientific innovators and developers of the future.

The physics material covered by the problems is that usually taught in advanced school courses; in the United Kingdom, and in countries where overseas UK examinations are sat, the topics covered are typically those included in an A-level syllabus. In the United States the relevant level is probably that of SAT examinations, whilst in Europe higher level physics in the International or French Baccalaureate would cover the necessary ground. In countries which adopt a less specialized school curriculum than that found in the UK, the relevant level may be that of junior college or the early years of university.

Whatever their school physics course, this book should provide valuable training for those making the transition to university-level study, as the principal aim of the problems is to foster an ability to think things through for oneself, as opposed to being guided through each step. For this reason, the book should also be of value to students already in their first year at university.

Most of the material covered is what is known as 'classical physics', although some 'modern' topics, such as quarks, ultrasound scanning, and the development of the Universe, are included. Several classical physics topics, among them polarized light, inductance, heat conduction, longitudinal waves and adiabatic changes, no longer appear in many school syllabuses, but I have included problems in some of these areas and provided the additional information needed to tackle them; being able to take in and use previously unknown information is a skill all would-be scientists should acquire.

The book is divided into three parts, the questions, the hints, and the full model solutions; this division will help a student to avoid 'accidentally' noticing a hint when reading the question, or the solution when seeking help from the hints! The Questions part of the book is divided into fifteen chapters, each centred on some particular area of physics, though, as always, physics is not compartmentalized, and the same ideas and principles can appear in many different places. The problems within any one chapter are not in any particular order of difficulty, though those marked with an * will probably be found to be the more challenging. Problems marked with a † require some mathematics that is not normally an integral part of school physics, but is likely to have been studied by any student who is also taking mathematics at an equivalent level.

The large majority of the problems posed call for quantitative answers, either as a formula or a numerical value. Seeking quantitative solutions makes it possible to provide intermediate answers and guidance in a reasonably compact form. When doing this, in the second part of the book, a sometimes difficult balance has had to be struck between being so helpful that there is nothing left to the problem, and being so oblique that the hint is merely one more baffling aspect; I hope that in the large majority of cases such a balance has been found. The Hints section, which is not divided into chapters, is comparatively short, and for the very best students will, hopefully, be little used.

Much the largest part of the book is the Solutions section, which aims to set out in explicit detail full solutions to every question. They are presented as one continuous set, since the solution numbering (in bold) provides at least as good a guide to the appropriate page as any division into chapters. As explained in the 'How to use this book' section, many of the model solutions start by setting out the sort of thoughts a student may, and should, have on reading the question.

Some of the problems are quite short, but where they are, they usually require some particular insight. The longer questions tend to be relatively straightforward but require a number of steps, and call for a clear plan and accurate use of the data given. The ability to decide, with little external guidance, what needs to be done, and in what order, is a major learning objective of all of the problems. Some could be solved more efficiently or elegantly by employing university-level methods, but I have tried wherever possible to use only approaches available after following a school-level course.

The symbols used for physical quantities are, in the vast majority of cases, those in standard use by all UK examination boards and by nearly all physics textbooks written in English. Where they are not, I have defined them within the individual problems. In order to make the book self-contained for its own purposes, a table of values for the standard constants is included on the very last page. I have not included a selection of formulae and relationships, partly because there is overlap, but no unanimity, amongst the major examination boards, but mainly because I feel that the practising scientist should, at least in the long run, have them as part of his or her vocabulary. If need be, the data sheet provided for their own course can be consulted.

Naturally, I take responsibility for all errors and ambiguities the questions, solutions and hints may contain, and would be most grateful to have them brought to my attention. Finally, I wish to place on record my appreciation of the help received from Simon Capelin and Esther Migueliz Obanos at CUP, and, although they may not expect it, from Helmut Kopka and Patrick W. Daly whose *A Guide to* LATEX I have consulted often over the last twenty years.

How To Use This Book

The principle aim of this book is to allow those advanced school students who seek a future in one of the physical sciences or engineering to develop the skill of being able to 'chart their own course' through a problem or investigation. And it is assumed that you are just such a person.

There is no particular aim to cover every part of some or all physics syllabuses, but rather to provide scenarios that offer the chance of logical self-organized thought, without the need for skills or knowledge outside of what you have already studied or is provided in the question. You may have to look up a few topics outside your own course, but they will be no more advanced than those you have already studied, and such research is, in itself, worthwhile training. Equally, you may need to look elsewhere for the values of some physical quantities. A table of standard physical constants is provided at the very end of the Solutions section; some others which will clearly be needed are stated in the question; but some appear only in the corresponding solution, as to include them in the question would provide too much of a clue as to the correct solution method.

As explained in the Preface, this book is divided into three parts, the questions, the hints, and the full model solutions. The Questions part of the book is divided into fifteen chapters, each centred on some particular area of physics, though, as always, physics is not compartmentalized, and the same ideas and principles can appear in many different places.

The large majority of the problems posed call for quantitative answers, either as a formula or a numerical value. That is not to imply that qualitative discussions of physical problems are not important; in some cases they are the only discussions that are possible. However, calling for a quantitative answer will usually ensure that the physics principles involved have been, not only understood, but correctly applied. Further, seeking quantitative solutions makes it possible to provide intermediate answers and guidance in a reasonably compact form. This is done in the second part of the book. There, each of the more than two hundred and twenty questions is provided with a hint, perhaps as a suggestion about an aspect of the problem that should be considered, perhaps as a signpost that should be passed on the way to a successful solution.

Much the largest part of the book is the Solutions section, which aims to provide in explicit detail full solutions to every question. Many of the solutions start by setting out the sort of thoughts a student may, and perhaps should, have on reading the question (but would not normally write down). They may include:

- What kind of processes are involved in this problem?
- Is the explicit data provided sufficient to yield the answer?
- If not, is there additional information implicit in the way the question has been posed?
- Or does everyday knowledge provide useful help, e.g. the length of a day gives the Earth's angular rotation speed.
- Do I need to introduce additional variables that are not mentioned in the question?
- Is there a self-consistency criterion to be satisfied?
- In what order should the various aspects of the problem be tackled?

Planning how issues such as these should be tackled is usually the key to a successful outcome. Some readers may find it helpful to begin each solution with a 'doodle' showing how they are going to approach the problem; a series of words or symbols joined by arrows, sometimes in the form of a closed loop, would suffice – many great inventions and ideas have started out that way!

The problems within any one chapter are not in any particular order of difficulty, though those marked with an * will probably be found to be the more challenging. Problems marked with a † require some mathematics that is not normally an integral part of school physics, but is likely to have been studied by any student who is also taking mathematics at an equivalent level. In some of the questions either too much or too little information appears to have been provided; try not to be distracted by red herrings, but do cast your net widely when thinking how any missing piece of the jigsaw might be provided.

Chapter 1

Dimensions
Stick to Your Own Kind!

Q1 Given Coulomb's law of electrostatic attraction, determine on dimensional grounds whether a quoted expression $E_i = 2\pi^2 m_e^2 k_e^2 e^4/h^2$ for the ionization energy of the hydrogen atom could be correct. If it could, obtain its value in eV; if not, suggest a corrected formula.

[Note: the Coulomb constant k_e can, alternatively, be written as $1/(4\pi\epsilon_0)$.]

Q2* According to the Wiedemann–Franz law, under certain conditions the ratio of a metal's thermal conductivity κ to its electrical conductivity σ is proportional to its absolute temperature T. Determine the dimensions of the constant of proportionality and, bearing in mind the processes to which it relates, suggest how it might be expressed in terms of fundamental constants.

Q3 Use the fact that the electrostatic force acting between two electric charges, q_1 and q_2, is given by $q_1 q_2/4\pi\epsilon_0 r^2$, where r is the distance between the charges and $\epsilon_0 = 8.8 \times 10^{-12}\,\mathrm{F\,m^{-1}}$, to express one farad in terms of the SI base units.

Q4* The following is a student's proposed formula for the energy flux S (the magnitude of the so-called Poynting vector) associated with an electromagnetic wave in a vacuum, the electric field strength of the wave being E and the associated magnetic flux density being B:

$$S = \frac{1}{2}\left[\left(\frac{\epsilon_0}{\mu_0}\right)^{1/2} E^2 + \left(\frac{\mu_0}{\epsilon_0}\right)^{1/2} B^2\right].$$

Assuming the normal definition of electric field strength, together with

  (i) Coulomb's law in the form $F = q_1 q_2/4\pi\epsilon_0 r^2$,
  (ii) the result that the magnetic field B inside a long empty solenoid of n turns per unit length, and carrying a current I, is given by $B = \mu_0 n I$,

1

(iii) the observation that the force acting on a rod of length ℓ that carries a current I at right angles to a field of magnetic flux density B is $BI\ell$,

determine whether the student's formula could be correct and, if not, locate the error as closely as possible.

Q5 If hens' eggs should be boiled for five minutes, for how long should an ostrich egg be boiled? If you have not previously studied heat conduction, see the introduction to Q 196 on page 68.

Q6* The radiation emitted per unit time by unit area of a 'black body' at temperature T is σT^4, where σ is the Stefan constant. Express σ in terms of the fundamental constants c, e, h, k and m_e, and then use the actual values of the quantities involved to evaluate any unknown constant.

Q7 Whilst they were dancing by the light of the full Moon, the Pussy-cat remarked to the Owl that it seemed almost as bright as day. The Owl first looked at the Moon, estimating that it subtends 9×10^{-3} rad, and then recalled from his pub-quiz days that its albedo is 7%. As they had only recently been married by the Turkey who lives on the hill, he thought it best not to spoil the moment by giving his own assessment of the actual situation. Which was what?

Q8* Whether or not an interstellar gas cloud undergoes gravitational collapse is determined by the relationship between the scale-length of its density perturbations and a characteristic length λ. The value of λ depends only on some or all of the following: the gravitational constant, the Boltzmann constant, the density of the gas, and the speed of sound in it.

Investigate the relationship between λ and the given quantities, (a) to determine whether the scale-length of the perturbations has to be noticeably greater than, or noticeably less than, λ to initiate collapse, and (b) to estimate the mass of a star formed from an atomic hydrogen cloud of density 10^{-17} kg m^{-3} and temperature 10 K, values typical of a so-called 'dense' cloud in interstellar space.

Q9 Three very different lengths that appear in quantum physics and cosmology are the Planck length ℓ_p, the Compton wavelength λ_m for a particle of mass m, and the Schwarzchild radius r_s. They are given by

$$\ell_p = \sqrt{\frac{hG}{2\pi c^n}}, \qquad \lambda_m = \frac{h}{mc}, \qquad r_s = \frac{2GM}{c^2}.$$

Here m and M are masses and c is the speed of light. Using only this information – that is, treating G and h as totally unknown quantities, in both magnitude and nature – deduce the value of n.

Q 10* The following table gives the observed values for the radius R of the fireball resulting from the explosion of an atomic bomb at sea-level, as a function of the time t since it was detonated.

$t\,(10^{-3}\,\text{s})$	0.24	0.66	1.22	4.61	15.0	53.0
$R\,(\text{m})$	19.9	31.9	41.0	67.3	106.5	175.0

Assuming that any numerical constants involved have values close to unity, estimate the energy E released in the explosion. (Ref. G.I. Taylor, *Proc. Roy. Soc. A*, **201**, p175, 1950)

Q 11 The relativistic Schrödinger equation for a spinless particle of mass m (the Klein–Gordon equation) is a (partial) differential equation describing how the relevant wave function, $\psi = \psi(x,\,y,\,z,\,t)$ develops in space and time. On one occasion, it was mistakenly written as

$$\frac{\partial^2 \psi}{\partial x^2} + \frac{\partial^2 \psi}{\partial y^2} + \frac{\partial^2 \psi}{\partial z^2} - \frac{1}{c^2}\frac{\partial^2 \psi}{\partial t^2} = \frac{(mc^2)^2}{\hbar^2}\psi, \qquad (*)$$

where $\hbar = h/2\pi$. Locate the likely error and suggest how it arose.

Q 12* When an incompressible fluid of density ρ flows through a small hole of diameter d in a thin plane sheet, the volume flow rate R depends upon ρ, d, η the viscosity of the fluid, and p the pressure difference between the two sides of the sheet. Measurements of the flow rate are made using a particular fluid. What can now be predicted for a second fluid with twice the density of the first, but only one-third of its viscosity? The size of the hole remains unaltered, but the pressure difference may have to be adjusted.

Chapter 2

Statics
Hold It There!

Q 13   A uniform square lamina ABCD, with centre O, has points P and Q as the mid-points of sides BC and CD respectively. If the square OPCQ is cut out and the remaining shape is suspended from a string attached at Q, what angle will the line OQ make with the vertical? Will the result change if a particle with one quarter of the mass of the original square is now attached at the mid-point of OB? If so, how?

Q 14*   An empty cylindrical beaker of mass m and internal radius r has its centre of gravity a distance h above the upper surface of its base. To what depth x_s should it be filled with a liquid of density ρ so as to make it as stable as possible? When it is, what is the height of G, the centre of gravity of the whole system?

Q 15   A light string of length $x + y$ is secured at its ends to two fixed points, a fixed distance apart and at the same height. A stone is tied to it at the point P, a distance x from one of the anchor points. The resulting tensions in the two parts of the string are T_1 in the part of length x, and T_2 in that of length y. Show that, as x and y are varied, the two tensions vary in such a way that

$$T_1^2 - T_2^2 \propto y^2 - x^2.$$

Q 16*   (*a*) The figure shows a pile of four identical books (treated as uniform solid blocks) and the amounts (in mm) by which each overhangs the book (or table) below it. Does the pile topple?

(*b*) Based on intuitive insight [ideal], or your experience with part (*a*) [probably more realistic], construct an algebraic criterion for the stability of a pile of N such books, in terms of their common length ℓ and the distances x_i, where x_n is the horizontal distance from the edge of the table of the right-hand side of the nth book.

4

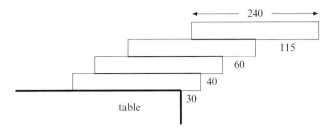

Q 17   A towel of length L hangs (unfolded) over a straight towel rail that has a circular cross-section of radius r. If the coefficient of friction between the two is μ, what is the maximum difference in length between the two vertical parts of the towel, if it is not to slip onto the floor?

Q 18***   The assembly instructions included in a kit for making a hanging mobile are as follows:

- This not-assembled pack has four light rods A, B, C and D, ten strings, six small animals with many weights, m, $2m$, $3m$, $\ldots$, and a hanging hook H with screw, for going through the roof. Please, to check yourself for any missing parts.
- When the hanging is all flat and all rods are level, the horizontal inch-positions of the strings, what they are hung from, and the loads on their bottoms, are in the following table:

String	Position	Hung from	Load
1	−2	B	$2m$
2	2	C	D
3	2	A	$3m$
4	−12	B	$3m$
5	10	D	m
6	0	H	C
7	−8	C	B
8	10	C	$4m$
9	0	D	A
10	−6	A	m

- Good luck!

What, if anything, has been lost in translation – apart from that of any sense of proper English?

Q 19

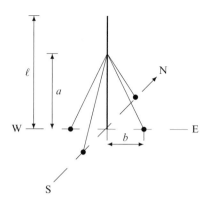

The figure shows a rigid radio mast of height ℓ that is pivoted by a ball joint at its base. It is secured against strong winds by four light but strong cables that are attached to it at a height a, and anchored at points that are b from its base, notionally in the N, E, S and W directions. The maximum safe tension in any one anchor cable is T_0, but when there is no wind, the actual tension is negligible.

By investigating the tensions in the cables when horizontal arctic winds, which exert a force of f per unit length on the mast, are blowing (a) from the North, and (b) from the North East, determine how a and b should be chosen so as to minimize the total length of cable needed, whilst maintaining the mast's stability against wind from any direction.

Q 20* A light rod of length a passes through a hole in a pivoted thin block, an end-on view of which is shown in the left-hand figure below. Because of friction between the block and the pivot's housing (at the surfaces shown by heavy lines) the pivot produces a torque that resists the block's movement.

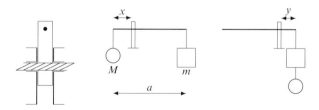

With the pivot a distance x from one end of the rod, and loads M and m on its two ends (see the middle figure), the system is in equilibrium. It is also in equilibrium if x is increased to $x + \frac{1}{4}a$, with the loads unchanged. The loads are now rearranged as shown in the right-hand figure. What is the maximum value of y for which equilibrium is guaranteed?

Q 21   A uniform ladder, 5 m long and of mass M, rests on a rough horizontal floor and leans against a smooth vertical wall just over 4 m high. If the foot of the ladder is more than 4 m from the wall it starts to slip. A man, carrying a load in both hands, uses the ladder, positioned with its foot 3 m from the wall, to access the top of the wall. If the man weighs $3Mg$, what is the maximum load he can carry safely?

Q 22*   A (rugged) hemispherical bowl of outer radius 10.0 cm is a shell of uniform thickness 2.0 cm. It is placed on a rough (no slipping can occur) plane inclined at an angle θ to the horizontal. What is the maximum value that θ can have if the situation is to be stable?

Q 23   A light rope carries a mass of 100 kg on one end and is coiled smoothly around a horizontal bar of circular cross-section, the coefficient of static friction between the two being 0.05. All parts of the wound rope lie on the surface of the cylinder, and not on any other part of the rope. Find the minimum number of turns required to prevent the 100 kg mass from falling, if the only available counterweight has a mass of 1 kg.

Q 24*   When a long wooden baton with square cross-section $2b \times 2b$ and density ρ' is placed in a certain oil, it drifts about aimlessly, sometimes on the surface, sometimes submerged. A pond of density $\rho \, (> \rho')$ is covered to a depth $\frac{1}{2}b$ by the oil, and the baton floats in the pond, with one of its long faces horizontal. Show that the baton is stable against small disturbances that tilt it about its long axis, provided $45\rho' > 37\rho$. Ignore all surface tension effects.

Q 25   A uniform cylinder of mass m hangs vertically from a spring balance A, of mass M_A, and is lowered slowly into a bowl of water, of mass M, until it rests totally submerged on the bottom of the bowl, which itself sits in the scale-pan of a second spring balance B, of mass M_B. This balance is, in turn, suspended from a third spring balance C, whose mass is M_C. On one diagram, show the variations of the scale readings of all three spring balances as a function of the vertical position x of the cylinder, measured downwards.

Q 26*   A thin uniform cocktail stirrer of length $2a$ has one of its ends inside an empty vertically-sided hiball tumbler of internal diameter d. The stirrer, which passes through the axis of the glass tumbler, makes an angle θ with the horizontal. Its points of contact with the glass are on the lip on one side, and with the inside surface on the other. It is on the point of slipping into the tumbler, and if the coefficient of friction μ (the same for both contact points) were only marginally less, it would do so. Show that

$$a\cos^2\theta\cos(\theta - 2\alpha) = d\cos^2\alpha,$$

where $\tan\alpha = \mu$.

Q 27   A chocolate orange consists of a sphere of smooth uniform chocolate of mass M and radius a, sliced into a number of segments by planes through its axis. It stands on a horizontal table with its axis vertical, and is held together only by a narrow ribbon around its equator.

Using spherical polar coordinates, find by direct integration an expression for the distance s from the orange's vertical axis of the centre of mass of a segment that subtends angle 2α at the axis. Check that your formula gives the expected values for a hemisphere ($3a/8$), and for a whole sphere (0); if not, have a look at the hint.

Find the minimum value that T, the tension in the ribbon, can have.

Q 28[*†]   This question requires the accurate integration of sinusoidal functions over non-standard intervals.

A rigid ping-pong ball of radius R and negligible mass rests on the rim of a round hole of diameter $\sqrt{3}R$ that has been cut in the base of a bucket. It is held in place by pressure from somebody's finger, whilst the bucket is slowly filled with water, none of which escapes. How deep does the water have to be before the finger can be withdrawn without the ball then floating upwards?

Q 29

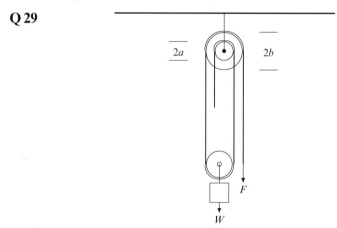

In the pulley system shown, the toothed pulleys of radii a and b are rigidly fixed together, but free to rotate about their fixed common axle. A load hangs from the axle of the freely suspended pulley, which is also toothed and may rotate about that axle. The total weight of this pulley and the load is W. The chain shown is of negligible weight and consists of many links that fit the teeth on all three pulleys; no relative slipping can take place. What force F is needed to keep the system in equilibrium?

If the right hand loop of the chain is now moved downwards with speed v, will the load move? If so, which way and how quickly?

Chapter 3

Dynamics
Keep It Moving!

Q 30   In a game of curling, the centre C of the *house* (the target) lies a distance $2L$ from the point O, from where the curler releases her curling stone (*rock*). Another stone, belonging to her team, lies halfway between O and C. If the coefficient of friction between a stone and the ice is μ, and the coefficient of restitution between two stones is e, with what speed U should she release her stone if the second one is to end up '*on the button*', i.e. at C?

Q 31[*]   Two rough blocks, each of mass M are joined by a light smooth inextensible string of length $4a$, at the centre of which a particle of mass m is attached. With the string taut and horizontal, the blocks are placed symmetrically on two rough tables separated by a distance $2a$, the string being perpendicular to both table edges. The coefficient of friction between the blocks and the tables is μ.

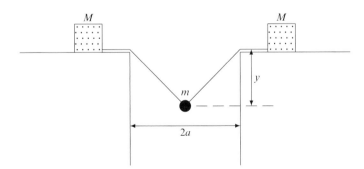

The system is released from rest, and at a subsequent time the blocks have moved through a distance x, whilst the particle has fallen through the vertical distance y shown in the figure. Find the inequality that must be satisfied if the blocks are to remain on the tables. If each block stops just at its table's edge, how far have they moved when the system has its largest kinetic energy?

Q 32   Without, if possible, looking up anything, which force would you say is the greater: that needed to prevent an open golf umbrella, held with its shaft horizontal, from blowing away in a 15 m s^{-1} wind, or the one needed to lift a golf bag containing a full set of clubs?

Q 33*   A tennis player serves from very close to the junction of the base and centre lines, and by making contact with the ball just after the racquet starts to move downwards, gives it a speed v at an angle α *below* the horizontal. The ball passes over the net at height h above the ground and hits the junction of the centre and service lines. The net, of height n, is a distance a from a base line and b from a service line. If the ball's height when struck is H, find expressions for $\tan \alpha$ and v in terms of the other relevant variables. Ignore air resistance and any sideways movement, spin or slice imparted to the ball.

In an actual tennis court, $a = 11.89$ m, $b = 6.40$ m and $n = 0.914$ m; check that your expression for $\tan \alpha$ agrees with

$$\tan \alpha = 0.1388\,H - 0.2404\,h.$$

The best players can serve at about 50 m s^{-1} with $H = 2.60$ m. Assuming that the ball's diameter d is 6.7 cm, and that $\tan^2 \alpha \ll 1$, by approximately how much does such a serve clear the net?

[For the numerical work involved, you should maintain at least 3 s.f. accuracy, as the subtraction of nearly-equal values is involved.]

Q 34   A liquid that flows through a narrow circular tube of internal radius r under a pressure gradient, does so at a volume flow rate of

$$\frac{\pi}{8}\,\frac{r^4}{\eta} \times \text{pressure gradient,}$$

where η is the viscosity of the liquid.

An ancient (and very primitive) water clock consisted of a horizontal capillary tube, 0.8 cubits (symbol q) long and internal diameter $\phi = 1.0$ mq, connected by a small right-angled bend to the bottom of a vertical thin-walled cylinder 0.4 q tall and 120 mq in diameter, which served as the water reservoir. The 'manufacturer's recommendation' was that the latter be fully refilled when the level in it fell to 40 mq, as readings became unreliable for levels lower than this. The dealers in the Bazaar marketed it as a 'five-day' clock. Were their customers delighted or disappointed with their purchases?

[The viscosity of water is 5×10^{-4} kg q^{-1} s^{-1}, and its density is 125 kg q^{-3}. $g = 19.6$ q s^{-2}.]

Q 35* A particle of mass m is projected vertically upwards with speed U in a medium that produces an opposing resistive force of magnitude kv^2, where v is the particle's speed. With what speed D will it return to the ground?

Q 36 Two steel balls are suspended by vertical strings (not necessarily of the same length), so that at rest they are just in contact, with their centres at the same height. One is pulled aside in the vertical plane defined by the strings and released when it is at a height H above its original position. Determine the maximum possible height to which the other can rise, and the condition for this to happen.

Q 37* On an unsatisfactory snooker table, two of the red balls are at rest and touching each other. The white cue ball, which has the same radius a and mass m as each of the reds, approaches them with velocity V along the perpendicular bisector of the line joining their centres. It is observed that after the collision the cue ball continues to move forward in its initial direction, and that, at a later time t, all three balls are in a straight line perpendicular to the path of the cue ball.

What can be said about e, the common coefficient of restitution between any two of the balls? Find an expression for t in terms of a, m, V and e.

Q 38† This problem is more of a mathematical curiosity than a practical aid to a cricket or baseball fielder, but it provides a good test of the accurate use of differential calculus and the manipulation of geometric identities.

A tallish batsman/hitter B strikes a ball towards a smallish fielder F, who is perfectly positioned to catch it. Assuming that the point where the bat struck the ball, the fielder's hands, and his face are all at the same vertical height, and the angle of elevation of the ball, as seen by the fielder, is ϕ, show that the rate of change of $\tan \phi$ is constant as he tracks the ball, and find an expression for that constant. Neglect air resistance.

Q 39* A spherical raindrop with mass m, radius r and downward speed v, falls through a cloud of water vapour, which moves upwards at speed U. The drop starts from rest with radius r_0 at time $t = 0$. As a result of condensation on it by the vapour, its mass grows at a rate equal to k times its surface area.

Establish the equation of motion of the raindrop, and find expressions for r as a function of time, and for v as a function of r. Use them to show that ultimately the drop's acceleration tends to a constant value, and determine that value.

Q 40 *Note* This will be a familiar problem to many, but it is included here as it calls for a clear understanding of the basic laws involved.

A fine chain of length ℓ and mass ρ per unit length is suspended so that it hangs vertically at rest with its lower end just touching the horizontal pan of an electronic set of scales. Its upper end is released at time $t=0$ and it falls freely, collapsing, without bouncing, onto the pan (see the figure).

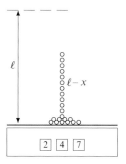

Determine the reading S of the scales (a) as a function of t, and (b) as a function of the length x of chain that has already fallen onto the pan. Make annotated sketch-graphs of these results, marking significant values.

Q 41* Two carriages A and B are moving freely in the same direction along smooth close parallel tracks, with speeds V and U respectively. Carriage B is of mass M, whilst carriage A, which carries a mailbag of mass m, has an unladen mass of $M - m$. Whilst the two carriages are alongside each other the mail bag is thrown from A to B, perpendicularly to the tracks as viewed from A. It is then returned in a similar manner, i.e. perpendicularly to the tracks as viewed from B. How much mechanical energy is lost as a result of the two transfers? And how is it accounted for physically?

Q 42 A straight river in a conservation research area has width a and flows in the positive y-direction. The velocity of the water at its surface is zero on the two banks, has a (horizontal) profile that is quadratic in x where $0 \leq x \leq a$, and has a maximum speed of $\frac{3}{4}u$ in mid-stream.

An otter crosses the river by swimming upstream, head above the water, at a steady speed u and at a fixed angle θ to the current, emerging on the other side directly opposite the starting point. Find the value of θ and sketch the otter's path as determined by its GPS collar, marking the value(s) of x at which it is furthest from the straight line joining the entry and exit points.

Q 43* An unusually-shaped and rather unsatisfactory 'billiard table' is in the form of an equilateral triangle ABC. The side cushions and table surface are smooth, and a billiard ball (to be treated as a point mass without any spin) that hits a cushion retains its velocity component parallel to it. Unfortunately, the component of its velocity perpendicular to the cushion, although reversed, is reduced in magnitude to one-third of its incoming value.

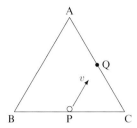

A billiard ball, initially at the mid-point P of the side BC, is cued with speed v towards Q, the mid-point of side AC. When it is next close to cushion BC, what will be its speed, and how far from P will the impact point be?

Q 44* A river of depth Y flows with uniform speed u_0. A suspected criminal leans over the side of an anchored boat on the river and surreptitiously releases 'a blunt object', which has a mass m and a density ρ greater than ρ_0, the density of water.

As it sinks, the blunt object experiences a directly-opposing resistive viscous force of magnitude β times its velocity relative to the water. How far downstream from the boat does the object come to rest on the river bed?

Q 45† Although the physics involved in this problem is no more difficult than that in most other questions, the integral calculus needed, whilst not particularly advanced, requires a great deal of care in its application.

A terrestrial rocket of total initial mass m_0 contains $\frac{1}{2}m_0$ of fuel, which it burns at a constant rate α, ejecting it backwards with speed u relative to the rocket. On ignition, it immediately launches itself vertically upwards. How high will it reach before it starts to fall to earth? Neglect air resistance and any variation of g with height.

Q 46 A golf ball was dropped from a height h of 2.00 m onto a concrete floor and stopped bouncing 12.1 s after its release. How high was its fifth bounce? Ignore air resistance.

Q 47* A trolley of length L is free to move on frictionless rails, and carries at one end a rapid-fire gun loaded with N bullets each of mass m; at the other end there is a target, thick enough to stop the bullets. The total mass of the trolley, including the gun, is M ($\gg Nm$), and the speed of each bullet when fired (at the target, of course) is v_0.

At time $t = 0$ the gun starts firing at a uniform rate f ($\gg v_0/L$). Ignoring air resistance, draw sketches of the (signed) variations with time of both the speed and position of the trolley, which starts from rest at $x = 0$. Quantify all significant values of time, speed and displacement. In particular, determine and comment on the values of the latter two well after the gun has run out of ammunition.

Q 48   A 'magician' has a collection of perfectly elastic balls, all of the same diameter, though weighted in their centres with a range of different masses; there are at least three balls of each mass. From the collection he selects three and loads them into a smooth vertical slotted tube.

The tube is fitted at height h ($\gg$ than the common diameter of the balls) with ganged horizontal inserts that ensure that the balls are closely spaced, but not touching, when at rest, and that they are all released at the same time when the inserts are withdrawn. The tube is there merely as a guide, and plays no part in determining the motion of the balls after release (i.e. no friction, air resistance, etc.).

Show that, with an appropriate selection for the three balls, the magician can, after loading and releasing them, amaze his audience by 'talking' the uppermost one into rising out of the guide tube which is some $20h$ high.

Q 49*   Both stages of a two-stage rocket are propelled by ejecting gas backwards with speed u relative to the rocket. The masses of the fuel-filled and empty stages are M_n and m_n, respectively ($n = 1, 2$). The empty first stage of mass m_1 is jettisoned before the second stage is fired, and the complete process takes place, starting from rest, in a field-free region.

Find an expression for the additional speed attained by the empty second stage, as compared with that achieved using a single rocket that is of mass $M_1 + M_2$ when full, and $m_1 + m_2$ when empty. Interpret your results qualitatively if m_1 is very much smaller than the other masses involved.

Q 50   The power theoretically extractable from the wind by a windmill whose sails sweep out an area A can be estimated using the following model. Suppose the wind is one-dimensional with speeds V and αV ($0 \le \alpha \le 1$) far up- and down-stream, and speed v at the mill itself. Calculate the power absorbed by the mill and then use 'power conservation' to find the ratio v/V. Determine the maximum fraction of the initial power in the wind that can be extracted on this model.

Q 51[†]   When a soap film is punctured, a rapidly-growing circular hole is formed, with, according to high-speed photography, the material already ruptured being concentrated on the rim of the hole, and uniformly distributed around it.

If a particular film has thickness w, density ρ and coefficient of surface tension γ, find the differential equation that describes the time dependence $r = r(t)$ of the radius of the hole as it enlarges under the surface tension produced by still-intact soap film. Solve it and determine the rate at which the hole's radius increases when it is already large.

Q 52   A long straight marble staircase has risers of height r and treads of width w. A golf ball bounces down the staircase in a totally repetitive way, one bounce on each step. If the coefficient of restitution is e, and any spin effects are negligible, how high above the level of each step does the ball rise following its bounce on that step? What is the angle between the horizontal and the direction of the ball's trajectory immediately following a bounce?

Q 53*†   A plane flies at a constant speed V relative to the air, and completes a level circular course in time T on a windless day. If a steady wind of speed kV, where $k \ll 1$, blows in a fixed horizontal direction, will the time for completing the same course be changed? If so, determine by how much (working to second order in k); if not, explain why this is the case.

Q 54   If I place a Rubik's cube of side $2b$ on a sloping rectangular tea tray, with one of its edges parallel to the tray's lower rim, it might slide down to that rim, which then presents a small but effective barrier to further translational motion. I find that sliding takes place if the slope θ of the tray exceeds $18°$. With the slope set at $20°$, and assuming that the static and dynamic coefficients of friction are equal, how far up the tray may I place the cube's front edge if the cube is to stop at the bottom rim without toppling over?

Chapter 4

Rotation
Round and Round We Go!

Q 55 A model aeroplane has mass m and always flies at a constant airspeed v. It is tethered to a fixed point by an inelastic chord and completes a horizontal circular orbit of radius R in time T_1. The original chord is now replaced by an elastic one that has force constant k and unstretched length R; the time for a horizontal circular orbit is now T_2. Show that the product $T_2(T_2 - T_1)$ is independent of both v and R, and find a value for it.

Q 56[†] The perpendicular axis theorem states that the moment of inertia of a plane body about an axis perpendicular to its plane is equal to the sum of the moments of inertia about two perpendicular axes lying in the plane; in an obvious notation

$$I_z = \int r^2 \, dm = \int (x^2 + y^2) \, dm = \int x^2 \, dm + \int y^2 \, dm = I_y + I_x.$$

(a) Use the theorem to find the moment of inertia of an elliptical disc, with major and minor semi-axes a and b respectively, about an axis through its centre C and perpendicular to its plane.
 [Should you need it, you may assume that $\int_0^{\pi/2} \sin^4 \theta \, d\theta = 3\pi/16$, and that the area of the ellipse is πab.][1]
(b) The elliptical disc is at rest on a flat rough table with its major axis vertical, but when very slightly disturbed it rolls sideways. What is its angular velocity when its minor axis is vertical?

Q 57 To practice being happy, despite having to go to work, Snow White's seven dwarves train on a compact, but infinitely long, road, in the form of a horizontal turntable – a uniform disc of mass 150 kg and diameter 5 m.

---

[1] For a proof of the first of these results, see the *Note* at end of the solution.

The dwarves, whose average weight is 300 N, sing as they march in Indian file around a circle painted on the turntable 0.5 m from its outer edge. Snow White herself, who weighs 500 N, stands at the disc's centre, watching the monitor for the GPS-fitted helmets that all miners are required to wear, whilst reprimanding anyone who is out of step, or out of tune.

The distance from their cottage to the mine, where the dwarves dig, dig, dig all day, is 1 km, and SW only calls a halt to the training when the monitor shows that the dwarves have covered this distance. On several occasions, Grumpy has complained that it had seemed more like three kilometres. Does he have a point?

Q 58   The body of a male skater of total mass 72 kg is represented by a uniform cylinder of radius 0.20 m, with two uniform thin 'arms' each of mass 4 kg and length 0.70 m, hinged at diametrically opposed points on its circumference. He is standing upright with his arms out-stretched and spinning about a vertical axis. He relaxes his arms and they fall gently to his sides without any effort on his part. What is his final rotational speed?

Q 59   Determine, in terms of the general properties of a door, the best distance from the line of its hinges at which to place a 'floor door-stop', so as to minimize the impulsive reaction at the hinges when an unlatched door is blown open by a gust of wind. Consider, in more detail, the specific case of a uniform solid door of width w.

Q 60***   A small block lies on the inside surface of a tall thin parabolic bowl whose shape is described by the equation $x^2 = 4ay$, where the y-axis is vertical. The coefficient of (static) friction between the two is μ, and the bowl can be rotated about its vertical axis with a frequency ω that can be varied. When the block is at a particular height y above the bottom of the bowl, there is a minimum value of ω, say ω_-, that is required to stop it slipping downwards; similarly, there is a maximum value ω_+ that must be observed, if it is not to slide upwards.

Obtain expressions for $\omega_\pm$ in terms of a, μ and x, and then show:

(a) that if $\mu \to 0$ then the block is stable at any height, but only if ω is exactly equal to ω_0 where $\omega_0^2 = g/2a$;
(b) that if $y = a$, then $\omega_- \omega_+ = \omega_0^2$, whatever the value of μ;
(c) that if $y = a\mu^2$, then there is no required minimum angular velocity, but for stability ω^2 must not exceed $2(1 - \mu^2)^{-1}\omega_0^2$. Explain the first of these observations in terms of a simpler situation;

(d) that if $y = a\mu^{-2}$, then there is no finite value of ω that can stop the block sliding upwards, and that to stop it slipping downwards requires ω^2 to be at least $\frac{1}{2}(1 - \mu^2)\,\omega_0^2$.

Q 61   When a particular small mass is suspended at rest from a fixed point P by a light elastic string of unstretched length ℓ_0 it produces an extension of length ℓ_1. Smart Alec claims that it is intuitively obvious that if the mass were made to move in a horizontal circular path, i.e. the system were to become a conical pendulum, the period of revolution would be $\tau = 2\pi[(\ell_0 + \ell_1)/g]^{1/2}$. Analyse the motion and determine whether or not Alec is really that smart.

Q 62*   A particular watermill has many buckets, and they are closely spaced around the perimeter of a vertical wheel of radius r. Water flows continuously, at a horizontal speed V, into the one that is uppermost at the time, and is not tipped out until that bucket reaches the bottom of the wheel. If the angular velocity of the wheel is set at its optimal value, find an expression for the efficiency of power extraction.

Q 63   A particle of mass m can slide smoothly inside a uniform tube, also of mass m, that can rotate freely in a horizontal plane about a pivot at one of its ends. The particle is at rest very close to, but not at, the pivot, when the tube is given an impulsive initial angular velocity. Will the particle leave the tube through the open end? If so, find the angle, relative to the axis of the tube, at which it does so. Why can your answer only be approximate?

Q 64*   A uniform cylinder of mass m and radius a rotates anticlockwise with angular velocity ω about its axis of symmetry in a fixed rough groove formed by two orthogonal planes, one of which makes an angle θ with the horizontal (as shown in the figure).

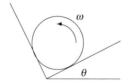

(a) If the coefficient of sliding friction at both points of contact is μ (< 1), and the cylinder's initial angular velocity is Ω, find how long it takes to come to rest.

(b) Your answer to part (a) should suggest that something 'goes wrong' if θ is large enough; in particular, if θ is replaced by $\pi/2 - \theta$, then a different stopping time would be obtained, even though this would only correspond to having the other plane make an angle θ with the horizontal.[2]

Why is the argument in the previous sentence invalid? Try to find the origin of the apparent discrepancy, determine the critical value of θ, describe what would actually happen, and explain why there is no corresponding change in behaviour when θ is small and the left-hand plane is nearly vertical.

Q 65   A uniform circular disc of mass M and radius a is placed on a smooth horizontal table and then set in motion by a tangential impulse applied to a point on its outer rim. About which point will the disc begin to rotate?

Q 66[*]   One end, A, of a thin uniform rod AB of length ℓ is placed on the edge of a very rough table, and then released from rest in an almost vertical position; it falls off, and at right angles to, the table's edge. The rod end that first hits the floor is end B, and it does so with BA vertical. How high is the table?

Q 67

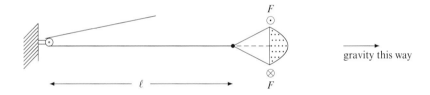

The figure (drawn sideways, for a change, and to save space) shows a conservatory hanging basket that has been lowered (for watering) by a light rope that passes round a pulley fixed to the conservatory ceiling. The basket, in the shape of a paraboloid of revolution with a depth of 0.2 m and a surface diameter of 0.4 m, is fastened to the end of the rope by three light, stiff rods that make angles of 60° with the horizontal and are attached at equal spacings around the rim of the basket.

If I displace the hanging basket sideways a little, it swings with a period of exactly three seconds. But if, instead, I rotate the rim of the basket through 20° by applying equal but opposite tangential forces F at diametrically opposite points on the rim, and then let go, the basket makes torsional oscillations with

---

[2] If your answer to (a) does not do this, consult the solution on page 170, reworking your answer if necessary, before proceeding with (or looking at the answer for) part (b) of this problem.

a period five times greater. Given that the density of moist soil is 1200 kg m^{-3}, find

(a) the length ℓ of the vertical part of the rope (to the nearest cm), and
(b) the magnitude of F.

Q 68[*]   In a misguided, and highly dangerous, attempt to measure the speed of a bullet from his newly-acquired rifle, a less-than-competent physicist decided to construct his own ballistic pendulum. He had available a light flexible wire of length 1.0 m and breaking strength 1000 N, which he used to suspend a spherical wooden block of mass 0.5 kg from a fixed horizontal beam B of circular cross-section. He fired the bullet, which weighed 10 g, into the centre of the block and expected to see the pendulum rise through an angle of about 45°.

However, what actually happened was that the block started to rapidly circle around the beam, progressively reducing the effective length of the wire – which snapped when it had been shortened to one half of its original length. Roughly, what sized wooden block should he have chosen?

Q 69   When scientists derive theoretical solutions to problems, usually in the form of a formula, it is important that they test it by considering any special cases in which one or more of the variables have or tend towards particular values, and for which the corresponding physical situations are easily understood. The problem below does not contain demanding physics, but has a theoretical solution that is somewhat difficult to visualize. Obtain your solution, and then test it for the special cases suggested.

A space station is in the form of a large wheel of radius R rotating with angular speed ω about its own centre. A small object to be delivered to the station's centre is projected inwards from the wheel's perimeter with speed V on a frictionless radial track. How long does it take to reach its destination?

For the following special cases, check that your solution predicts the behaviour expected on physical grounds: (i) $V \to \omega R$; (ii) $V \to \infty$; (iii) $V < \omega R$; (iv) $\omega \to 0$.

Q 70[*]   A compact mass is suspended by a light inextensible (and unbreakable) string of length ℓ from a point P on a horizontal bar that has clear space both above and below it. A second and parallel bar, Q, of very small diameter, is fixed a vertical distance $\lambda \ell$ below P, where λ can be varied over the range $0 < \lambda < 1$. The mass is drawn aside in the vertical plane containing P and Q, so that the string is both horizontal and taut. It is then either released from rest or given an initial downward velocity V.

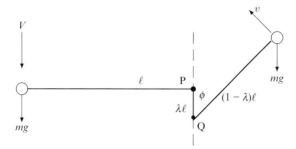

(a) Describe qualitatively, but in as much detail as you can, what you think would then happen. Consider whether the behaviour changes according to the various ranges in which λ might lie.

(b) Make a physics analysis of the situation using the notation indicated by the above figure, finding, in particular, the condition(s) for the taut string to become vertical again after it first impinges on Q.

(c) Review your answer to (a) in the light of your answer to (b), and either congratulate yourself or take note of something for future use.

(d) Whether or not you did well in part (c), look at the experimental data recorded in the Solutions section and compare it with your predictions.

(e) Do the experiment for yourself; it requires only a minimum of apparatus, and some clear space around you. And it will certainly make a change from observing *vanishingly small* swings of a pendulum – as well as being a lot more exciting!

Chapter 5

Gravity
To Be Taken Seriously!

In this chapter assume that all orbits are circular and coplanar, unless the question states otherwise.

Q 71   Astronauts travel (in a straight line for the purposes of this question) from the Earth to the Moon, but before they take off they note that it subtends an angle of 9.05×10^{-3} rad. When they get there, they gaze back at the Blue Planet and see that it subtends an angle of 33.5×10^{-3} rad. From experiments with golf balls, they measure the Moon's gravitational acceleration to be 1.62 m s^{-2}.

What fraction of their journey had they completed when the Moon's gravity took over from that of Earth?

Q 72   An unpowered spacecraft is to be landed, as gently as possible, on the airless moon of a planet. The moon has radius R and mass M, and is in an orbit of radius $10R$ around the planet, whose mass is $9M$. If the motion of the planet during the period of the flight can be neglected, and the flight path is along the line joining the planet's and moon's centres, what is the minimum possible impact velocity when the spacecraft reaches the moon's surface?

Q 73   Three geostationary satellites are positioned, one (G) above the Galapagos Islands, a second (V) over Lake Victoria, and the third (N) above New Ireland, which is located $35°$ due North of Canberra.

A signal emitted by V is detected and immediately retransmitted by G to N, which in turn immediately retransmits it back to V. What is the time interval between the emission and reception of the signal at V?

[The radius R_E of the Earth is 6.4×10^3 km. Ignore any possible relativistic effects.]

Q 74 The expansion of the Universe is thought to be governed by an equation of the form

$$\left(\frac{dR}{dt}\right)^2 = \frac{a^2}{R} + bR^2 - c,$$

where R is its 'size' and t is time, with a, b and c real constants. Obtain an expression for d^2R/dt^2 and then, knowing that the Universe is observed to be expanding at the present time, say what you can about its past history and future behaviour, if

(i) b is negative, and
(ii) b is positive and c is negative.

Do not attempt to solve any differential equations; base your arguments on the possible shapes of a graph of R as a function of t.

Q 75 An astronomer posted to an observatory on Mars, where the year is 687 Earth-days long, measured the distance to the Andromeda galaxy, using the parallax method. Forgetting that 'he was on another planet' he reported the distance as 0.51 Mpc. Assuming that his measurements were up to scratch, what value should he have reported?

Q 76 Without worrying about the insuperable practical difficulties of constructing and installing it, or the unsustainable strains within it, find how long a uniform cable that rises vertically along its entire length should be, if the load on its one anchor point, located on the Equator, is to be minimized.

Q 77 If the Earth were suddenly stopped in its orbit around the Sun, how many days would it take for it to fall into the Sun? You may assume that any movement of the Sun towards the Earth is negligible, and the mathematical result

$$\int_0^1 (x^{-1} - 1)^{-1/2}\, dx = \tfrac{1}{2}\pi.$$

Q 78 According to Einstein's theory of general relativity, a small correction of magnitude $6GMmv^2/(r^2c^2)$ should be added to the Newtonian gravitational force between a star of mass M and a planet of mass m moving with speed v in a circular orbit of radius r around the star. Estimate the advance per century of the planetary motion of Mercury due to this correction.[1]

[The mass of the sun is 2.0×10^{30} kg and the radius of Mercury's orbit, assumed circular, is 5.8×10^{10} m.]

---

[1] Known as its perihelion precession; its prediction and measurement were crucial in establishing Einstein's theory.

Q 79   Forces that follow an inverse cube law do not seem to occur in nature, at least when viewed from an 'inertial' frame (typically the 'laboratory frame'). Nevertheless, a study of their properties can be of value in analysing other systems.

For example, when a particle of mass m is orbiting in a plane under the influence of a *central* attractive force, its angular momentum L cannot change – in plane polar coordinates (r, θ), this means that $L = mr^2\dot{\theta}$ is a constant. If the system is described in a co-rotating frame, there appears to be a (centrifugal) force $mr\dot{\theta}^2$ present, in addition to those apparent in an inertial frame. Since $\dot{\theta}$ appears in both L and the fictitious force, it can be eliminated from the equation of motion which becomes one-dimensional and of the form

$$m\ddot{r} = F(r) + \frac{L^2}{mr^3},$$

i.e. the fictitious force is an inverse cube one.

Now for the actual question. Even if they don't appear naturally, systems operating under an inverse cube force law, have a few interesting, but easily derived, properties. Investigate the energy and angular momentum associated with a particle of mass m moving in a circular orbit under an attractive central force given in magnitude by k/r^3. Draw any qualitative conclusions you can from your results.

Q 80*   A satellite S is in a circular orbit of radius r about the Earth, which has radius R and angular velocity Ω. The orbit lies in the equatorial plane and is traversed in the direction of the Earth's rotation. An observatory on the Equator records S as being directly overhead at 12 hour intervals. For how long during each pass is the satellite observable?

Express your answer in terms of the given quantities, together with g, substitute actual values, and check whether your calculated value is qualitatively reasonable.

Q 81   Seen from the Moon, the Earth has 3.6 times the angular diameter of the Sun. Estimate the relative density of the Sun and Earth.

Q 82*   Because of the presence of interplanetary gas, a particle of mass m orbiting the Sun in a circular orbit with angular frequency ω, experiences a small resistive force that is $-k$ times its velocity, in both magnitude and direction. Assuming that $k \ll m\omega$, answer the following, making clear the sign/direction of any changes:

(a) Does the particle's angular momentum J about the Sun remain constant? If so, prove that it does. If not, find its variation with time.
(b) By what fraction does the radius R of the particle's orbit change per revolution?

(c) By what fraction does the particle's tangential speed v change per revolution?

Q 83 Two similar spacecraft, each of mass m, are to be launched from a space station that has speed v_S relative to the Earth, and orbits above the Equator in the same sense as the Earth orbits the Sun. The rockets burn for only a few minutes, but the launches aim to use the minimum fuel for the intended missions.

Spacecraft A is to be launched in such a way that it just escapes from the solar system, whilst spacecraft B is designed to fall straight into the centre of the Sun immediately after its rocket has finished burning. Which spacecraft requires the more powerful rocket? Find an expression in terms of known quantities for the difference between the impulses required in the two cases, ignoring the Earth's gravitational field.

Q 84* Consider the following three methods of visiting the antipodal (diametrically opposite) point on the Earth's surface and then returning to the starting point:

(a) Using a low-orbit satellite, just above the surface of an Earth that has uniform density ρ_0.
(b) By drilling a (hypothetical) diametric hole through an Earth that has uniform density ρ_0, and then falling into it.
(c) By falling into a (hypothetically) drilled diametric hole through an Earth that has a density that varies inversely with distance from its centre, but has the same radius and total mass as both the Earths in (a) and (b).

Before you start any calculations, try forecasting the relative order of the times T_a, T_b and T_c for the round trips. Then find an expression for each in terms of quantities that allow them to be numerically compared.

(d) In the light of your answers to (a), (b) and (c), would you expect the time for a return journey via a diametric hole through an Earth with an iron core (more like the real one) to be greater or less than that for a low-orbit satellite?

Q 85 Light of wavelength $\lambda = 500$ nm is emitted from the surface of the Sun, and eight minutes and twenty seconds later arrives on Earth. Using only data known to non-scientists (assumed to include $c = 3.0 \times 10^8$ m s^{-1}) and the fact that the Sun subtends an angle of $0.53°$ when viewed from Earth, determine the wavelength at which it is received.

Q 86*   A uniform spherical dust cloud of mass M expands or contracts in such a way as to remain both uniform and spherical. Show that, as a function of time t, the cloud radius R satisfies the equation

$$\dot{R}^2 = 2GM \left(\frac{1}{R} - \frac{1}{R_0} \right), \tag{*}$$

stating the particular quality that characterizes R_0 (assumed > 0).

Verify that a particular solution of $(*)$ is given parametrically by

$$R = a \sin^2 \phi, \quad \text{and} \quad t = b(\phi - \sin \phi \cos \phi),$$

where a and b are constants that can be expressed in terms of G, M and R_0.

Show that this solution describes a cloud that expands from infinite density at $t = 0$, and later collapses back to infinite density at a time T which is to be determined.

Disclaimer. Any similarity between this scenario and a Big Bang theory – past, present or future – is purely coincidental.

Chapter 6

SHM
Here We Go Again!

Q 87   A clock pendulum consists of a uniform rod AB of length 2ℓ and mass m together with a uniform circular disc of radius a and mass M centred on B. The disc lies in the vertical plane in which the pendulum swings. Determine the period T of small swings of such a pendulum.

For a particular pendulum, in which $\ell = 3a$, find the value of $\lambda = m/M$ which will make its period equal to that of an idealized pendulum consisting of a weightless rod of length 2ℓ carrying a point mass at its lower end.

Q 88*   A particle of mass m moves (in one dimension) on the inner surface of an upright parabolic bowl that has the form $y = bx^2$. The inside of the bowl is smooth except for a small region of length 2ℓ centred on the origin; there, the mass experiences a frictional retarding force that can be assumed to have a constant magnitude μmg. The mass starts from rest at the point (x_0, bx_0^2) with $x_0 \gg \ell$.

(a) Estimate the time that elapses before the mass comes to rest near the bottom of the bowl.
(b) Refine your estimate by giving separate consideration to the final part of the particle's motion, i.e. the stage during which it is confined to the region $|x| < \ell$.

Q 89   A particle is executing SHM of period π s and it is observed that its velocity at a particular time is 4.0 cm s^{-1}, whilst half a second later it is -3.0 cm s^{-1}. What is the amplitude of the vibration, and how long after the particle had reached the preceding maximum positive displacement was the first observation made?

Q 90* A 'baby-bouncer' can be represented by the following model. A particle is attached to the mid-point of a light elastic string of unstretched length 2ℓ, the ends of which are attached to two points $2b$ apart and at the same horizontal level. As shown in the figure, the particle is in equilibrium when hanging a distance a below that level.

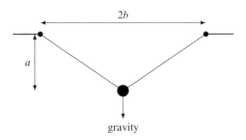

The baby/particle is pulled downwards through a small distance and then released. Show that the period of the resulting small oscillations is $2\pi/\omega$, where

$$\omega^2 = \frac{g}{a}\left(\frac{1-(b^2\ell/c^3)}{1-\ell/c}\right) \quad \text{and} \quad c^2 = a^2 + b^2.$$

Q 91 In a school physics laboratory, a uniform wooden disc is supported in a horizontal plane by a number of light vertical threads attached at equally-spaced points around its circumference.

When the class was asked which factors would affect the period τ of small torsional oscillations of the disc in its own plane, many combinations of the following were suggested: the disc's mass M, density ρ, surface density σ, and radius a; the strings' number n, length ℓ, tension T, and Young modulus E; gravitational acceleration g and the vertical component of the Earth's magnetic field B_V.

Which students had done their homework? Without calculating anything, write down your own prediction and then analyse the situation, producing the complete formula for τ in terms of some or all of the proposed factors.

Q 92* A light elastic string of spring constant k is fixed at its upper end and carries at its lower end an initially stationary scale pan of mass M. A lump of putty of mass m falls onto the pan from a height h and sticks to it. In the subsequent motion will the pan rise above its original position? If so, how far above? If not, by how much does it fail to do so?

Q 93 The figure shows a simple pendulum of length ℓ set up at a distance $a\,(\ll \ell)$ from a vertical wall, with P marking its equilibrium position.

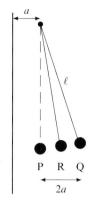

The bob is pulled back a distance $2a$ to Q and then released at time $t = 0$; when it hits the wall it loses half of its kinetic energy and then returns along the original path as far as R. How far is R from P, and at what time does the bob get there on its return swing?

Q 94* A particle of mass m rests on a smooth table and is connected to four fixed points P, Q, R and S by four identical springs of unstretched length a and spring constant k, as shown in the figure.

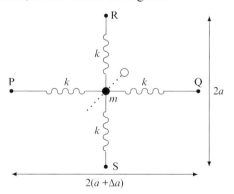

The separation of R and S is $2a$, but that between P and Q is $2a + 2\Delta a$, where $\Delta a \ll a$; the 'x-springs' in the figure are therefore slightly stretched when the system is in equilibrium. The mass is now displaced slightly in a direction making $45°$ with both PQ and RS, and then released. By finding the angular frequencies of small oscillations both parallel and perpendicular to PQ, obtain an approximate expression for the time that will elapse before the mass is again moving along the direction of its initial displacement.

[As noted in the solution, for transverse oscillations, changes in the spring tensions can be ignored if the oscillation's amplitude is small.]

Q 95   A rough uniform sphere of radius a is released from rest at a point that is on the inside surface of, and near the bottom of, a rough spherical bowl of (internal) radius na. Approximately how long will it be before the sphere returns to its original position?

Q 96*   Although in practice a considerable amount of damping is deliberately introduced, one model of a car and its suspension is a spring of spring constant k carrying a load of mass M, as shown in the picture.

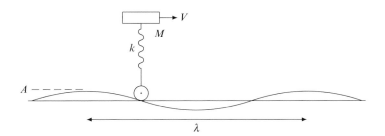

A car represented by this model travels at speed V along a road that has sinusoidal ripples in its surface of amplitude A and wavelength λ. As a consequence, a baby asleep in the car is subjected to a vertical velocity that varies in magnitude between 0 and v; the baby wakes up and cries if v exceeds v_0. Within what speed range(s) should the car driver aim to maintain V? Make a qualitative sketch showing how v varies with V, and specify the 'allowed' region(s).

Assuming values of $k = 5000$ N m^{-1} and $M = 500$ kg for the car, and that $v_0 = 0.2$ m s^{-1} for the baby, show that on a road that has peak-to-trough sinusoidal variations of 10 cm, with the peaks appearing every 10 m, the driver should try to avoid maintaining road speeds in the range 3.4 to 7.4 m s^{-1} for any significant length of time.

Q 97   A piece of apparatus in a school physics laboratory that was designed to illustrate spring-controlled simple harmonic motion, consisted of a trolley whose frictionless wheels ran on a smooth horizontal track. As shown in the figure, it was connected to rigid end walls by a pair of light springs with spring constants λ and 3λ. When the trolley was in its equilibrium position, indicated by $x = 0$ in the figure, there was no tension or compression in either spring.

If the trolley was displaced a distance X to the right and then released, it oscillated with period T.

Later, after the apparatus had somehow become damaged – and, of course, nobody had any idea how it could have happened – the connection at A between the trolley and the right-hand spring was found to be broken, though

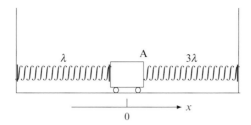

they still just touched in the equilibrium position. A repeat initial displacement and release then produced oscillations of period T'. What is the numerical relationship between T and T'?

Q 98*† Three identical light springs, each having spring constant k, are connected in a line and anchored to two fixed points on a smooth horizontal table; the points are sufficiently far apart that in equilibrium the springs are all stretched. Attached at the two spring–spring junctions are small bodies X and Y, both of mass m, whose displacements from their equilibrium positions, measured in the same direction, are denoted by x and y respectively (see the figure).

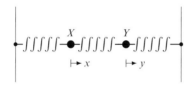

At time $t = 0$, with the system at rest, X is given an impulse mV towards Y. At what time will the (first) maximum displacement of X occur?

[This question needs slightly more advanced mathematics than usual, and if you need help with solving the differential equations involved, consult the hint.]

Q 99 A certain type of linear molecule can be modelled as three atoms, A, B and C, with masses m, M and m respectively, connected by two light springs of spring constant k, as shown in the figure.

Such a system can vibrate along the line of centres in two ways (*modes*); in each mode all moving atoms oscillate with a common frequency, whilst the overall centre of gravity of the molecule remains stationary. In mode 1, B does not move; in mode 2, A and C move equally in the same direction. Find the angular frequencies w_1 and w_2 of the two modes.

For a fixed value of M, make an annotated sketch of the variations of w_1^2 and w_2^2 with m.

Q 100* A turntable for 'vinyl' records, with a variable rotation speed Ω, sits on a fixed horizontal table top and a pendulum clock has been placed on top of the turntable. The pendulum can swing only in one particular vertical plane, fixed with respect to the clock case, and at rest is directly above the centre C of the turntable. With the turntable stationary, the angular frequency of small oscillations of the pendulum is w_0. This frequency changes as Ω changes and beyond a certain frequency, Ω_0, the pendulum ceases to swing.

In three separate sketches, show the paths of the pendulum bob, *relative to the fixed table top*, when Ω takes the values 0, w_0, and $\Omega_0/\sqrt{2}$, justifying your sketches with quantitative analysis as far as possible.

Q 101*† A framework consisting of four freely-jointed light rods, AB, BC, CD and DA, is suspended from A in a vertical plane and supports a mass m attached at C. The rods are all of length a, and a light spring of natural length $\sqrt{2}a$ joins B and D; in the equilibrium configuration it has length a. Find the period of small vertical oscillations of the mass.

Q 102* A particle of mass m moves in a potential well $V(x)$. Show, by means of a sketch-graph and notes, how you would expect the frequency f of oscillations to depend upon their amplitude A for the following forms of $V(x)$:

(a) $V(x) = \frac{1}{2}k\dfrac{x^2}{4 + x^2}$.

(b) $V(x) = k(1 - \cos x)$.

(c) $V(x) = k(\cosh x - 1)$.

(d) $V(x) = k(\cosh x - \cos x)$.

(e) $V(x) = \frac{1}{2}k(x^2 - 2ax)$.

(f) $V(x) = \frac{1}{2}k(x^2 - x^3)$.

Chapter 7

Waves
We're Not Phased!

In this chapter, and perhaps elsewhere, the following two trigonometric identities may prove useful; the second is explicitly proved in solution S 118 on page 236.

$$\cos 2\alpha + \cos 2\beta = 2\cos(\alpha + \beta)\cos(\alpha - \beta),$$
$$\cos 2\alpha - \cos 2\beta = 2\sin(\alpha + \beta)\sin(\beta - \alpha).$$

Q 103 A plank of wood about one metre long and a small cork are floating on a pond across which a wave-train is travelling. The ends of the plank, whose length happens to be parallel to the direction of the wave-train, rise and fall as the waves pass, and the cork bobs up and down once every 2.5 s.

The speed v of water-surface waves of wavelength λ is given by

$$v^2 = \frac{g\lambda}{2\pi} + \frac{2\pi\gamma}{\lambda\rho},$$

where ρ is the density of water and γ is its surface tension, equal to 7.0×10^{-2} N m^{-1}.

What is the wavelength of the waves?

Q 104 Twins Peter and Paula were given one 'Young Scientist's DIY Optical Kit' as a joint birthday present, rather than the one each they had hoped for. They were further disappointed to find it contained only three lenses, two convergent with focal lengths of 30 and 40 mm, and one divergent of focal length 45 mm. There were also a number of 150 mm tubes into which the lenses fitted, and that could be screwed together if necessary.

Peter, a would-be plant biologist, wanted to make a 150 mm long compound microscope to examine seed structure, whilst Paula, a would-be astronaut, had hoped to build an astronomical telescope to study the Moon. After some experimenting, they both found ways to construct what they wanted, though not at the same time. With viewing in normal adjustment (the eye focussed at infinity), the magnifications they achieved were equal. What is Peter's least distance of distinct vision?

Q 105 The radiation from two transmitters 3.00 m apart, and emitting equal amplitudes of 30 mm microwaves in a horizontal direction normal to the line joining them, is detected 100 m away. Due to a small difference in their construction, the left-hand transmitter has a frequency that is 1 Hz greater than that of the right-hand one. What is the nature and behaviour of the observed interference pattern?

Q 106 Although line-of-sight Doppler effects are quite often mentioned in qualitative descriptions of emergency-vehicle sirens, and used in simplified quantitative form in areas such as astrophysics, proof of more general results is largely absent from school physics syllabuses. The following problem aims to develop an understanding sufficient to make quantitative investigations possible.

It simulates a radar speed gun, but using sound rather than electromagnetic waves. In fact, the sound analogy is somewhat the more challenging of the two, as car (and tennis ball) speeds are not negligible relative to v, the speed of sound, but are when compared to c.

Obtain formulae connecting the frequency f_r recorded by the observer to that of the source f_s:

(a) When the source is fixed but the observer is moving towards it with speed u; consider the detected frequency directly.
(b) When the source is moving towards the fixed observer with speed u; consider the detected wavelength before converting it into a frequency.

Then use your formulae to answer the following question.

(c) A car travels towards a stationary source of sound waves of frequency 1.00 kHz. Waves reflected from the car return to the source and are used to produce beats with the original waves, the beats having a frequency of 121 Hz. What is the speed u of the car? Take the speed of sound in air as $v = 350$ m s^{-1}.

Q 107 One of the strings of a concert piano is 1.5 m long and is made of steel, which has a density of 7.7×10^3 kg m^{-3} and a Young modulus of 2.2×10^{11} N m^{-2}; after tuning, it has an internal elastic strain of 1%. One of the strings of the viola player's instrument is 40 cm in length and tuned to A (440 Hz). Finding the piano lid open, she attempts to make the piano string vibrate by bowing the A string on her viola. By how much should she shorten it (by fingering) so as to have the best chance of success?

Q 108[†]   A generalized form of Snell's law ($n_1 \sin\theta_1 = n_2 \sin\theta_2$) that accommodates continuously variable media is as follows:

> 'For a light ray in an x-y plane in which the refractive index n depends only on y, $n\sin\phi$ is constant along the ray, where ϕ is the angle between the tangent to the ray and the y-axis.'

A man, whose eyes are 1.7 m above the ground, stands on a long flat concrete runway above which the temperature of the air falls uniformly with height, producing a variation in its refractive index $n(x, y) = n_0(1 + \alpha y)$ for all x, with $|\alpha| = 1.5 \times 10^{-6}$ m^{-1}. As a result, he cannot see the runway beyond a certain point D, a distance d away.

(a) If the equation of the light ray from D which enters his eye is $y = y(x)$, find the differential equation satisfied by $y(x)$.

(b) The mathematics needed for the full solution of this equation may not have been studied in a physics course, but, assuming that $|\alpha x| < |\alpha d| \ll 1$, try a solution of the form $y = \beta x^2$, and so find an approximate value for d.

Q 109   A radio receiver R receives, simultaneously, two signals from a transmitter T that is broadcasting on a frequency of 100 MHz from 500 km away. One signal follows a path along the (flat) surface of the Earth, whilst the other is the result of reflection from a portion of the Heaviside layer, which is a perfect horizontal reflector situated about 200 km above the Earth's surface. The combined signal strength varies from maximum to minimum and then back to maximum eight times per minute. With what vertical speed is the Heaviside layer moving?

Q 110[†]   The Doppler red shift of the light from a star or galaxy receding with speed v, relative to the observer, is given non-relativistically by $z = v/c$. But this form is only valid for $v \ll c$, and, according to special relativity, the correct form is

$$z = \sqrt{\frac{1 + \beta}{1 - \beta}} - 1,$$

where $\beta = v/c$.

(a) Show that the two forms agree when $v \ll c$, and, either by experiment with a calculator, or (preferably) by algebraic methods, determine which of the two forms gives the greater value for z over the range $0 \le v < c$.

(b) How far away must a galaxy be for the difference between the relativistic and non-relativistic expressions for z to equal 10% of the latter?

Q 111   In the horizontal x-y plane $z = 0$, the line $x = 0$ is interrupted by its intersections with four (long) vertical slits situated at $(0, \pm a)$ and $(0, \pm 2a)$. The slits are identical except that the inner pair are covered by thin films that reverse the phase of the parallel light of wavelength λ, that is incident upon them from the region $x < 0$. The observation screen is the plane $x = L$, where $L \gg a$.

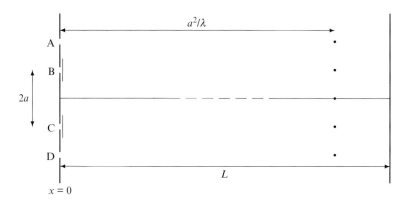

Construct an equation, valid for $L \gg a$, that gives the relative phases of the contributions of the four sources to the screen image at (L, Y), and use it to answer the following questions.

(a) What is the general formula for the values of L at which the light intensity shows a local maximum at $(L, 0)$?
(b) When $L = a^2/\lambda$, which of the following are positions of local maximum intensity: $(L, 0)$, $(L, \pm a)$ and $(L, \pm 2a)$?
(c) What is the ratio of light intensities at $(2a^2/\lambda, 0)$ and $(3a^2/\lambda, 0)$?

Q 112*   A man with a pole stands in front of one of the vertical rods that form a long iron railing, and at a distance D from it. The rods are spaced a distance a apart, and effectively stretch to infinity in both directions (see the figure).

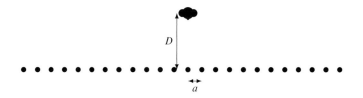

He bangs the pole on the concrete ground, and listens for the echoes from the railing. Ignoring the attenuation of sound as it travels through air, do the following:

(a) Without doing any calculations, make an annotated sketch of the echo frequency $f(t)$ you expect him to hear as a function of time t, taking $t = 0$ to be the moment he bangs the pole.
(b) Find a formula for that frequency variation, identify its main features, and compare them with your predictions.

Q 113 When a light spectrum is examined using a diffraction grating at normal incidence, the following very close groupings of lines are observed to occur:

(i) A red-blue doublet at an angle of 24.5°.
(ii) A red-green-violet triplet at 38.4°.
(iii) A blue-yellow doublet at 43.6°.

The NASA guide to the visible spectrum gives the following approximate equivalences between 'color' and wavelength:

Color	Violet	Blue	Green	Yellow	Orange	Red
Wavelength (nm)	400	475	510	570	590	650

The green line is known from other data to correspond to a wavelength of 518 nm. What are the wavelengths of the other detected colours?

Q 114*† When describing the use of ultrasound for medical scanning of the human body, the following formula is often quoted:

$$\frac{I_r}{I_i} = \left(\frac{Z_2 - Z_1}{Z_2 + Z_1} \right)^2 .$$

Identify the physical meaning of the various symbols and then consider scanning a region that contains layers of three different materials, characterised by Z_1, Z_2 and Z_3, in that order. Taking Z_1 and Z_3 as fixed, and ignoring differences in attenuation factors, find (a) the relationship between these quantities for the ultrasound intensity in the third layer to be the maximum possible, and (b) an expression for that maximal intensity relative to that in the first layer.

The following table gives some data for five different types of human body material. Determine the order and type of materials that come closest to replicating the situation analysed in the previous paragraph. Do not worry about whether such a combination can be found in practice!

Material	Blood	Fat	Lung	Muscle	Water
Density $(kg\ m^{-3})$	1060	920	400	1070	1000
Speed of sound $(m\ s^{-1})$	1566	1446	650	1598	1480

Q 115   The screen in a Young double-slit experiment is replaced by a centrally-located vernier microscope, which is then used to view the transmitted light. When a thin converging lens of focal length 210 mm is moved along the central line joining the slits and the microscope, it is found that there are two positions for which sharp images of the slits are formed on the microscope's cross-wires; the corresponding observed slit separations are 0.490 mm and 0.250 mm.

With the lens removed, a horizontal traverse of 30.1 mm by the microscope resulted in 22 complete fringes being counted. What colour is the light?

Q 116*   Light emitted by a hot vapour is analysed using a normal-incidence diffraction grating with 5×10^5 lines m^{-1}. In the line spectrum so formed, lines *A–E* are found at diffraction angles

$$\theta_A = 8.34°,\ \theta_B = 11.16°,\ \theta_C = 11.48°,\ \theta_D = 18.48°,\ \theta_E = 35.49°.$$

There is also some evidence for a further line *F* at an angle in the range 24°–25°.

(a) Determine the smallest set of atomic energy levels that will account for the established lines.
(b) Show that line *F* is consistent with this set, and determine a more accurate value for θ_F.
(c) Determine the angle θ_G, in the range 10°–12°, at which a further line should be found.

Q 117   A telephoto lens combination is used to gain a magnified image of a distant object without unduly increasing the physical length of a camera; starting from the standard (thin) lens formula, $u^{-1} + v^{-1} = f^{-1}$, its principles can be illustrated as follows.

(a) Show that the size of the image of a distant object formed by a simple converging lens is proportional to the focal length of the lens.

(b) Draw a ray diagram showing how a telephoto lens, which consists of a converging objective lens and a diverging lens a distance d behind it, can be used to produce an effective focal length f' that is greater than the length L of the camera.

(c) For converging and diverging lenses with focal lengths of $f_1 = 76$ mm and $f_2 = 25$ mm respectively, and a 60 mm gap between them, determine from the geometric properties of your diagram the effective focal length of the combination, the length of the camera, and the gain in magnification as compared to a simple converging lens in the same camera body.

(d) Check your answer against that obtained by 'automated substitution' in the standard formula

$$\frac{1}{f'} = \frac{1}{f_1} + \frac{1}{f_2} - \frac{d}{f_1 f_2}.$$

Q 118*† As a prototype for a device to cancel unwanted sound from a noisy piece of machinery, two small identical point sources, a distance d apart, produce spherical waves at a frequency of 100 Hz, but with a phase difference of π between them. Far away, in a direction making an angle $\theta = 45°$ with the normal to the line joining the sources, the sound pressure amplitude is one-tenth of that recorded when only one of the sources is operating.

Given that the speed of sound is 340 m s^{-1}, find the range of sound pressure amplitudes experienced in far-away positions as a function of θ.

Q 119 A long row of identical radio transmitters, spaced a distance d apart along a line that runs North–South, are operating at a wavelength λ. The signal from each aerial is in phase with that from the next-but-one transmitter, but the two groups can be out of phase with each other by an adjustable phase angle ϕ. Show that the strongest possible signals, in some particular direction(s), are only obtainable if ϕ is a multiple of π.

If ϕ is set at π, determine the relationship between λ and d if there are to be *exactly* six directions in which a strong signal is broadcast and identify those directions.

Q 120*† This question, essentially a 'reverse engineering' investigation of a travelling wave, involves a significant amount of manipulation of trigonometric quantities, and it is suggested that four or five figure accuracy be maintained throughout to avoid serious (possibly misleading) rounding errors. Some parameters of the problem have been chosen so as to make the solution path easier to discover.

The transverse displacement (in cm) of a taut string carrying a sinusoidal wave is measured at points x along its length that are close together as compared to the wavelength λ of the wave, and at times t that are separated by intervals that are small compared to its period T. The results (which are given here to an unrealistic accuracy) are as follows.

t/s	x/m	6	7	8
3		a	13.508	b
4		11.340	9.059	6.687
5		c	4.249	d

Determine the values of a, b, c, and d, the speed and direction of the wave, and its amplitude.

Q 121 When light from an unpolarized source is incident on a sheet of polaroid, it is found that the emergent light has an intensity that is 32% of the incident one. A second identical sheet is used to block completely the light transmitted by the first one. A third identical sheet is now placed between the two, and initially orientated so that it transmits no light at all. It is next rotated in its own plane through an angle θ. What is the intensity of any light that is transmitted through the whole system?

Chapter 8

Solids and Liquids All Together Now!

Q 122   A thin ring of radius R is made of a material that has density ρ and Young modulus E. It is rotated about its centre, and in its own plane, with angular velocity ω. By how much, if at all, does its radius change?

Q 123   Thin-walled flexible garden hose is often strengthened by two helical windings (one clockwise, the other anticlockwise) of nylon string embedded in its wall. If string of diameter 0.5 mm with a breaking stress of $1.0 \times 10^8 \, \mathrm{N\,m^{-2}}$ is used to strengthen a hose of radius 10 mm, and designed to withstand 4 atmospheres of excess water pressure, estimate how much nylon string will be needed for each metre of hose? Verify, in retrospect, that any approximations that you have made in arriving at your estimate are justified.

Q 124   A washing machine has an automatic rinsing cycle in which clean water is admitted at a rate F that can be controlled manually. When the machine contains a volume V of water, a pump is switched on. Although water continues to enter, the pump removes it at a rate $P \, (> F)$ until there is no water left, at which point the inlet valve closes, the pump switches off, and the cycle is complete.

How, after observing a single cycle, can you determine whether to increase or decrease F so as to reduce the time for the next rinse cycle? What is the best (i.e. shortest) cycle time that can be achieved?

Q 125   A straight rainwater channel that runs alongside a high solid brick wall, has a rectangular cross-section of depth c and width d. The normal height of water in the channel is c or less, but, when exceptionally heavy rains are expected, a triangular concrete storm barrier is installed on one edge, as shown in the figure.

The height b of the barrier is chosen to be just greater than the maximum expected flood depth; naturally, its base length a is decided by reference to

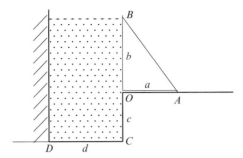

page 473 of the Health & Safety Manual (Waterways). The latter states that the barrier must be over-engineered so as to remove any chance of (a) it being pushed sideways along the ground, and (b) it being tipped over about point A.

If the density of concrete ρ is 2,400 kg m^{-3}, and the coefficient of friction between concrete and the ground is $\mu = 0.7$, what is the minimum allowed value for a?

Q 126   The compression of a certain material is proportional to the applied pressure up to its yield point, at which the compression is 8% and the pressure is 10^7 N m^{-2}; thereafter the compression continues indefinitely, without any change in the pressure. If the pressure is subsequently reduced, the compression, as a function of pressure, falls at the same rate as it rose.

A mass of 16 kg, with a light block of this material mounted in front of it as a buffer, collides with a rigidly-mounted similar block; each block is 20 mm thick and has an area of 0.2 m^2. Determine how the rebound velocity v_r of the mass varies with its impact velocity v_i.

Q 127   A 'test cylinder' for measuring the elastic properties of threaded rods, consists of a hollow cylinder of inner and outer radius $\sqrt{2}r$ and $2r$, two rigid washers, two nuts, and the test rod, which has radius r. Some heat insulating material is also incorporated, but this can be ignored in the analysis. The assembly is shown in cross-section in the figure. The test rod is stressed by heating the outer cylinder until the latter's expansion causes the rod to reach its yield stress.

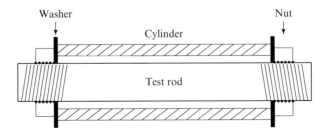

To calibrate the device, a rod of known yield stress σ_y, matching that of the cylinder, is used. Both are perfectly elastic up to that stress, but the Young's modulus is E for the cylinder, whilst for the rod it is $2E$.

Initially the nuts are tightened and there is a stress of $\frac{1}{2}\sigma_y$ in the rod. The cylinder (but not the rod) is then heated and when the cylinder's temperature has been raised by T the rod fails and starts to extend plastically. What is the linear coefficient of expansion α of the cylinder material?

Q 128 The density of a low viscosity liquid increases linearly with depth, being ρ_0 at its surface and $2\rho_0$ at depth D. A small rigid sphere of density $3\rho_0$ is released from rest at a depth of $\frac{1}{2}D$. Assuming that there are no fluid pressure changes as a result of the sphere's motion, describe quantitatively what subsequently happens, and establish the greatest depth to which the sphere sinks.

Q 129* A man in a boat on a pond of area A has with him blocks of stone and timber, the former more dense, the latter less dense, than water. What happens to the level of the water in the pond if he throws overboard (a) the stone, (b) the timber, and (c) the two tied together, which then sink to the bottom of the pond?

Q 130 The velocity v of longitudinal (compressional) waves in a material of density ρ and Young modulus E is $(E/\rho)^{1/2}$. A vertical rod, fastened at its upper end, is extended elastically by 0.4 mm when a mass one thousand times its own is attached to its lower end. If the rod is now clamped horizontally at its mid-point, what is the lowest possible frequency of longitudinal standing waves within it?

Q 131*† This question requires the accurate manipulation of fractional powers of both symbols and numbers, and an elementary application of a Taylor expansion.

A long-chain molecule is composed of identical atoms evenly spaced. The potential energy of interaction between nearest neighbours is given by

$$V(x) = -\frac{A}{x^6} + \frac{B}{x^{12}},$$

where x is the distance between the two atoms, and A and B are positive quantities; other interactions are negligible. Calculate the modulus of elasticity and breaking strain of the molecule.

Q 132 For a change, this question is about cables that are *not* 'light'.

(a) A helicopter capable of lifting a maximum load of weight W is converted to a rescue helicopter by the addition of a winch cable of length ℓ, density ρ and breaking stress σ_0. What is the maximum mass M that can be supported on the end of the fully extended cable?

(b) A fixed volume V of a material that has breaking stress σ_0 and Young modulus E is to be made into a cable freely suspended from one end. What is the maximum elastic energy that can be stored in it?

Q 133* A rectangular framework has, as two of its opposite sides, metal rods, each 20 mm long, rigidly held 200 mm apart. The other sides are light rubber threads, straight but initially unstretched, that obey Hooke's law with force constants 15 N m^{-1}. When a soap film is formed in the framework, it is found that the threads almost touch at their centres. How large is γ, the surface tension of the film?

Q 134 Steady heat flow in one dimension has many similarities with the flow of d.c. electricity through resistors and capacitors. The heat flux (energy flow per unit area per unit time) through a uniform medium is equal to $-k \times$ the temperature gradient, where k is the thermal conductivity of the material, and the minus sign indicates that the flow is from a higher temperature to a lower one. The heat transfer coefficient H of a complete structure, which may be of composite construction, is the ratio of the heat flux through it to the temperature difference between its two extremities.

Find the heat transfer coefficient H_2 for a structure consisting of two slabs of equal thickness but differing thermal conductivities, k_1 and k_2, that are placed in contact; give your answer as an expression for $1/H_2$. Using an appropriate analogy with either resistors or capacitors, write down an expression for H_3, the heat transfer coefficient for a structure consisting of equal-thickness slabs of three different materials that have thermal conductivities k_1, $\frac{1}{2}(k_1 + k_2)$ and k_2.

By substituting trial values, or (very much to be preferred) by proving it algebraically, determine which of the two composite slabs provides the better heat insulation, if the overall thickness of the structure is predetermined.

Q 135* A lock, that separates a saltwater harbour from a freshwater canal, consists of two pairs of hinged full-height gates which, when closed, form V-shaped barriers, with the 'V's pointing towards the harbour (quasi-symbolically, H $<$ L $<$ C). The two water heights, measured from the horizontal canal bed are s in the harbour and f ($< s$) in the canal, with corresponding densities of ρ_1 and ρ_0.

After a boat has passed through the lock, from the harbour into the canal, and both pairs of gates are closed, a small one-way valve in a by-pass pipe connecting the harbour to the lock is opened. Some time later the harbour pair of gates open of their own accord. Explain what has happened, and deduce what you can about the height h of the valve above the canal bed.

Q 136   Two identical rods of length L, have cross-sectional area A, Young modulus E_1 and (positive) coefficient of expansion α_1. A third rod also has length L, but its cross-sectional area, Young modulus and (positive) coefficient of expansion are $3A$, E_3 and α_3, respectively. The three are joined end-to-end with the thicker rod sandwiched between the other two. The composite rod so formed is rigidly clamped at its two ends in a horizontal position, as shown in the figure.

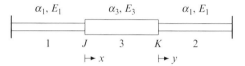

The set-up is then subjected to a small temperature rise of T. How far, x and y, from their original positions do the junctions J and K move? What is the stress in (a) one of the end rods, and (b) the central rod?

Q 137[*†]   If a (so-called Newtonian) viscous fluid is in motion, and its speed varies with position, then a force parallel to the direction of flow is needed to maintain that motion. The coefficient of viscosity η is defined as the ratio of the required shearing stress to the (local) velocity gradient in the moving fluid; it is in several respects a dynamic analogue of the Young modulus E which gives the ratio of the required longitudinal stress to the displacement gradient (strain) in a static solid.

Oil of viscosity η and density ρ flows steadily downhill in a flat shallow channel of width W that is sloped at an angle θ to the horizontal. Ignoring viscous effects at the side walls, find the volume flow rate V if the oil is everywhere of depth D (relative to the base of the channel). Use axes that are parallel and perpendicular to the channel base, and assume that the flow velocity immediately above the latter is zero.

Q 138   A rotating blade in an aero engine may be modelled by a uniform elastic rod rotating about one end with a constant angular velocity. The blade, of unstressed length 90 cm, is made from a material of density 4510 kg m^{-3} and Young modulus $E = 117$ GPa, and, for safety reasons, the maximum tensile stress in it must not exceed 400 MPa. Determine the maximum allowed r.p.m. for the engine, and the corresponding lengthening of the blade.

Chapter 9

Electrical Circuits
Go With the Flow!

Q 139 Two physics students, A and B, living in neighbouring college rooms, decided to economize by connecting their 100 W ceiling lights in series and paying equal shares of the electricity bill. However, both secretly thought that they could get better lighting at the other's expense; A installed a 200 W bulb and B installed a 50 W bulb.

Which of the two was the brighter student, and by what factor, in terms of cost per unit of lighting received? How did what actually happened measure up against their probable expectations? Assume (unrealistically) that the bulb resistances are independent of the power level at which they are working.

Q 140 A battery of open circuit emf V and internal resistance r is to deliver power to a variable external resistor R. Show how R should be chosen to maximize that power.

The battery terminals are now joined by a fixed resistance S. How should the variable resistor be used so as to maximize the total power dissipated in the external resistances? You should consider both $S > r$ and $S < r$ situations, and calculate the power dissipated in S in each case.

Q 141 Points A and B are joined by an underground twin-core cable, which is uniform and 7 km long. Some way along it, a fault in the insulation makes a resistive connection between its two cores. With the other end open-circuited, the cable's resistance when measured at A is 64 Ω; when measured at B it is 70 Ω. If 16 V is applied across the terminals at A, the reading of a high impedance voltmeter connected across those at B is 15 V. What percentage of power supplied at A would be dissipated in a 50 Ω load connected at B?

Q 142 A conducting balloon, of radius a, and charged to a potential V, is connected to earth through a resistance R at the same time as a valve is opened to deflate it. Show that it is possible, with a particular time dependence for the decrease of the balloon's radius, to maintain the potential at V, and find the heat then dissipated in the resistance.

Q 143 In the circuit below, how much energy is stored in the whole capacitor system?

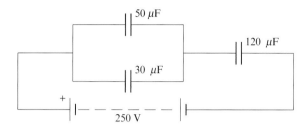

And how much in the 50 μF capacitor?

Q 144 In school physics courses, electrical resistors are nearly always assumed to have a fixed resistance (defined as voltage/current), whatever their roles in a circuit. However, in practice, this is seldom the case, and the resistance of most devices increases if the current (and hence the heating) in them increases; for example, a 100 W light bulb must have a working resistance of about $V^2/P = (240)^2/100 \approx 580\,\Omega$, but when measured cold it is only about one-tenth of that.

Two non-linear devices A and B are connected in parallel and a third such device C is connected in series with the combination. A voltage source V is then connected across the whole system as shown in the figure.

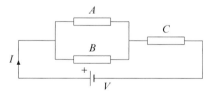

The resistance of device A, still defined as voltage/current, is given by aI_A, where I_A is the current through A; the resistances of B and C are similarly defined. Find the power P drawn from the source.

By taking suitable limiting values for some or all of a, b and c, and comparing the results with $P = V^2/R$, establish the formulae for finding the equivalent resistances of series and parallel combinations of two such devices.

Q 145 What is the largest current that can be carried by a copper fuse wire of radius $r = 1$ mm? Ignore heat conducted away by the supply circuit. Some of the following data on copper may be useful: atomic mass, 63.5 u; melting point,

1084 °C; density, 8.96×10^3 kg m^{-3}; thermal conductivity, 400 W m^{-1} K^{-1}; resistivity 1.68×10^{-8} Ω m.

Q 146   The figure shows a nominal 2.5 mF capacitor under test.

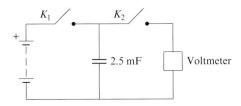

The following observations were made:

(a) With both switches closed, the voltmeter reading was 16.0 V.
(b) K_2 was opened, then K_1 was opened for 30 s, after which K_2 was closed and an immediate reading of the voltmeter showed 14.1 V.
(c) Using other apparatus, it was found that it takes a steady current of 120 μA to keep the capacitor charged to 15 V.

Explain why these observations could be self-consistent, and, on the assumption that they are, determine values that should be used when designing circuits that incorporate the capacitor.

Q 147   The right-hand figure below shows the voltage-current characteristic of the element Z in the circuit on the left.

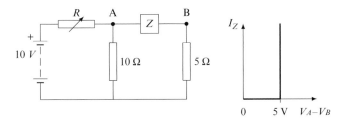

Make an annotated plot of the current drawn from the (ideal) battery as a function of R, for $0 \le R \le 20\,\Omega$.

Q 148†   This problem involves the multiplication of 2×2 matrices. It can be solved without them, but is likely to become tedious and error-prone.

The figure below shows a network of three equal resistors and defines the voltages and currents at the two terminal pairs 1 and 2.
If we define a 'column vector' $\mathbf{x}_n$ as

$$\mathbf{x}_n = \begin{bmatrix} V_n \\ I_n \end{bmatrix},$$

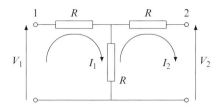

then the connection between quantities measured at the pairs of terminals 1 and 2 can be written in the matrix form

$$\begin{bmatrix} V_1 \\ I_1 \end{bmatrix} = \begin{bmatrix} a & b \\ c & d \end{bmatrix} \begin{bmatrix} V_2 \\ I_2 \end{bmatrix},$$

or, more compactly, $\mathbf{x}_1 = M\mathbf{x}_2$, where M is a 2×2 matrix.

(a) Determine the values of a, b, c and d.
(b) Three such networks are joined together in cascade (terminals 1 of the second unit being connected to terminals 2 of the first, etc.) and terminals 2 of the third unit are then joined by a further resistance R. What would be the resistance measured across terminals 1 of the first unit?

Q 149 By considering, on the basis of the model described below, the power loss in a wire with length L and cross-sectional area A, estimate the resistivity ρ of copper, which has a relative atomic mass of 63.5 and a density of 8.9×10^3 kg m^{-3}.

One model of ohmic loss is that the free electrons undergo collisions with the atoms of the lattice, as a result of which they lose all their kinetic energy, which then appears as heat. They are then uniformly accelerated from rest until they suffer their next collision, which, on average, occurs after $\tau = 4 \times 10^{-14}$ s. Assume that each copper atom contributes one free electron to the drift current.

Q 150[*†] This question introduces a topic that does not normally appear in school syllabuses, but as being able to assimilate and apply new ideas is a valuable skill, it is included here.

An emf is generated if the flux linkage in a circuit changes, with the magnitude of the emf proportional to the rate of change. The converse of this is that if the flux linkage is *made* to change, an emf is required to make it do so.

One way to change the flux linking a coil is to alter the current that produces it, and so, even in a coil with negligible resistance, a p.d. across the coil is needed to counteract the emf generated by the changing flux linkage. Quantitatively, the voltage drop required (V_L) is given by

$$V_L = L \frac{di}{dt},$$

where t denotes time and L, whose value depends upon the physical construction of the coil, is known as its inductance (measured in Henrys).

(a) Show that inductances combine in the same way as resistances when in
   series or parallel combinations.
(b) For the circuit shown in the figure, assume that the charge on the capacitor is given by $Q = Q_0 \cos \omega t$, where Q_0 may depend upon ω, and by
   deriving the potential differences around the circuit from it, show that
   there is current resonance when $\omega = \omega_0 = 1/\sqrt{LC}$.

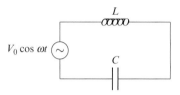

(c) Using at most two equal inductances L and at most two equal capacitances C, draw circuit diagrams for four arrangements that will exhibit
   current resonance at frequencies other than ω_0, marking each with its
   actual resonance frequency.

Q 151 This problem contains no difficult physics, but does require an orderly
approach, an attribute aspiring science students should aim to acquire.

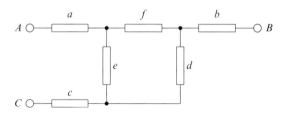

Six resistors with resistances $1\,\Omega$, $2\,\Omega$,..., $6\,\Omega$ are connected as shown in the
figure. The resistances measured between the three pairs of terminals are: AB,
$7\frac{3}{13}\,\Omega$; AC, $6\frac{9}{13}\,\Omega$; BC, $10\frac{1}{13}\,\Omega$. Which resistor is which?

Q 152*† A diode, which can be assumed to have negligible resistance when
conducting in the forward direction, is used in the circuit shown to rectify a
20 V peak-to-peak a.c supply of frequency $f = 50$ Hz.

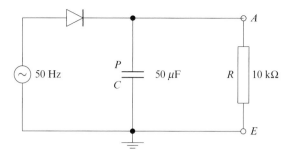

Explain in general terms the time dependence of the voltage measured across AE, make a rough sketch of it, and estimate its peak-to-peak ripple. Obtain equations whose solution gives the interval δ in each cycle during which the diode is conducting. Verify that $\delta \approx 0.045f^{-1}$ satisfies them, either by direct substitution or, as a mathematical challenge, by assuming that $\delta f \ll 1$ and making some appropriate function expansions.

Q 153

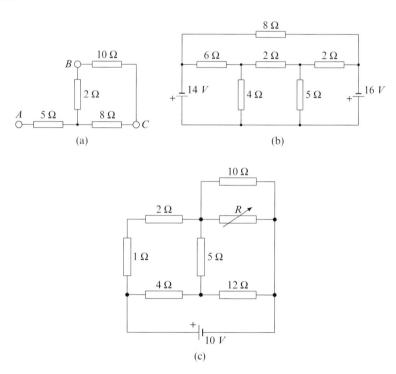

(a) For the network shown in figure (a), determine the resistance between the following pairs of terminals (i) A and B, (ii) B and C, (iii) B and C when terminals A and C are shorted.

(b) Find the current in the $4\,\Omega$ resistor in network (b).

(c) At what value should the variable resistor R in network (c) be set so as to minimize the heat generated in the $5\,\Omega$ resistor?

Q 154*

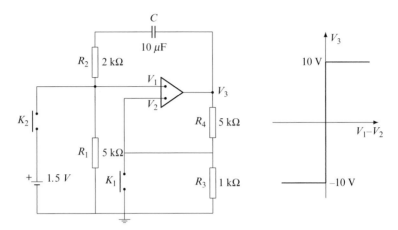

The left hand figure above shows a circuit incorporating a powered earthed (differential) amplifier whose output V_3 is either $+10$ V or -10 V, depending upon which of its two inputs, V_1 and V_2, is at a higher potential (as shown in the right-hand figure). Both inputs take negligible current.

Initially the capacitor C is uncharged and key K_1 is closed. Key K_2 is momentarily closed and then left open. Several seconds later K_1 is opened.

(a) What is the time variation of the current taken from the output of the amplifier before K_1 is opened?

(b) Show that after K_1 is opened the system becomes an oscillator and, without a detailed mathematical analysis, make an annotated sketch of the time variation of V_1.

Q 155   The time-base circuitry for an oscilloscope has to generate a periodic voltage signal that increases from zero at a uniform rate for the (time) period $0 \le t \le T$. One possible circuit is shown in the figure below. It consists of a resistance R and a capacitance C connected in series with a low impedance source of emf V_0. Across the capacitor is connected a device D that has the property that, if initially non-conducting, it remains so until the voltage across it rises to V_1 ($< V_0$), at which point it very rapidly discharges the capacitor, and then returns to the non-conducting state.

Sketch the voltage waveform V such a circuit will produce and establish the relationship between T and the circuit parameters.

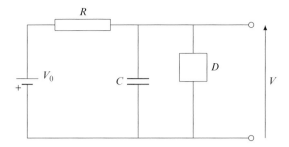

If the time-base is to trigger 100 times per second and be linear (in time) to within 1%, what conditions must be imposed on those parameters?

Chapter 10

Electrostatics
Reach Your Potential!

Q 156   A battery consists of N identical cells, each of emf $\mathcal{E}$, connected in series. A capacitor of capacitance C is charged through a resistor R either (a) by connecting it across the entire battery, or (b) by connecting it first across a single cell, then across two cells, and so on, until it is again fully charged. Is there any difference between the two methods with regard to the energy wasted in the resistor?

Q 157   (a) An isolated spherical conductor (in vacuo) carries a uniform surface charge density σ. By considering the change in stored energy produced by a small change in its radius, the total surface charge being held constant, find an expression for the outward pressure p exerted on the sphere.

(b) A thin-walled metal sphere of radius a and wall thickness t contains a gas at excess internal pressure p. By considering the tensile forces acting on a (small) part of its surface, relate p to the tensile stress T in the metal.

(c) A sphere of radius 0.5 m and wall thickness 1.0 μm is connected to a voltage source that is steadily increased. Using the results from (a) and (b), find the minimum breaking stress the metal must have if the sphere is not to rupture before it discharges by ionizing the surrounding air, which has a breakdown field intensity of $E_0 = 3.0$ MV m^{-1}.

Q 158   (a) An air-filled capacitor consists of three parallel metal plates each of area 1.0×10^{-2} m^2, with the spacings between the inner plate and the two outer ones being 1.0 mm and 2.0 mm. Initially the outer plates are both earthed and the central one is charged to 2000 V. All three plates are now insulated, and this is the situation shown in the left-hand figure.

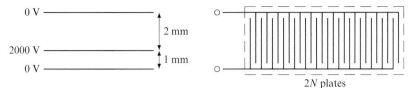

2N plates

The central plate is now removed. What is the final potential difference between the two remaining plates?

(b) The right-hand figure shows $2N$ identical thin conducting plates arranged parallel to one of the faces of a rectangular box of volume v. Alternate plates are electrically connected together and the spaces between the plates are air-filled. If the breakdown electric field strength for air is E_0, find the maximum electrostatic energy that can be stored in the box.

Q 159 A copper ribbon, of width $2a = 2.0 \times 10^{-2}$ m in the y-direction and carrying a current in the positive x-direction, is placed in a magnetic field of flux density 0.75 T applied in the positive z-direction. It is found that a voltage of 3.0 μV is generated across the width of the ribbon; this phenomenon is known as the Hall effect, after its discoverer. Deduce the drift velocity of the electrons, assuming that they are the only carriers. Which edge of the ribbon is at a higher potential?

Q 160 In a way roughly analogous to the production of electric fields by static charges, moving charges, i.e. currents, produce magnetic fields. One particular device based on this is a solenoid in which a magnetic field B is produced inside a (long) cylinder by wrapping it with (many) turns of an insulated current-carrying wire; the field is directed parallel to the cylinder's axis.

The strength of the field (measured in teslas) is given by $B = \mu_0 nI$, where I is the current, n the number of turns per unit length of the cylinder, and μ_0 is a universal constant, analogous to ϵ_0, and called the permeability of free space.[1]

The turns of a particular solenoid are wound to fill the space between two concentric cylinders of fixed radii. For a given field strength inside the inner cylinder, determine how the heat dissipated in the winding depends upon the (cross-sectional) radius of the wire used.

Q 161* It can be shown, as a general result, that the electric field close to a conductor carrying a charge density σ is σ/ϵ_0. It is also the case that the capacitance of a pair of concentric spheres with radii a and b is given by $4\pi\epsilon_0 ab/(b-a)$, where b is the radius of the larger sphere.[2]

---

[1] It is related to other fundamental constants through the identity $\mu_0 \epsilon_0 c^2 = 1$.

[2] The capacitance of an isolated sphere, $4\pi\epsilon_0 a$, is a special case of this, when $b \to \infty$.

Assuming both of these results, consider a spherical capacitor in which b is fixed, but a can be chosen. The space between the spheres is filled with air, which has breakdown electric field strength E_0. Which values of a and V, the potential difference between the spheres, should be chosen so as to maximize the energy stored in the capacitor?

Q 162[†]   There have been suggestions that a parallel-plate capacitor might be constructed in which the plate separation d varies with the charge Q on the plate according to the formula $d = d_0 - \alpha Q$, where d_0 and α are constants.

Suppose that such a capacitor carries a total charge made up of a fixed charge Q_0 and a variable charge $Q_1 \cos \omega t$ supplied by an external circuit. Show how to choose Q_0 so as to remove any component of frequency ω from the voltage V across the capacitor. When this has been done, what is the time dependence of V?

Q 163[*]   The figure shows four identical very thin plates A, B, C and D forming two parallel-plate capacitors; their plate separations and charges are as shown, and the left-hand one has a capacitance of 2.0×10^{-11} F.

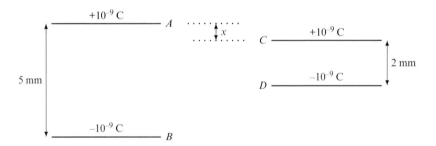

The right-hand pair of plates is now slid between the left-hand ones without touching them, and at a height such that plates A and C are separated by a uniform distance x. What are now the p.d.s V_{AB} and V_{CD}? Was the second capacitor 'sucked in', or did it have to be pushed, or neither? How much external work, if any, was involved in the process?

Q 164   The two parallel conducting plates of a capacitor, each of area A and with separation d, are connected to a source of constant voltage V. The plates are allowed to gradually approach each other until their separation is $\frac{1}{3}d$. The source is then disconnected and the plate separation x is slowly restored to its original value.

(a) What difference, if any, is there between the initial and final stored electrostatic energies?

(b) How does the force needed to separate the plates vary with x?

Q 165*   *Note* This problem is an example of the so-called *Method of Images*, which is normally beyond school physics, but is here adapted so as to be accessible at that level.

(a) Show, by manipulating it, that the following mathematical result is valid.

$$\frac{b}{[(b - a\cos\theta)^2 + (a\sin\theta)^2]^{1/2}} = \frac{a}{[(a\cos\theta - a^2/b)^2 + (a\sin\theta)^2]^{1/2}}.$$

(b) Relate this result to the physical situation shown in the figure, in which there is a charge q at B, and a charge $-qa/b$ at A, where $OB = b$ and $OA = a^2/b$.

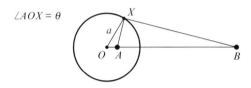

Show, in particular, that the potential on a sphere centred on O and of radius a is *everywhere* the same.

(c) It can be proved that in any given region of space, the forces acting upon any charge located within it are *uniquely* determined by any other charges in that space and the potentials on its boundaries. Use this 'theorem' and the result of (b) to deduce the force F with which an earthed conducting sphere of radius a attracts a point charge q placed a distance b $(> a)$ away from its centre.

Q 166   The electric field (in vacuo) at a radial distance r from a long line charge of uniform linear density σ is $E(r) = \sigma/(2\pi\epsilon_0 r)$. Assuming this result, consider two concentric cylinders with the space between them filled with a material of relative permittivity ϵ_r and electrical resistivity ρ. If the inner cylinder is given a charge, and the outer one earthed, show that the time constant for the subsequent decay is $\epsilon_r\epsilon_0\rho$.

Q 167*   The figure shows the active part of an high voltage variable capacitor, in which cylinder A fits into coaxial hollow cylinder B, and is separated from it by insulating oil of relative permittivity 4.5; the relevant diameters are as indicated.

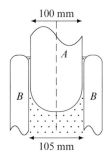

The capacitor is charged to 30 kV, but then isolated from the voltage supply. By considering what happens if cylinder *A* is withdrawn by a small amount Δx, find the magnitude and sense of the axial force between the two cylinders.

Chapter 11

Magnetic Fields
Stay in the Loop!

Q 168   A wire loop in the form of an equilateral triangle of side a carries a current I and is suspended from one of its vertices, with the opposite side lying parallel to a uniform horizontal magnetic field of flux density B. What couple, if any, acts upon the triangle?

Q 169   This question introduces the notion of currents as the origin of magnetic fields, and is not normally included in school physics. It should be treated as an exercise in absorbing and applying new material.

It can be shown that the magnetic field produced at a point P by a long straight wire carrying a current I is azimuthal (i.e. perpendicular to both the wire and the shortest line joining P to the wire). Its strength B depends only on the distance r of P from the wire, and is given by

$$B = \frac{\mu_0 I}{2\pi r},$$

where μ_0 is a universal constant, with an S.I. value of $4\pi \times 10^{-7}$ H m^{-1} (and is analogous to ϵ_0 in electrostatics).[1] The magnetic field lines therefore form closed circles centred on the wire. The sense of the magnetic field is clockwise when viewed along the direction of the current in the wire.

In the arrangement of a (long) straight wire and a wire rectangle shown in the figure, both conductors carry currents, with magnitudes and directions as indicated.

(a) What is the strength and direction of the resultant force on the rectangle?

(b) What force, if any, due to the current in the straight wire is experienced by side AB of the rectangle?

---

[1]   H is the symbol for the Henry, the S.I. unit of inductance.

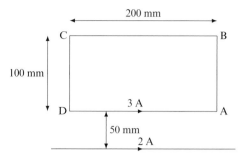

Q 170   Electrons of a particular speed v ($\ll c$) are injected at a given point P into a uniform magnetic field, making a spread of small angles θ with the field's direction. Show that they will all pass close to a specified downstream point Q.

Q 171   A current-carrying ring can be floated above a horizontal supercon-ducting plane without any mechanical support. In the figure, P is such a ring, floating above the superconducting plane $z = 0$.

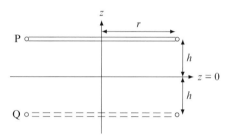

Following a method similar to that of image charges in electrostatics (see Q 165), it can be shown that the magnetic flux density in the region $z \geq 0$, above and on the plane (but not below it), is the same as it would be if the plane were removed and an identical 'image ring' Q, carrying the same current, were placed the same distance below the plane as the real ring is above it.

If the ring radius r is much greater than the ring-plane separation h, the two rings can be treated as if they were long parallel straight wires. Bearing in mind the state of the real ring P, determine the relative sense of the currents in P and Q, the equilibrium height h_0, and the frequency of the ensuing oscillations if the ring is displaced slightly in the vertical direction and then released.

Q 172   A 90 cm long tube of rectangular cross-section, 10 mm $\times$ 15 mm, is sealed at its lower end and contains mercury to a depth h. It is held ver-tically in a horizontal magnetic field of flux density $B = 2.0$ T in the positive y-direction, which is parallel to the 15 mm dimension. The 15 mm $\times$ 90 cm walls are conducting, but the other two walls and the end seal are not. If a current of $I = 1.0 \times 10^3$ A is passed through it in the x-direction, what will happen?

Q 173† A small square loop of side a is suspended with its sides vertical at the centre of a large fixed circular loop of radius R by a thread of torsion constant k; currents I_1 and I_2 flow in the two loops in opposite senses. The angle between the planes of the loops is denoted by θ. Given that the flux density of the field at the centre of a circular loop of radius r that carries a current I is $\mu_0 I/2r$, find k_0, the minimum value of k that is needed to ensure that the position $\theta = 0$ is one of stable equilibrium.

What happens if (i) $k = (2/\pi)k_0$, and (ii) $k = \frac{1}{2}k_0$?

Q 174 *Warning.* This problem requires the careful manipulation of very large and very small numbers; one slip would lead to a ridiculously wrong answer!

When a relativistic charged particle is moving with speed v in a magnetic field, the parameters of its motion are the same as those for a non-relativistic particle, provided the particle's mass m is replaced by its relativistic counterpart $m_0(1 - v^2/c^2)^{-1/2}$, where m_0 is the particle's (fixed) *rest mass*.

A μ^- meson is a particle with the same charge as an electron, but 207 times its rest mass. Such a meson (which has a lifetime of about 2.2 μs) is projected horizontally into a region in which there is a uniform vertical magnetic field of flux density $B = 2.00$ T, and subsequently describes a circular orbit of radius $r = 0.198$ m. How long, τ, does it take to complete one orbit?

Q 175* A uniform conducting wire of length 0.5 m and density 9.0×10^3 kg m^{-3} is supported at its ends under tension, and is perpendicular to a uniform horizontal magnetic flux density of 1.0×10^{-2} T. The slight sag in the wire due to its weight is removed if a voltage $V_0 = 88$ mV is applied across its ends. What is the resistivity of the wire material?

The wire is now formed into a rigid rectangular loop and laid flat on a rough wooden table placed in the magnetic field. The loop's short sides are half the length of the longer ones, and lie parallel to the field. Determine the behaviour of the loop as the voltage V across its ends is slowly increased from 0 to $2V_0$.

Q 176 Do not attempt this problem until you have studied Q 169 and its solution – unless you are already familiar with the description of magnetic fields arising from current-carrying conductors.

Direct current power is supplied at 33 kV to a distant load by means of two parallel cables having a capacitance between them of 1.00×10^{-11} F m^{-1}. Such cables, when separated by a distance d and loaded with a charge of σ per unit length, each experience a force per unit length of $F = \sigma^2/2\pi\epsilon_0 d$. At what (non-zero) electrical load level is there no net force between the two supply cables?

Q 177* A narrow beam of positive particles consists of singly-charged ions of two isotopes of neon, ^{20}Ne and ^{22}Ne with a limited, but non-zero, range of kinetic energies. Initially the beam travels in the y-direction, but moves into a region of length ℓ in which there are uniform electric and magnetic fields, E and B, both being parallel to the z-axis. It then passes through a field-free region of length L before striking a fluorescent screen, whose plane is perpendicular to the y-axis. The set-up is shown in the figure.

Assuming that the deflections caused by the fields are all small, find and sketch the pattern of fluorescence on the screen. On the same sketch indicate how the pattern would change if (i) the direction of the magnetic field were reversed, and (ii) the mean energy of the initial beam were increased significantly.

Chapter 12

Electromagnetism
A State of Flux!

Q 178   A generator consists of a disc of radius 1.2 m and thickness 10 mm made from metal with a resistivity of 1.0×10^{-7} Ω m, and is supported by a vertical axle of radius 20 mm. It is rotated in a vertical uniform flux density of 10 T at a frequency of 5 Hz. Fixed leads are connected to the axle, which has negligible electrical resistance, and uniformly to the rim of the disc. How much current flows in the axle when the two leads are joined together?

Q 179

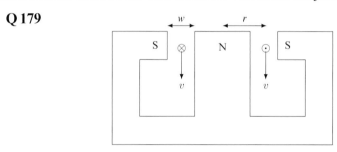

The figure shows a central cross-section of a permanent magnet similar to that used in loudspeakers. It also shows two sections of an aluminium ring that is concentric with the magnet, and is dropped into the air gap between the poles. It is observed that, shortly after it is dropped, the ring moves downwards with uniform velocity v. Ignoring air viscosity, and using a minimum selection from the data below, estimate this velocity.

Radius of the ring, $r = 2.0$ cm; Cross-sectional area of the ring, $A = 9.0 \times 10^{-6}$ m^2; Radius of the central (N) pole, $R = 1.5$ cm; Width of the air gap, $w = 1.0$ cm; Radial magnetic flux density at radius 2 cm, $B = 0.8$ T; Density of aluminium, $\rho = 2.7 \times 10^3$ kg m^{-3}; Resistivity of aluminium, $\rho_e = 2.8 \times 10^{-8}$ Ω m; Specific heat of aluminium, $s = 9.10 \times 10^2$ J kg^{-1} K^{-1}.

Q 180*   A uniform magnetic flux density B in the z-direction occupies the region $x > 0$. A rectangular loop of wire with its sides of length a parallel to the x-axis and those of length b parallel to the y-axis, is moved with constant velocity v through the range $-L < x < L$, where $L \gg a$. Make an annotated sketch of the time dependence of the force required to do this, and show that the total work done is equal to the heat dissipated in the wire.

Q 181* Two meshed brass gear wheels of radii 10 and 50 mm are mounted in a metal frame and placed in, and with their axles parallel to, a uniform horizontal flux density of 0.5 T. A mass of 10 g is supported by a light thread wound round the 2 mm diameter shaft of the larger gear wheel. If the points of contact of the two wheels are electrically separated, the resistance between them (via the wheels, the axles and the metal frame) is 10 mΩ. Ignoring frictional losses, find the terminal speed of the mass as it falls under gravity.

Q 182 A radio set, which behaves as a resistance of 600 Ω, uses as an aerial a large planar coil of negligible resistance, with 100 turns, each of area 0.25 m^2. The coil is orientated so that its normal is parallel to the magnetic field of a 100 kHz radio wave. If the sensitivity of the set is limited by random noise of power 10^{-12} W, determine the minimum magnetic flux density that is detectable.

Q 183* Two long wires are placed on a pair of parallel rails, which are a distance d apart and perpendicular to the wires. Both the wires and the rails have resistance ρ per unit length. A magnetic flux density B is applied perpendicularly to the plane of the rectangle so formed, and one wire is held stationary whilst the other is moved along the rails with uniform speed v. The force on the fixed wire vanishes when the wire separation x takes a particular value x_0. Show that x_0 satisfies an equation of the form

$$x_0^2 + dx_0 - \lambda d = 0,$$

where λ is a constant to be determined.

Q 184 Charged particles moving (non-relativistically) in a circular orbit of radius r, oriented perpendicular to a magnetic field, may be accelerated by the emf that arises when the magnetic flux enclosed by the orbit changes. Find the relationship between the magnetic flux density at the orbit and the average flux density within it, if r is to remain unchanged.

Q 185* A uniform thin wire of length $2\pi a$ and resistance ρ per unit length has its ends joined to form a circle. A small voltmeter V of resistance R is connected, in two different configurations, to points on the circle with angular separation θ, as shown in figures (i) and (ii). The connecting leads have negligible resistance.

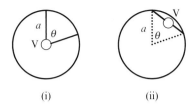

(i)   (ii)

A uniform magnetic flux density perpendicular to the plane of the circle is changing at a rate $\dot{B}$. What will the voltmeter read in each case?

Q 186 The Earth's magnetic flux density is 3×10^{-5} T at the Equator, 6×10^{-5} T at the North Pole, and 5×10^{-5} T over London, where its angle of dip is $66°$. Find the directions of the emfs that could be detected in the following situations, and estimate their magnitudes.

(a) A voltmeter is connected between the banks of the Thames at Tower Bridge, where the river is 250 m wide and flows at 2.0 m s^{-1}.

(b) Sensing devices are attached to the surface of a jumbo jet of wing span 80 m, length 60 m and height 8 m that is flying horizontally at 720 km h^{-1} (i) over the North Pole, (ii) northwards over the Equator, (iii) eastwards along the Equator, and (iv) northwest over London.

Q 187* A stream of mercury of resistivity ρ fills a horizontal rectangular pipe of length ℓ. Two opposite walls of the pipe, a distance a apart, and of height b, are conducting and shorted together; the other two walls are insulating. A magnetic field of flux density B is applied parallel to the conducting walls and perpendicular to the length of the pipe, whilst a pressure difference P is applied along its length.

If the speed v of the mercury along the length of the pipe is proportional to the total external force acting upon it, find the relationship between v and v_0, where the latter is its speed when there is no magnetic field present.

Q 188 The figure shows a simplified version of the ignition system used in the 1948 Citroën 2CV car. The car had only two cylinders, in each of which was a spark plug with a spark gap of 0.75 mm. The rotating cam, which caused the contact C to open and close, was driven at a speed matching the engine revolution speed.

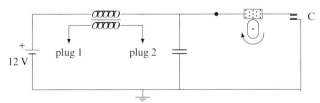

From this information and the figure, describe qualitatively what you can deduce about the physics of spark generation in each cylinder, and, given (a) that the breakdown electric field strength for air is about 3×10^6 V m^{-1}, and (b) that reliable ignition of the air-petrol vapour mixture required about 50 mJ per spark, estimate the winding ratio of the ignition coil and the current taken from the battery (for this purpose) when the engine was running at 2000 r.p.m.

Chapter 13

Heat Transfer
Some Like It Hot!

Q 189 The latent heat needed to evaporate water is about 2.3 MJ kg^{-1}. Estimate the temperature at which an airing cupboard should be kept in order to dry washing ten times faster than it would at $T_0 = 27°$C.

Q 190 A convex lens of diameter $d = 100$ mm and focal length $f = 500$ mm is used to produce a focussed image of the Sun on a thin matt black disc the same size as the image. What is the highest temperature to which the disc can be raised if the Sun has a black body temperature of 6000 K?

Q 191 A steam power station delivers 100 MW to the national grid, whilst waste heat is removed by a river at $10°$ at the same rate. In what range does the working temperature of the boilers lie?

Q 192 A very large black sheet, maintained at $100°$C, is above and parallel to a similar sheet which is kept at $10°$C by a stream of water, the space between them being evacuated. Three further sheets of black metal, thermally insulated from the first two and from each other, are now inserted between them and allowed to come into equilibrium.

(a) What temperatures will they take up?
(b) If n sheets, rather than three, were inserted, how, if at all, will the heat carried away by the water be affected?

Q 193[†] Newton's law of cooling states 'that the rate of heat loss of a hot body to its surroundings is proportional to the excess of its temperature over that of the surroundings'.

A body of heat capacity C, when warmed by a heater of power P, attains a maximum temperature of T_1 when placed in surroundings at temperature T_0. Find the time τ it takes for the body's temperature to fall to $\frac{1}{2}(T_1 + T_0)$ if the heater is then switched off.

Q 194   In a thermos flask the gas density is so low that the molecules hit the walls much more often than they hit each other. The walls may be treated as parallel and the temperature difference across them as small. Obtain an approximate expression for the heat flux ϕ in terms of the pressure p, the average temperature T, and the mass m of a molecule.

[Consider only the radial components of the molecular velocities; they are responsible both for the heat flux and the pressure on the walls. Assume that on impact with the walls the molecules immediately come into thermal equilibrium, and that $\bar{c} = (8/3\pi)^{1/2}c_{\text{rms}}$.]

Q 195[†]   This problem involves the reasoned use of mathematical approximation to solve an equation that has no routine solution method.

A uniform metal bar of length ℓ and thermal conductivity λ has one of its end faces maintained at temperature T_1 and its sides lagged so that no significant heat is lost through them. The face on the other end of the bar radiates heat to its surroundings, which have temperature T_0; both this face and the surroundings behave as if they were black bodies. Assuming that it is not too different from T_0, show that the temperature T of this face is given approximately by

$$T \approx \frac{\lambda T_1 + 4\sigma \ell T_0^4}{\lambda + 4\sigma \ell T_0^3}.$$

Q 196   Although heat conduction is not taught in many school physics courses, the following problem can be tackled using an electrical analogy.

The above figure shows analogous one-dimensional situations drawn from two different areas of physics. On the left, charge flows through a medium of resistivity ρ under the influence of a potential difference between its bounding surfaces; on the right, heat flows through a medium of thermal conductivity λ under the influence of a temperature difference between its bounding surfaces. The bounding surfaces all have area A.

The analogous governing equations are (writing j for the current density and ϕ for the heat flux density, and using the standard equations $\nabla V = RI$ and $R = \rho L/A$)

$$j = \frac{I}{A} = \frac{1}{A}\frac{dQ}{dt} = \frac{1}{\rho}\frac{V_1 - V_2}{L} \quad \text{and} \quad \phi = \frac{1}{A}\frac{dH}{dt} = \lambda \frac{\theta_1 - \theta_2}{L}.$$

The quantity λ is the thermal analogue of ρ^{-1} [itself sometimes called the (electrical) conductance, and denoted by σ] and is measured in W m^{-1} K^{-1}.[1]

On the basis of the above analogy, consider the following.

The air above a large lake containing water at $0°$C has a constant temperature of $-2°$C. Assuming that only thermal conduction is important, and using relevant data selected from that given below, find how long it would take for a layer of ice 10 cm thick to form on the lake's surface.

Thermal conductivity of water $= 0.56$ W m^{-1} K^{-1}; Thermal conductivity of ice $= 2.3$ W m^{-1} K^{-1}; Specific latent heat of fusion of ice $= 3.3 \times 10^5$ J kg^{-1}; Density of water $= 1000$ kg m^{-3}; Density of ice $= 917$ kg m^{-3}.

Q 197 The Sun's diameter subtends an angle of $0.35°$ when seen from Mars. Taking the Sun as a black body with a surface temperature of 6000 K, estimate the surface temperature of Mars.

Q 198* This question is only accessible if *entropy* has been studied in some depth.

Two identical bodies, each of constant heat capacity C, are at temperatures T_1 and T_2. What is the maximum amount of work that can be extracted by allowing them to reach equilibrium with each other?

Q 199† Unless you have studied heat conduction as a full topic, do not attempt this question without first reading Q 196.

As a result of atomic fission, heat is generated uniformly throughout the volume of a uranium sphere of radius $a = 100$ mm at a rate of $H = 5.5 \times 10^3$ W m^{-3}. If the thermal conductivity of uranium is 46 W m^{-1} K^{-1}, what is the temperature difference between the centre and the outside of the sphere in the steady state?

Q 200*† In a so-called *adiabatic* change, in which the pressure p and volume V of a fixed mass of an ideal gas are changed, but without any heat being added or taken away, the product pV^γ remains constant. The quantity $\gamma\,(>1)$ is characteristic of the particular gas involved.[2]

A simplified description of the cyclic changes undergone by the air in a cylinder of a petrol engine is as follows.

1. Adiabatic compression from volume V_1 and atmospheric pressure p_1 to a smaller volume $V_2 = sV_1$ and pressure p_2.
2. Rapid heating at constant volume V_2 by burning fuel, the air's pressure rising to p_3.

---

[1] Verify for yourself that the units are dimensionally correct.
[2] For a proof of this result see Q 210.

3. Adiabatic expansion to the original volume V_1, but with pressure p_4.
4. Rapid reduction of the pressure from p_4 to p_1 by opening a valve, the volume remaining at V_1.

Assuming that the air behaves as an ideal gas, draw these changes on a p-V diagram [plotting V along the 'x-axis']. Show that the work done by the air is

$$W = (p_3 - p_2)\, V_2 \, \frac{1 - s^{\gamma-1}}{\gamma - 1},$$

and so find the efficiency η of the engine.

Chapter 14

Gases
Little Things Matter!

Q 201 A balloon, fitted with an inlet valve, is filled with air at atmospheric pressure p_0 and temperature T_0. A small, but non-zero, excess pressure is needed to stretch its very light envelope. Determine, in quantitative terms, and from the point of view of subsequently lifting a load, whether it is better to have the valve open or closed whilst the air is being heated to a working temperature of $T\,(>T_0)$.

Q 202 On a particular day the temperature of the atmosphere, which decreased linearly with height, was $15°C$ at sea-level and $7.5°C$ at a height of 1.15 km. Treating air as a perfect gas of mean relative molecular mass 28.8, find the height x at which the atmospheric pressure was one half of that at sea-level.

Q 203 Two cylinders, A and B, each of volume V and negligible thermal capacity are in thermal contact and connected by a small closed valve. Cylinder B is sealed and empty, but A is fitted with a piston, initially fully withdrawn, and is filled with a perfect monatomic gas at temperature T.

The valve is slightly opened and the gas driven as far as it will go into B by pushing the piston at such a rate that the pressure p in A remains constant. Find the final temperature T_f of the system.

Q 204 A sealed central heating system includes, at its lowest point, an expansion bottle containing trapped air. As the system heats up and the water in it expands, it does so by entering the bottle, and so compresses the air.

The system is filled initially by isolating the bottle, completely filling the rest of the system with water at $5°C$, sealing the top, and then opening the isolating valve. If the maximum safe operating pressure is 3.0 atmospheres above atmospheric pressure, use the information below to find the minimum volume the bottle must have.

Height of the system, $h = 15$ m; Volume of the system, $V = 0.20$ m^3; Operating range, $5°C$ to $100°C$; Cubic expansivity of water, $\alpha = 2.00 \times 10^{-5}$ K^{-1}.

Q 205   A balloon filled with 1.00×10^{-4} kg of hydrogen at a pressure of 1.05 atmospheres (1.05×10^5 Pa) is allowed to rise to the ceiling of a room. Neglecting the weight of the fabric of the balloon, and taking the mean relative molecular mass of air as 28.8, determine the area A of the balloon in contact with the ceiling.

Q 206* Bearing in mind the assumptions made in deriving the ideal gas equation, examine the following measurements of gas pressure p, which were made with a constant volume gas thermometer at three different (monatomic) gas densities. Explaining your reasoning, determine the unknown temperature T as accurately as the data will allow.

p at $T_0 = 273.15$ K (atm)	p at temperature T (atm)
1.013	1.3795
0.6243	0.8494
0.2021	0.2749

Q 207   As part of an industrial process, 1000 ball-bearings, each of mass 1 g, are collected in a level square tray of side $L = 2$ m, and have a total translational energy of $E = 100$ J. On the basis that their collisions are effectively elastic, and that one half of the ball-bearings are moving parallel to any particular side, whilst the other half move perpendicularly to it, answer the following:

(a) What is the average force exerted on one side of the tray?
(b) If an opening $d = 10$ mm wide were made in one side of the tray, estimate the time taken for the number of balls to fall to (i) 900, and (ii) 50.
(c) Ignoring any effects due to statistical fluctuations, for each of your three answers in (a) and (b), state, with reasons, whether you consider it to be 'exact' or an approximation.

Q 208*† A gas molecule of mass m and initial speed v_0 bounces perpendicularly back and forth between two walls which are initially a distance x_0 apart.

One of the walls is now moved towards the other at a speed V that is very much less than v_0.

   (a) Show that, if all molecule-wall collisions are elastic, then the product of the molecule's speed v and the wall separation x is essentially constant.
   (b) Find the temperature rise of the 'one molecule gas' when the wall separation has been reduced to x.

Q 209   A thermally insulated container initially holds N_0 molecules of an ideal monatomic gas at temperature T_0. Molecules escape from the container through small holes in the walls, and it can be shown that in such a process at temperature T the average kinetic energy of the escaping molecules is $2kT$. How many molecules remain in the container when their temperature has fallen to $\frac{1}{2}T_0$?

Q 210*†   Specific heats of ideal gases do not normally appear in school physics, but, if quantum effects are ignored, they can be studied using only material that does normally appear. This problem is a guided set of steps in that direction, finishing with the adiabatic equation quoted in Q 200 on page 69. The mathematics required is a little more advanced than that associated with most other problems in this book.

To simplify the notation, we will work with one mole of an ideal gas (not necessarily monatomic). When its temperature is raised by ΔT, the heat that it takes in is $\Delta Q = C\Delta T$, thereby defining the (molar) specific heat capacity C of the gas. However, there are two particular situations for which C has a specific name: if the volume remains constant then C is denoted by C_V; if the pressure remains constant then C is denoted by C_p.

   (a) C_V and C_p are not equal, since the latter has to include the energy needed for the gas to expand – necessary in order to keep the pressure from rising. Write this as a formula, and then use the ideal gas equation to express this additional energy in terms of ΔT. Deduce that

$$C_p = C_V + R.$$

   (b) Now consider a change of state, in which all of p, V and T change, that is characterized by no heat being added or subtracted from the system (an *adiabatic* change).
       The general heat requirement for a change, normally written $dQ = dU + p\,dV$, is the sum of the change in internal energy plus the work done [essentially as in part (a)]. By the definition of adiabatic, $dQ = 0$,

whilst $dU = C_V dT$. From the ideal gas equation for one mole, remembering that all three dynamic variables may change, we can write

$$T = \frac{pV}{R} \quad \Rightarrow \quad dT = \frac{V\,dp + p\,dV}{R}.$$

Use these results, and that from part (a), to obtain an equation for an adiabatic change that contains only p, V and the ratio $\gamma = C_p/C_V$.

(c) Compare your answer to (b) with $pV^\gamma = \text{constant}$.

Chapter 15

Particles and Atoms
All Bits and Pieces!

Q 211* In an investigation of the photoelectric effect, a clean sodium surface was irradiated with light of wavelength $\lambda = 450$ nm; the resulting photoelectric current could be suppressed by a stopping potential of $V_s = 0.50$ V. Would such a current flow for the following combinations of λ and V_s: (i) $\lambda = 300$ nm, $V_s = 1.50$ V, (ii) $\lambda = 500$ nm, $V_s = 0.25$ V, and (iii) $\lambda = 750$ nm, $V_s = -1.00$ V?

Q 212† This question requires some 'delicate' mathematics, but gives practice in deciding whether a proposed general formula gives the expected results in particular, but more easily resolved, situations.

A nucleus X decays with decay constant λ_1 into nucleus Y, which itself decays with decay constant λ_2. Initially, at time $t = 0$, there are N nuclei of type X, and none of type Y. Denoting the number of nuclei of type Z by $z(t)$, construct the relevant equations and associated conditions, and, by substitution, show that

$$y(t) = \frac{\lambda_1 N}{\lambda_2 - \lambda_1} \left(e^{-\lambda_1 t} - e^{-\lambda_2 t} \right).$$

For each of the following conditions, obtain a simplified approximate form for $y(t)$ and give a qualitative description of how it is to be interpreted physically.

(a) The half-life of X is very much smaller than that of Y.
(b) The half-life of Y is very much smaller than that of X.
(c) The half-life of Y is equal to that of X.

Q 213 An unusual form of atom is the so-called π-mesic atom. In such an atom, a π-meson, which has mass $m = 2.4 \times 10^{-28}$ kg and the same charge as the electron, is in a circular orbit of radius r about a nucleus, with an orbital angular momentum of $h/2\pi$. If the radius of a nucleus of atomic number Z is given by $R = R_0 Z^{1/3}$, where $R_0 = 1.7 \times 10^{-15}$ m, estimate the limiting value of Z for which π-mesic atoms might exist. Is it an upper or a lower limit?

Q 214* In an experiment very similar to that undertaken in 1909 by Millikan and Fletcher to determine the charge on the electron, oil is sprayed into an air-filled region to which an adjustable vertical electric field can be applied. With the field switched off the terminal speed of fall of an oil drop is measured. The field is then switched on, and, if the drop responds by accelerating or decelerating, the voltage generating the field is adjusted to the value V_1 needed to hold the drop stationary.

A sample believed to contain unusual particles called 'quarks', which are known to have masses much less than that of a typical oil drop, is then introduced, and if the drop then starts to move, the voltage V_2 needed to restore it to rest is noted. Below are the entries for six of the drops followed, recording, for each, its mass m and the corresponding two voltages V_1 and V_2.

Drop	$m/10^{-15}$ kg	V_1/V	V_2/V
a	3	459	690
b	1	306	230
c	3	307	230
d	2	612	1836
e	2	305	204
f	4	613	525

(i) Explain, with relevant formulae, how the very small masses of the drops could be deduced from the measurements made, and estimate the order of magnitude of the terminal velocities to be measured. The viscosity of air is approximately 1.6×10^{-5} kg m^{-1} s^{-1}.

(ii) From the data given, what can you conclude about the likely charges of quarks? How do your deductions compare with their currently accepted properties?

Q 215 Antiparticles have the same masses as their corresponding particles.

(a) An electron and a positron, each with a speed $v = 0.8c$, collide head-on and annihilate each other according to the equation

$$e^- + e^+ \to \gamma + \gamma.$$

Explaining your reasoning carefully, determine the wavelengths of the resulting γ-rays.

(b) Two protons with equal speeds v are made to collide head-on. What is the minimum value of v required, if the aim is to produce antiprotons?

Q 216* An alternative notation for a nuclear reaction produced by particle a incident upon a target nucleus A and producing particle b and a different nucleus B, i.e.

$$a + A \longrightarrow b + B,$$

is $A(a, b)B$. The energy associated with this reaction is called its 'Q value', with positive Q values implying that the energy is released (rather than absorbed). Since Q values are simply signed numbers, they and their associated notations can be added and subtracted (in a kind of algebra). Clearly, the Q value for $B(b, a)A$ is the negative of that for $A(a, b)B$.

Given below are some Q values for reactions involving protons (p), neutrons (n), deuterons (d = $^2_1$H), α-particles ($\alpha = ^4_2$He), isotopes of silicon ($^{**}_{14}$Si), isotopes of phosphorus ($^{**}_{15}$P), and gamma rays.

	Reaction	Q value/MeV
(a)	$^{31}P(\gamma, n)^{30}P$	-12.313
(b)	$^{31}P(p, \alpha)^{28}Si$	1.916
(c)	$^{28}Si(d, p)^{29}Si$	6.251
(d)	$p(n, \gamma)d$	2.226
(e)	$d(d, \gamma)\alpha$	23.847

Deduce Q values for (i) $^{30}P(d, p)^{31}P$, (ii) $^{28}Si(\alpha, n)^{31}P$, and (iii) $^{29}Si(d, n)^{30}P$.

Q 217 Part of the radioactive series that starts with $^{238}_{92}U$ and ends with $^{206}_{82}Pb$ is given below, together with the half-life τ of each isotope.

$$^{238}_{92}U \rightarrow \cdots \rightarrow ^{234}_{92}U \rightarrow ^{230}_{90}Th \rightarrow ^{226}_{88}Ra \rightarrow ^{222}_{86}Rn \rightarrow ^{218}_{84}Po \rightarrow \cdots \rightarrow ^{206}_{82}Pb$$

| 4.5×10^9 y | 2.5×10^5 y | 7.5×10^4 y | 1620 y | 3.83 d | 3.05 m | stable |

(a) In the first half of the twentieth century the dials of many watches and clocks were made luminous by painting them with a mix of a zinc sulphide phosphor and an α-particle emitter. What mass of radium was needed to produce an average of ten α-particles per second for this purpose?

(b) A sample of ore contains all of the isotopes in the series given above. If radioactive equilibrium has been established, and there is 1 kg of $^{238}_{92}\text{U}$ in the sample, how many $^{226}_{88}\text{Ra}$ nuclei would you expect to find?

(c) If the abundances of $^{238}_{92}\text{U}$ and $^{206}_{82}\text{Pb}$ in a sample of ore are A_U and A_Pb, what is the approximate age of the Earth?

Q 218* In a so-called Franck–Hertz type of experiment, electrons from a heated cathode pass through a low density sodium gas towards a metal grid which is at a variable positive potential relative to the cathode. The electrons that pass through the grid are collected by an anode that is slightly negative with respect to the grid, and the current they produce is measured by an ammeter. The set-up is shown schematically in the figure.

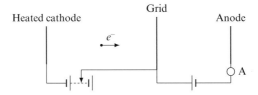

In such an experiment, it was found that the measured current generally increased as the voltage was increased, but showed relatively sharp reductions at voltages of 2.10 V, 3.18 V, 3.75 V, and 4.34 V.

When sodium gas is heated and its emission spectrum examined, spectral lines are found at the following wavelengths: 2.853×10^{-7} m, 3.303×10^{-7} m, 5.893×10^{-7} m, 1.139×10^{-6} m, and 2.207×10^{-6} m.

Reconcile these observations and suggest other wavelengths which should be observed in the emission spectrum.

Q 219† The *relativistic* relationship between the *total* energy E of a particle, its momentum p, and its rest mass m, is $E^2 = p^2c^2 + m^2c^4$, where c is the speed of light in a vacuum.

A photon of frequency f collides with an electron initially at rest, and is scattered through an angle of $90°$. Show that its new frequency is

$$f' = \frac{m_\text{e}c^2}{hf + m_\text{e}c^2} f.$$

Q 220* It is believed that one of the reaction sequences taking place inside suitably hot second-generation red-giant stars is as follows:

1. In a process known as radiative proton capture, $^{12}_{6}\text{C}$ captures a proton to form nucleus A and a γ-ray.

2. *A* undergoes β^+-decay to produce *B*.
3. *B* transforms to *D* by radiative proton capture.
4. *D* also undergoes radiative proton capture to produce nucleus *E*.
5. *E* transforms to *F* through β^+-decay.
6. *F* captures a proton to yield *either* (a) *G* and an α-particle, *or* (b) *H* and a γ-ray.

In case (b), the process continues as follows:

7. *H* captures a proton radiatively to form *I*.
8. *I* decays by β^+-emission to *J*.
9. *J* captures a proton to yield *K* and an α-particle.

(a) Identify the ten nuclei *A*, *B*, *D*, ..., *K*, showing, in particular, that *D* and *K* are the same.

(b) Show that two closed cycles are possible, establish their basic reactions, and identify those nuclei that are 'merely catalysts'.

(c) How much energy is released when 1 g of hydrogen is converted to its final products? Use the data from Q 216 on page 77 and the fact that the Q value for the decay p $\longrightarrow$ n + e$^+$ + ν_e is -1.805 MeV.

Q 221 The existence of strange particles was first reported when the interaction

$$\pi^- + p \rightarrow \Lambda + K^0$$

was detected. The K^0 is a meson with strangeness of $S=+1$, whilst the π-meson has strangeness zero. Mesons, which have zero baryon number, each contain two quark elements. Construct the quark representation of the above equation, and so determine the quark structure of the Λ-particle.

The decay of strange particles can be considered as the decay of one of its quarks. The strange quark decays according to

$$s \rightarrow u + W,$$

where *W* is a charged boson. The decay of the positively charged *W*-boson is

$$W^+ \rightarrow e^+ + \nu_e.$$

Use this information to determine one of the decay modes of the Λ.

Hints

H1 Using the notation that $[X]$ stands for the dimensions of X, simplify the 'algebra' by working in terms of the combination of base dimensions appropriate to energy, namely $[J] = ML^2T^{-2}$. The (arbitrary) symbol J has been chosen because energies are usually measured in joules.

H2 Express the dimensions of both κ and σ in terms of those for power (or energy).

H3 Use another symbol, say P, for the force, to avoid any confusion with the symbol F for farad.

H4 Show that $[\epsilon_0] = M^{-1}L^{-3}T^4I^2$ and $[\mu_0] = MLT^{-2}I^{-2}$, and then check the compatibility of the two terms.

H5 Consider how the relevant physical properties scale with linear size.

H6 Establish, for as many of the fundamental constants as possible, any explicit parts they could play in the expression for σ.

H7 Note that the origin of moonlight is reflected sunlight, and that the intensity of direct sunlight is the same for both the Moon and the Earth.

H8 Use dimensional analysis to determine which of the four given quantities are actually involved in determining λ. In a sparse gas the speed of sound is comparable to the speed of the atoms.

H9 Use the implications of dimensional consistency to remove the relevances of h and G.

H10 Consider the implication of the phrase 'at sea-level' being included in the question.

H 11   Convince yourself that the dimensions of ψ, [actually $L^{-3/2}$] are irrelevant, and that the dimensions of a derivative such as dy/dx are $[y]/[x]$.

H 12   If q_i are dimensionless combinations of the relevant variables, the equation describing the flow takes the form $f(q_1, q_2, \ldots) = 0$ for some (unknown) function f.

H 13   The suspended lamina may not provide the easiest coordinate system with which to work. Use the coordinate system that is the most convenient for determining the direction of the line joining Q to the modified lamina's centre of gravity.

H 14   This problem can be solved either (a) the 'easy way' by giving a little thought to the condition that the centre of gravity of the partially-filled beaker must satisfy when maximum stability has been attained, or (b) the 'hard way' through the use of differential calculus.

H 15   Find four simultaneous equations involving x, y, T_1, T_2 and the angles that the strings make with the horizontal. Then eliminate the angles.

H 16   Start by considering the 'plight' of the uppermost book, and then work downwards through the pile.

H 17   Determine the difference in tensions between the two ends of an infinitesimal length of towel that subtends an (infinitesimal) angle $d\theta$ at the centre of the towel rail. Then use elementary integral calculus to find the result for the whole of that part of the towel that is in contact with the rail.

H 18   It may help to use graph paper when constructing the layout of the mobile.

H 19   The ratio of a/b that minimizes the total cable length needed can be determined by inspection of the expression for the cable tension, without the need for calculus.

H 20   Take moments about the pivot, but, in all three cases, include a torque of unknown magnitude and direction. Then consider the minimum properties that it must have, given the recorded observations.

H 21   Determine the coefficient of friction between the ladder and the floor from the slipping, and then consider which part of the man's ascent is even more hazardous than the rest of it!

H 22   When finding the centre of mass of the bowl, consider it as a solid hemisphere containing a second hemisphere of negative density.

H 23   Determine the difference in tensions between the two ends of an infinitesimal length of rope that subtends an (infinitesimal) angle $d\theta$ at the centre of the cylinder. Then use elementary integral calculus to find the result for the finite rope in the question. Or simply take a (selective) peep at the solution to a related problem; it appears on page 111.

H 24   Determine how much of the baton lies below the oil, and the distances below the oil surface of the centres of gravity and buoyancy/upthrust. Then find the various couples acting when the upper surface of the baton is tilted through a (very) small angle.

H 25   Beware of red herrings – there are several swimming about!

H 26   Choose as a point about which to take moments, one that keeps the number of forces involved to a minimum.

H 27   In spherical polars, $dV = r^2 \sin\theta \, dr \, d\theta \, d\phi$ and $x = r \sin\theta \cos\phi$; use these to find $s = \bar{x} = 3\pi a \sin\alpha / 16\alpha$. Then take moments about a line which passes through any unknown force.

H 28   For an assumed depth of water in the bucket, obtain an integral giving the downward component of the net hydrostatic force acting on the part of the ball that is in the water.

H 29   Take moments about the axle of the compound pulley for the static situation, and then consider the actual rate of working when F moves downward with speed v.

H 30   Determine the speed the second stone must have immediately after the collision, if it is to end up on the button. Then work backwards from there.

H 31   Use energy conservation to determine the kinetic energy of the system, and simple geometry to relate the factors contributing to it.

H 32   The density of air is 1.22 kg m^{-3}. Airlines allow golf bags of up to about 23 kg as check-in luggage.

H 33   Treating the ball's trajectory as two separate flights, obtain expressions for $H - h$ and H and then rearrange them to give separate ones for v^2 and $\tan\alpha$.

H 34   Produce a simple differential equation for the depth of water left in the reservoir, showing that it varies exponentially with time.

H 35   Consider which one quantity is common to the rise and fall of the particle, and adjust the equation of motion to take advantage of the fact that it is common.

H 36   Consider how the relationships between the various velocities depend on the mass ratio of the two balls.

H 37   Note that e relates to relative velocities *along the line of contact*, and that when the balls collide the cue ball's centre is *not* in the line containing those of the reds.

H 38   Show that

$$\tan \phi = \frac{t\,V\sin\theta - \frac{1}{2}g\,t^2}{(V^2/g)\sin 2\theta - t\,V\cos\theta},$$

where V is the speed with which the ball leaves the bat, and θ is the angle between its initial direction of motion and the horizontal.

H 39   Establish that the raindrop's radius grows linearly with time, and later use that result to cast $\int m\,dt$ into a more manageable form.

H 40   Remember that the force on the scale pan is made up of two components.

H 41   When the mailbag is thrown as described, there is no change in the *speed* of the delivering carriage. However, it does change its momentum parallel to the tracks, since the bag carries some of it away.

H 42   Consider what happens in a small time interval dt, and so form an expression for dy/dx, which can then be integrated.

H 43   As an intermediate result show that the ball rebounds along the normal to cushion AB, and then starts off as if to retrace its path.

H 44   Obtain simple first-order differential equations for the horizontal velocity component u, and, separately, for the vertical component v. Integrate the equations, and then express X and Y as integrals involving u and v.

H 45   As intermediate steps, show that, whilst the fuel is burning, the rocket's speed v and mass m vary with time t according to the equation

$$m\frac{dv}{dt} = \alpha u - mg,$$

and that it's speed when the burn is complete is

$$u \ln 2 - \frac{m_0 g}{2\alpha}.$$

H 46   Find an expression for the total time T in terms of h, g and e, the coefficient of restitution between the ball and the concrete floor. Use this to find e and then consider the energy changes at each bounce.

H 47   As an initial step consider what sort of motion is taking place at a typical time t in the range $0 < t < L/v_0$, and then determine how the situation changes when $t > L/v_0$, and again when all the bullets have been fired.

H 48   Set up the momentum conservation equations for the subsequent collisions, and express the perfect elasticity as relationships between the velocities involved.

H 49   Show that the speed gained during any particular stage of the acceleration process is proportional to the logarithm of the initial mass/final mass ratio.

H 50   Determine the force that acts on the sails as a result of its slowing down the wind, and then convert this to the power the mill absorbs.

H 51   Apply Newton's law, but note that if dF is the surface tension force acting on a length $r\, d\theta$ of the hole's rim of mass dm, the required equation is *not* $dF = \ddot{r}\, dm$.

H 52   Note that the value of e does not affect the horizontal component of the ball's velocity. Follow the various stages of each bounce, and then impose the 'repetitive condition'.

H 53   Using plane polar coordinates, draw a careful two-dimensional vector diagram of the velocities to be taken into account, and then consider the conditions they must satisfy.

H 54   Consider the interplay between kinetic and gravitational potential energy when the cube reaches the bottom of the tray.

H 55   Suppose that the chord stretches by x, and then balance the centrifugal force with the tension in the chord.

H 56   (a) Use the standard equation for an ellipse to obtain the appropriate limits for the first stage of the double integration, then make a substitution of the form $u = A\cos\theta$. (b) Use energy conservation involving gravitational, linear kinetic, and rotational kinetic energies.

H 57   Consider the implications of the conservation of angular momentum – and also what a GPS system actually measures.

H 58   Calculate the skater's moment of inertia, both before and after his arms are lowered, and then apply the conservation of both angular momentum and energy, noting that his arms fall gently.

H 59   Consider which impulsive velocity increments the hinges would receive, if they were free to move horizontally. There are two.

H 60   Deduce, and then use, the fact that the tangent plane to the bowl at the point $(x, x^2/(4a))$ has slope $x/(2a)$.

H 61   Establish as many equations as you can that connect the explicit and implicit variables, and then eliminate those about which you have no information.

H 62   There are two factors that contribute to the torque experienced by the wheel – and two sources that contribute to the energy available.

H 63   Apply both angular momentum and total energy conservation. The moment of inertia of a rod of mass M and length ℓ about an axis through one of its ends is $\frac{1}{3}M\ell^2$.

H 64   Resolve forces parallel to the two planes, rather than vertically and horizontally (to ease the algebra). Note how the two normal reactions vary as θ is increased.

H 65   Apply the principles of conservation of linear momentum and rotational angular momentum, whilst remembering the nature of an impulse.

H 66   Identify the condition to be satisfied when the rod first loses contact with the table, and use conservation of energy to find the angle it then makes with the vertical. The rod is then a constantly rotating object in free fall.

H 67   Find a suitable equation for the basket's shape, and then use it to calculate any of its properties needed for the analysis of the oscillations.

H 68   As only an approximate answer is needed, any effects due to air resistance or gravity can be ignored when analysing the rapid motion of the block and the shortening of the wire. Consider which conservation laws are applicable.

H 69   Form a simple differential equation for the radial position r as a function of time t, and look for solutions involving exponential functions.

H 70   Establish the energy conservation equation for the system, and then impose the requirement that the string stays taut, so obtaining an expression for the minimum required value of V. For the more general analysis, keep in mind the finite length of the string.

H 71   Express the required answer in terms of ratios before inserting numerical data. There is no need to determine the masses or radii of the Earth and the Moon.

H 72 Find the point in the flight where the spacecraft is moving most slowly, and then consider the subsequent change in its gravitational potential energy. It may help to sketch the gravitational equipotentials in the space between the centres of the moon and planet.

H 73 Establish that V, G and N form an equilateral triangle that contains the Earth's centre C, and determine how far from C each one is.

H 74 Pay particular attention to the curvature of the graph, deciding whether it is/could be positive, zero or negative.

H 75 Use the length of the year on Mars to determine the radius of its solar orbit.

H 76 Consider the contrasting variations with distance of gravitational and centrifugal forces.

H 77 Multiply the equation of motion through by $\dot{r}$ and so gain an explicit expression for $\dot{r}$, or use conservation of energy.

H 78 Show that there is an increase of $3GM/(rc^2) \times \omega_0$ in the angular frequency ω_0 of the orbit.

H 79 Remember that both the kinetic and the potential energies contribute to the total.

H 80 Obtain an expression for the ratio r/R, and then use a frame of reference in which the Earth is stationary.

H 81 Establish a formula of the form $GM = r^3\omega^2$ and then consider the ratios of relevant quantities.

H 82 For relating small changes, differentiating the equilibrium conditions can prove useful, but careful attention to the signs of the changes is needed.

H 83 Consider how each rocket would be launched relative to the space station's orbit, and note that in any stable gravitational orbit the magnitude of the kinetic energy is one half that of the potential energy.

H 84 Show that in (b) the traveller executes simple harmonic motion, but in (c) is subjected to uniform accelerations and decelerations. All the answers can be expressed in terms of G and ρ_0.

H 85 This is *not* a Doppler effect, as, to first order, the Earth is not moving towards or away from the Sun. Note the chapter title.

H 86 Consider a particle on the cloud's surface, and either construct and integrate its equation of motion, or use the conservation of total energy. Obtain an equation of the form $\alpha \cos^2 \phi = \beta - \gamma \sin^2 \phi$ and consider how this can be made consistent for all ϕ.

H 87   *T* is *not* given by the period of a simple pendulum whose length is equal to the distance between A and the actual pendulum's centre of mass.

H 88   Use the rate of frictional energy loss, averaged over complete cycles, to find a first approximation. For part (b), use a similar approach to determine how the amplitude of the oscillations is reduced to zero.

H 89   Consider which of the variables describing the situation can be chosen arbitrarily, and which cannot.

H 90   Establish the relationship between the mass m of the particle and a, and then determine the change induced in a by a small change in m.

H 91   Consider carefully the relationship between the angle of rotation of the disc and the angle a typical string makes with the vertical.

H 92   The pan and putty will perform SHM about a new equilibrium position, with the moment immediately following the impact providing a 'snap-shot' of it.

H 93   Express the total energy of each part of the motion in terms of its amplitude and angular velocity.

H 94   Note that for oscillations in the PQ direction, the 'y'-springs play no part, as they are not under tension.

H 95   Use the fact that, because there will be no slipping, the total energy of the system does not change with time.

H 96   Construct the equation governing the deviation of the car's (centre of) mass from its equilibrium position, showing that it represents an undamped harmonic oscillator driven by a source whose frequency is related to V. Note that the phase of the response (relative to that of the driving force) reverses at a certain value of V.

H 97   Consider the forces involved during each distinct phase of a single oscillation by the damaged trolley, and how long each phase lasts.

H 98   Obtain a pair of coupled differential equations, and then set $x + y = u$ and $x - y = v$. Later, note that if $\cos \alpha = - \cos \beta$ then one possibility is that $\alpha + \beta = \pi$.

H 99   For mode 1 establish the equation of motion of A; for mode 2 consider that of B.

H 100   Determine the conditions under which the centrifugal force matches the gravitational one. In the final part, show that in polar coordinates $r = R\cos\theta$, and then move the origin to $(x, y) = (\frac{1}{2}R, 0)$.

H 101   Use the equilibrium information to determine the spring constant, and introduce a vertical position coordinate for the mass.

H 102   Compare each of the potentials with what you expect for the reference simple-harmonic potential $V(x) = \frac{1}{2}\alpha x^2$.

H 103   Use the qualitative information provided by the plank to give a good approximation to the equation satisfied by λ.

H 104   The single lens formula is $u^{-1} + v^{-1} = f^{-1}$ and the effective focal length of a two-lens combination is given by $f^{-1} = f_1^{-1} + f_2^{-1} - d/f_1 f_2$, where d is the distance between the lenses.

H 105   Break the analysis into three parts: with the transmitters in phase and with exactly the same frequency; when they have the same frequency but the left-hand one is, say, $\frac{1}{2}\pi$ ahead in phase; the question as posed.

H 106   (a) $f_r = [(v + u)/v] f_s$,   (b) $f_r = [v/(v - u)] f_s$.

H 107   Show that the fundamental frequency of the piano string is 178 Hz, and then consider what might be possible using the viola string.

H 108   Consider the implication of the temperature gradient for the sign of α, and so sketch the general shape of the ray. Relate its tangent to Snell's law as stated in the question.

H 109   Obtain the phase difference as a function of the height of the Heaviside layer, and then differentiate it with respect to time.

H 110   Use Hubble's law to convert the required galactic speed into the required galactic distance.

H 111   Show that, with the given conditions, the distance from $(0, y)$ to (L, Y) is greater than that from $(0, y)$ to (L, y) by approximately $(Y - y)^2/2L$.

H 112   Show that if t_n is the time of the nth echo, then $f(t_n) = dn/dt_n$ for n significantly greater than 1.

H 113   Consider how lines of different colour can appear at the same angle, and what is constant, even if unknown, for all the observations.

H 114   For plane wave ultrasound the impedance Z of a material is equal to the product of density and sound speed. Note that the transmitted intensity is equal to $I_i - I_r$.

H 115 With the lens in place and sharp images formed, the sum of the object and image distances must be equal to the slit-microscope separation. Further information about these distances is derivable from the two observed slit separations.

H 116 Remember that the observed lines are not necessarily first-order spectra.

H 117 Use similar triangles involving h and h', where the former is the distance from the symmetry axis that a ray, initially parallel to that axis, enters the objective, and the latter is the corresponding distance when it enters the diverging lens.

H 118 For the sounds received at any particular time from the two sources, determine how the phase difference between them depends upon θ. Then convert the *sum* of their two amplitudes into a *product* of factors.

H 119 Show that $\phi = (2n - m)\pi$, where n and m are both integers. Consider the implication of *exactly* six (not a multiple of four) strong signal directions.

H 120 Construct quantities that allow you to make use of the first identity given at the start of this chapter, and so establish values for the angular frequency and wave number of the wave.

H 121 Consider the implication of the intensity reduction of the unpolarized light to 32%, rather than to 50%, and then express your conclusions in terms of the relevant amplitudes.

H 122 Consider the forces acting upon a part of the ring that subtends a small angle 2θ at its centre.

H 123 Assume that you may consider the string as being everywhere perpendicular to the axis of the hose. Check on this later.

H 124 Show that the optimum value for F is $\frac{1}{2}P$.

H 125 Consider how, if at all, the channel dimensions are involved in determining the minimum value for a. Show that the force on the barrier is $\frac{1}{2}\rho_0 g b^2$.

H 126 Sketch a plot of 'the volume of material that has been compressed' as a function of the pressure, and then relate certain regions of it to particular aspects of the collision.

H 127 Consider the additional forces that act on the washers when the cylinder is heated.

H 128   Apply Archimedes' principle and consider at what depth the sphere will ultimately float.

H 129   Consider what, if anything, happens to the boat in each of the three cases.

H 130   Show that E/ρ is proportional to the square of the length of the rod.

H 131   First determine the equilibrium spacing x_0 of the atoms, and then use Taylor's theorem to evaluate dV/dx at $x = x_0 + y$.

H 132   Identify the point of greatest stress and so determine what the cable cross-section needs to be.

H 133   Establish the relationship between γ, the tension T in the threads, and r their radius of curvature. Then use geometric properties.

H 134   For the two-slab composite,

$$\frac{1}{H_2} \propto \left(\frac{1}{k_1} + \frac{1}{k_2} \right).$$

Use an analogy with either capacitors in series or resistors in parallel.

H 135   Consider, on physical grounds, the conditions necessary (a) for water to flow through the valve, and (b) for the harbour gates to open; then combine your conclusions.

H 136   Use the symmetry of the situation, and consider whether the larger cross-section of rod 3 has any significance.

H 137   With x measured normal to the base of the channel, and $v(x)$ the velocity profile, show that

$$\eta \frac{d^2 v}{dx^2} + \rho g \sin \theta = 0.$$

H 138   In a standard notation, show that the tension T in the blade obeys

$$\frac{dT}{dr} = -\rho A \omega^2 r,$$

and consider where the maximum stress occurs.

H 139   Determine the relevant resistances using $P = V^2/R$, and the relevant power dissipations using $P = RI^2$.

H 140   Determine whether or not there is a value of R that produces a turning point in the expression for the power dissipated in the relevant resistor(s).

H 141   Determine the fault resistance, its location, and the resistance per unit length of a core, before attempting the power calculation.

H 142   Consider how the charge on the balloon must change if certain conditions involving the balloon and the resistor are to be met. Note that the capacitance of an isolated sphere of radius r is $4\pi\epsilon_0 r$.

H 143   Either calculate the charges on each of the plates, or treat the system as a 'potential divider'.

H 144   Work from the basic principles of (i) the conservation of current, and (ii) equal potential differences across alternative paths connecting the same two points.

H 145   Consider under what circumstances the mechanism for heat dispersion can 'no longer cope'.

H 146   Consider why the voltmeter reading in (b) was not 16.0 V, and devise a more realistic representation of the capacitor.

H 147   Find the value of R corresponding to Z just starting to conduct, and then divide your analysis into two regimes.

H 148   The net voltage drop around a closed loop must be zero; write equations expressing this for the obvious loops in the network, and then rearrange them in the form indicated. As a check, you should find that $ad - bc$ (the determinant of M) has unit value, whatever the value of R.

H 149   Note that, just before it collides, an electron has a speed that is twice that of the drift current. Equate the heat generated in the specimen to $(\rho L / A) I^2$.

H 150   (a) and (b). Equate the voltage drops around the external circuit to the voltage supplied by the source. Obtain expressions for the current and its rate of change by differentiating that for the charge on the capacitor.

H 151   Establish that $d + e + f = 13$, and that the general form of a measured resistance is $n_1 + n_2 + n_3(13 - n_3)/13$.

H 152   Note that the capacitor discharge lasts for just under 0.02 s, but that the time constant for it is 0.5 s. Show that $\delta = f^{-1} - T_1$, where T_1 is the solution of

$$e^{-T_1/RC} = \cos(2\pi f T_1).$$

H 153   (a) Consider, in each case, which resistors are in parallel and which are in series.

(b) The 8 Ω resistor is irrelevant. Analyse the circuit in terms of three circulating currents, one in each of the three (obvious) loops.

(c) Find out what a 'Wheatstone's bridge' is, and consider the given circuit analogously.

H 154   Show that the time constant for charging the capacitor C is 0.07 s. Note the roles of the two potential dividers in the circuit.

H 155   To test the linearity, for $t = T$ expand the exponential function in your equation as a power series, and then examine the relative sizes of the linear and quadratic terms in T.

H 156   For case (b), calculate the energy wasted as the difference between that supplied by the battery and that stored in the capacitor. Note that $\sum_1^N r = \frac{1}{2}N(N+1)$.

H 157   (a) $p = \sigma^2/2\epsilon_0$.   (b) $p = 2tT/a$.

H 158   (a) Consider what the p.d. would/must be if the two spacings were equal.
(b) Note the relationship between the volume of the box and the area, number, and spacing of the plates.

H 159   Use Fleming's left hand rule to determine the direction of the electromagnetic force.

H 160   Relate both the number of turns used and the total length of the wire to its radius.

H 161   Show that given a, the voltage V should be set at $a(b - a)E_0/b$.

H 162   Show that $V(t)$ contains a component of frequency 2ω.

H 163   Note that the *net* charge on each plate must remain constant, as they are not connected to anything else or to each other.

H 164   For each of the two stages of the procedure, consider which physical variable is constant.

H 165   Show that *on* and *outside* the sphere, the conditions are exactly the same whether we are considering the actual physical situation or one containing only the two charges (and no earthed sphere).

H 166   Assume values a and b for the cylinder radii $(b > a)$, and L for its length. Then determine (separately) the capacitance of, and the resistance between, the two cylinders.

H 167 Find the *change* in the capacitance of the system corresponding to Δx, and the resulting change in the stored energy. Then consider the external work done to bring that change about.

H 168 The couple varies with height above the base of the triangle, and a simple integration is required.

H 169 Use Fleming's left hand rule to establish the direction of any electromagnetic induced force. In part (b), a simple integration is required because the field strength varies with position along AB.

H 170 Resolve an electron's velocity into components parallel and perpendicular to the field and analyse them separately, before combining the results.

H 171 Use Fleming's left hand rule to determine the required current directions, and the formula given in Q 169 to determine the magnetic forces involved.

H 172 Compare the weight of the mercury with the electromagnetic force acting upon it.

H 173 Make a careful sketch of the situation for non-zero θ, looking vertically downwards on the loops, and so establish the sense of the forces acting on the vertical sides of the square.

H 174 Show that $r = mv/Be$, substitute for m, and then solve for v.

H 175 In both parts, the effect of the electromagnetic force has to be balanced against that of gravity.

H 176 Establish that the electric and magnetic forces have opposite signs, but the same d^{-1} dependence.

H 177 Find expressions for the x- and z-deviations separately, and then combine them by eliminating the ion's velocity V.

H 178 Find the resistance of the disc by considering an annulus of radius x and radial depth dx.

H 179 Work in symbols and show that, in addition to g, the only data needed are ρ, ρ_e and B.

H 180 Determine during which interval(s) there is a current flowing in the wire, and use the information as the basis for both the mechanical and electrical aspects of the problem.

H 181   Show that the electric fields generated in the two wheels act in the *same* direction at the contact point. In the terminal state, equate the electrical and gravitational power losses.

H 182   Remember that power is determined by the r.m.s. value of the applied voltage, and not by its peak value.

H 183   Find an expression for the current induced in the wires, and then consider the interaction of the fixed wire with (i) the moving one, and (ii) the applied field.

H 184   When the particle velocity changes by dv, find expressions for the force(s) needed to keep it in the same orbit, and for those arising from a change in the magnetic field strength.

H 185   For two different closed circuits, equate '$\Sigma R_k I_k$' to the rate of change of flux through the circuit. Use a consistent convention for current directions and the sense of circuit traversal.

H 186   Resolve the Earth's field vertically and northwards, and the jet's velocity northwards and eastwards. Then contort the first two fingers and the thumb of your *right* hand into a multitude of orientations.

H 187   Show that the current that flows is $Bvb\ell/\rho$, and then consider the effect of the consequential force. There is a self-consistency condition to be satisfied.

H 188   Note that an ideal transformer does not change the power available, and that in this engine there are four sparks generated per revolution, even though only two are useful.

H 189   Determine the energy needed to evaporate a single water molecule and then consider the Boltzmann distribution.

H 190   Note that the angular diameters of the object and image are equal.

H 191   For an ideal system working between T_1 and T_2, the maximum possible efficiency is $(T_1 - T_2)/T_1$.

H 192   Establish that $T_{i+1}^4 - T_i^4$ is independent of i for $i = 0$, 1, 2, 3.

H 193   Use the maximum temperature information to determine the constant of proportionality in Newton's law.

H 194   Note that the frequency of collisions with the walls depends upon $\bar{c}$, but that the pressure depends upon c_{rms}.

H 195   To solve the quartic equation, set $T - T_0 = \Delta T$ and use the binomial theorem to expand the expression for T^4, keeping only the two leading terms.

H 196   The temperature gradient will get less as the thickness x of the ice layer increases; consider the small change dx that occurs in time dt.

H 197   Treat Mars as a black body, both as a receiver and as an emitter of radiation.

H 198   Note that the temperature at which each small amount of heat is gained or lost varies throughout the process; an integration is necessary.

H 199   *Either*
Equate the heat flow out of a sphere of radius r to the total heat generated within it,
or
Equate the difference between the heat flow out of a thin shell of thickness dr and the heat flow into it, to the heat generated within it.

H 200   Express, and then evaluate, W as the difference between two integrals of the form $\int p\,dV$. Use the increase in internal energy of the air in step 2, to determine the external energy supplied.

H 201   Note that the basic equation is 'lift = upthrust − weight'. Determine how each of these factors varies for the two heating modes.

H 202   Establish a quantitative relationship between the pressure and density of an ideal gas, and then apply it to the equilibrium of a layer of air of depth dx.

H 203   Find two equations linking the final temperature with the fraction of cylinder A that is traversed by the piston, noting that the pressure remains constant.

H 204   Find two equations connecting the hot and cold volumes of the air in the bottle; one from its own properties as a gas, the other from the expansion of the water.

H 205   Show that for any ideal gas, $\rho = N_A mp/RT$.

H 206   Retaining four decimal place accuracy, compute the ratios $p(T)/p(T_0)$ and interpret any trends you observe.

H 207   For each answer, consider on physical grounds the relevant subset of the ball-bearings and the average velocity associated with it.

H 208   For (a), find iteration formulas for v and x, and for the product vx.

H 209   Consider very carefully the signs of the small changes that occur when dN molecules escape.

H 210   In part (b), after substituting for dT, group the dV and dp terms separately, and then divide through by $C_V pV$.

H 211   Use the original observation to determine the work function for sodium.

H 212   Recall that half-lives are inversely related to decay constants. In (c), take particular care when evaluating expressions that take the form $0/0$ on direct substitution.

H 213   Establish the relationship $Ze^2 = \epsilon_0 h^2/(\pi mr)$.

H 214   Show that the number of elementary charges on a drop is proportional to m/V_1, and that, for the six cases given, these numbers are in whole number ratios. The drops acquire terminal speeds when the viscous drag of the air through which each falls is equal to its weight.

H 215   Note that the speeds involved in both parts are relativistic, and that overall lepton and baryon numbers are conserved in interactions.

H 216   (i) Consider adding the equations for reactions (a) and (d).
(ii) Beware!
(iii) Extend the method used in (i).

H 217   For (b), show that in equilibrium the relative abundances of the isotopes are in proportion to their half-lives.

H 218   Show that each of the emission lines corresponds to the difference between two of the energy levels implied by the experimental results.

H 219   Use total energy conservation and momentum conservation in two perpendicular directions, all treated relativistically.

H 220   For part (b), note that G is $^{12}_{6}C$. For part (c), show that $4p \longrightarrow \alpha + 2e^+ + 2\nu_e$ has a Q value of 24.7 MeV.

H 221   Keep careful track of particle–antiparticle distinctions, and make sure your quark representations reflect them.

Solutions

S 1 Although all the quantities involved could be expressed in terms of the basic dimensions, M, L, T and I, as the formula claims to be for an energy and Coulomb's law gives a simply related quantity, a force, it should simplify the algebra to express as much as possible in terms of an energy. Since energies are usually measured in Joules, we will denote their basic dimension by $[J]$ ($\equiv ML^2T^{-2}$).

Coulomb's law states that the force between charges q_1 and q_2, when they are separated by a distance r, is given by $F = k_e q_1 q_2 / r^2$. So, $k_e e^2$ has the same dimensions as force $\times$ length^2, or, equivalently, energy $\times$ length, i.e. $[J]\,L$.

So, for the quoted formula,

$$
\begin{aligned}
[E_i] &= [m_e]^2\,[J]^2 L^2\,[h]^{-2} \\
&= M^2\,[J]^2\,L^2\,([\text{joule} \times \text{second}])^{-2} \\
&= M^2\,[J]^2\,L^2\,[J]^{-2}\,T^{-2} \\
&= M\,(ML^2T^{-2}) = M\,[J].
\end{aligned}
$$

Since E_i, being an energy, should have dimensions $[J]$, the quoted formula for it is clearly wrong; it has one mass dimension too many. This strongly suggests that m_e, rather than m_e^2, should appear in the formula. When this amendment has been made, the formula is of the correct form – though, of course, our analysis cannot determine whether $2\pi^2$ is the appropriate numerical factor. But it is.

S 2 As stated, the Wiedemann–Franz law takes the form

$$
\frac{\kappa}{\sigma} = aT,
$$

where a is the constant of proportionality to be investigated.

So far as the thermal conduction is concerned, the defining equation for κ takes the form

$$\frac{\text{power}}{\text{area}} = \kappa \times \text{temperature gradient},$$

which in dimensional form reads

$$\frac{[P]}{L^2} = \frac{[\kappa][T]}{L} \quad \Rightarrow \quad [\kappa] = \frac{[P]}{L\Theta},$$

where T is temperature, and not to be confused with the basic dimension of time.

Similarly, for electrical conductivity,

$$\frac{\text{current}}{\text{area}} = \sigma \times \text{voltage gradient},$$

leading to

$$\frac{I}{L^2} = \frac{[\sigma][\text{voltage}]}{L} = \frac{[\sigma]}{L} \times \frac{[P]}{I} \quad \Rightarrow \quad [\sigma] = \frac{I^2}{L[P]}.$$

Substituting these results into the defining equation for a, we obtain

$$[a] = \frac{[\kappa]}{[\sigma]\Theta} = \frac{[P]^2}{\Theta^2 I^2} = \frac{[J]^2}{\Theta^2 I^2 T^2}.$$

Now, $[J]/\Theta$ would usually be measured in units of joules per kelvin, i.e. exactly the same as the fundamental Boltzmann constant k – perhaps not surprising given the involvement of thermal conductivity. More straightforwardly, IT has the same dimensions as charge, and the fundamental electronic charge e is immediately suggested – in line with the mechanism of electrical conduction.

Thus the strong indication is that the constant of proportionality has the form

$$a = \text{numerical constant} \times \left(\frac{k}{e}\right)^2.$$

In the Wiedemann–Franz law, the numerical constant is $\pi^2/3$, and a is known as the Lorentz number, usually denoted by L. Despite its name, L is not dimensionless; its value is $2.44 \times 10^{-8}\ W\Omega K^{-2}$.

S3   Using the symbol P for the force, the given equation can be rearranged:

$$P = \frac{q_1 q_2}{4\pi\epsilon_0 r^2} \quad \Rightarrow \quad \epsilon_0 = \frac{q_1 q_2}{4\pi P r^2}.$$

Taking dimensions yields

$$[\epsilon_0] = \frac{(IT)\,(IT)}{(MLT^{-2})\,L^2} = M^{-1}L^{-3}T^4 I^2.$$

The numerical value of ϵ_0 is irrelevant for our present purposes and only its dimensions matter. From these we see that

$$[\text{farad}] = M^{-1}L^{-3}T^4I^2 \times L = M^{-1}L^{-2}T^4I^2,$$

and it follows that $1\,\text{F} = 1\,\text{A}^2\,\text{s}^4\,\text{kg}^{-1}\,\text{m}^{-2}$.

S4 We first need to check that each of the terms in the proposed expression has the appropriate dimensions for an energy flux, and this requires us to establish the dimensions of each of ϵ_0, μ_0, E and B. Directly from (i) and (iii) we have

$$[\epsilon_0] = \frac{IT\,IT}{MLT^{-2}\,L^2} = M^{-1}L^{-3}T^4I^2 \text{ and } [B] = \left[\frac{F}{I\ell}\right] = \frac{MLT^{-2}}{IL} = MT^{-2}I^{-1}.$$

From (ii) we then deduce that $[\mu_0] = (L^{-1}I)^{-1}[B] = MLT^{-2}I^{-2}$. Finally, $[E] = [F/q] = MLT^{-2}/IT = MLT^{-3}I^{-1}$.

Energy flux has dimensions $(ML^2T^{-2})L^{-2}T^{-1} = MT^{-3}$. For the E^2 term in the proposed formula, the dimensions are

$$\left(\frac{M^{-1}L^{-3}T^4I^2}{MLT^{-2}I^{-2}}\right)^{1/2} (MLT^{-3}I^{-1})^2 = \frac{T^3I^2}{ML^2} \times \frac{M^2L^2}{T^6I^2} = MT^{-3}.$$

These are the same as those of S, and so the suggested E^2 term is at least plausible.

The corresponding calculation for the B^2 term yields

$$\left(\frac{MLT^{-2}I^{-2}}{M^{-1}L^{-3}T^4I^2}\right)^{1/2} (MT^{-2}I^{-1})^2 = \frac{ML^2}{T^3I^2} \times \frac{M^2}{T^4I^2} = L^2M^3T^{-7}I^{-4}.$$

This is clearly incompatible with representing a contribution to an energy flux, and so invalidates the student's proposed formula.

A little further consideration, coupled with the notion that energy expressions tend to contain the squares of dynamic variables – and hence that a B^2 term is probably correct, to 'balance' the E^2 term – can lead us to the correct formulation. In order that the calculated dimensions of the supposed B^2 term can be converted to those of an energy flux, they need to be multiplied by T^4I^4/M^2L^2. Looking over our earlier results, we see that is exactly the dimensional combination of $1/\mu_0^2$, strongly indicating that the correct formula is

$$S = \frac{1}{2}\left[\left(\frac{\epsilon_0}{\mu_0}\right)^{1/2} E^2 + \frac{1}{\mu_0^2}\left(\frac{\mu_0}{\epsilon_0}\right)^{1/2} B^2\right].$$

More symmetrically, and showing explicitly the 'balance' between electric and magnetic fields, this can be rewritten as

$$S = \frac{1}{2(\epsilon_0\mu_0)^{1/2}}(\epsilon_0 E^2 + \mu_0^{-1}B^2). \qquad (*)$$

This is, in fact, the correct expression for the energy flux. Of course, the 'correction factor' could have contained a non-unit numerical factor – and our analysis would not reveal this – but the symmetry of $(*)$ suggests that this is not the case.

[*Note*: The factor $1/(\epsilon_0\mu_0)^{1/2}$ can be more simply written as c, the speed of light in a vacuum, and the final expression for the flux considered as the product of the energy density of an electromagnetic field and the speed at which it moves.]

S5   For an egg to be described as 'boiled', its various parts have to be brought to particular temperatures. We can assume that certain properties, such as specific heat, density, shape and thermal conductivity, are essentially the same in the two types of egg, and work in terms of one that clearly isn't, namely linear size.

The heating proceeds via thermal conduction, and we have assumed that the thermal conductivity λ is the same for both eggs. As it is expressed in watts per unit area per unit temperature gradient, the total rate of heat flow scales as L^2 (for the area factor) and as L^{-1} (for the temperature gradient), i.e. as L overall. Clearly the total heat needed scales as the egg's volume, i.e. as L^3. Thus the time needed to boil the egg scales as $L^3/L = L^2$. Since the mass m of an egg is proportional to the cube of its linear dimension, the required time $\propto (M^{1/3})^2 = M^{2/3}$.

A typical medium-sized hen's egg weighs about 50 grams and the weight of an average ostrich egg is about 1.5 kg, and so the time required to boil the latter is approximately

$$t_o = t_h\left(\frac{m_o}{m_h}\right)^{2/3} \approx 5 \times \left(\frac{1.5}{0.05}\right)^{2/3} = 48 \text{ minutes.}$$

S6   Five fundamental constants are listed and σ is to be expressed in terms of them. Since there are five basic dimensions that could be involved (M, L, T, I and Θ), all five constants could be required. To investigate this we set out the dimensions of all six quantities:

$$[c] = LT^{-1} \qquad [e] = IT \qquad [h] = [\text{J}]T = ML^2T^{-1}$$

$$[k] = [\text{J}]\Theta^{-1} = ML^2T^{-2}\Theta^{-1} \qquad [m_e] = M,$$

whilst those of σ are

$$[\sigma] = [J]T^{-1}L^{-2}\Theta^{-4} = MT^{-3}\Theta^{-4}.$$

We note that, among the base dimensions, I appears *only* in $[e]$, and it is not present in $[\sigma]$. Therefore e cannot be involved in the required expression for σ. Further, among the base dimensions, only k contains Θ; the presence of Θ^{-4} in $[\sigma]$ thus demands a factor of k^4 in the expression for σ.

The presence or absence of c, h, and m_e in the required expression have to be determined by considering the base dimensions M, L and T simultaneously. Let

$$\sigma = \mu k^4 c^\alpha h^\beta m_e^\gamma, \qquad (*)$$

where α, β, γ and μ are (currently) unknown numerical constants. Now, equating the three remaining dimensions on each side of $(*)$ gives

$$
\begin{aligned}
M: &\quad 1 = 4 + \beta + \gamma, \\
L: &\quad 0 = 8 + \alpha + 2\beta, \\
T: &\quad -3 = -8 - \alpha - \beta.
\end{aligned}
$$

These simultaneous equations have the (unique) solution $\alpha = -2$, $\beta = -3$, $\gamma = 0$. The immediate conclusion from $\gamma = 0$ is that the electronic mass is not involved in the Stefan constant – and, on physical grounds, it is hard to see how it could be!

Gathering together the other results, we see that σ takes the form

$$\sigma = \mu k^4 c^{-2} h^{-3}.$$

The actual physical values of the quantities involved are

$$k = 1.38 \times 10^{-23}\,\mathrm{J\,K^{-1}}, \qquad c = 3.0 \times 10^8\,\mathrm{m\,s^{-1}}, \qquad h = 6.6 \times 10^{-34}\,\mathrm{J\,s},$$

$$\sigma = 5.7 \times 10^{-8}\,\mathrm{W\,m^{-2}\,K^{-4}}.$$

It follows that the (strictly) numerical factor μ is given by

$$\mu = \frac{5.7 \times 10^{-8}\,(3.0 \times 10^8)^2\,(6.6 \times 10^{-34})^3}{(1.38 \times 10^{-23})^4} = 40.7,$$

and that the full expression for σ is

$$\sigma = 40.7\frac{k^4}{c^2 h^3}.$$

The true value of μ is $2\pi^5/15 = 40.8$; the small discrepancy is accounted for by rounding errors in the quoted experimental values.

S7 This question contains only two numerical pieces of data, and, for example, does not indicate the Moon's size or its distance from Earth; we will have to assign them symbolic values and hope/expect their actual values not to be required for the final answer.

If the Moon's distance from Earth is R, then its radius r is given by $2r = 9 \times 10^{-3}R$. The light diffusely reflected from the illuminated half of its surface is $P = 0.07 \times \pi r^2 \times I$, where I is the intensity of sunlight falling on it – the same as that which reaches Earth. This diffusely reflected light is emitted into a solid angle of 2π, and so the intensity M of the resulting moonlight, as seen from Earth, is $P/2\pi R^2$. Thus,

$$\frac{M}{I} = \frac{0.07 \times \pi r^2}{2\pi R^2} = 0.035 \frac{r^2}{R^2} = 0.035 \times (4.5 \times 10^{-3})^2 = 7 \times 10^{-7}.$$

So, in fact, moonlight is only about one-millionth as bright as sunlight, but, being a wise old Owl, he kept that to himself – it might also be useful as a physics problem, one day!

S8 To see whether we can eliminate any of the four given quantities from the expression for λ, we begin by listing their dimensions:

$$[G] = M^{-1}L^3T^{-2}, \quad [k] = ML^2T^{-2}\Theta^{-1}, \quad [\rho] = ML^{-3}, \quad [V] = LT^{-1}.$$

As Θ appears *only* in the Boltzmann constant k, and therefore cannot be absent from any product containing k, the latter cannot be involved in an expression for a length. Of the remaining three, only G and ρ contain M, and consequently they must appear in the expression for λ as the combination $G\rho$. Similarly, to eliminate any T-dependence of λ, which is a 'pure' length, G and V must appear in the combination G/V^2.

We therefore examine the dimensions of the combination $G\rho/V^2$:

$$\left[\frac{G\rho}{V^2}\right] = \frac{M^{-1}L^3T^{-2}\ ML^{-3}}{L^2T^{-2}} = L^{-2}.$$

As expected/contrived, this contains only dimension L, and λ must, within a dimensionless constant μ, be given by

$$\lambda = \mu\sqrt{\frac{V^2}{G\rho}} = \frac{\mu V}{\sqrt{G\rho}}. \tag{*}$$

(a) Since G appears in the denominator of this expression, and we would naturally expect gravitational collapse to be more likely if G were larger, and hence λ smaller, the expression (*) must represent a lower limit for collapse, i.e. the scale-length of the density perturbations must *exceed* λ for the gas to coalesce under its own gravity.

(b) Although numerical factors between 0.1 and 10 hardly matter in estimations such as this, and we do not know the value of μ in any case, we will take μ as unity and keep any other known numerical factors at their actual values. To compute a numerical value for λ, we need one for V. This can be estimated as being equal to the speed of the atoms in the cloud, and, although the Boltzmann constant does not appear in the expression for λ, it can be used to connect V with the temperature T of the atomic hydrogen cloud:

$$\frac{1}{2}m_{\mathrm{p}}V^2 = \frac{3}{2}kT \quad \Rightarrow \quad V = \sqrt{\frac{3kT}{m_{\mathrm{p}}}} = \sqrt{\frac{3 \times 1.4 \times 10^{-23}\ 10}{1.7 \times 10^{-27}}} = 500\ \mathrm{m\,s^{-1}}.$$

This gives a value for λ of

$$\lambda = \frac{500}{\sqrt{6.7 \times 10^{-11} \times 10^{-17}}} = 1.93 \times 10^{16}\ \mathrm{m}.$$

The mass M_* of a star formed by gravitational collapse, following a density perturbation whose scale exceeds λ in linear dimensions, will be the mass of all the atomic hydrogen within a sphere of radius $\lambda/2$, i.e.

$$M_* = \frac{4\pi}{3}\left(\frac{\lambda}{2}\right)^3 \rho \sim 4 \times 10^{31}\ \mathrm{kg}.$$

S9 We need to eliminate the involvement of the unknown (for this problem) physical constants h and G, and it will be slightly less complicated if we work with ℓ_{p}^2, rather than ℓ_{p} itself:

$$\ell_{\mathrm{p}}^2 = \frac{hG}{2\pi c^n} = mc\lambda_{\mathrm{m}} \times \frac{c^2 r_{\mathrm{s}}}{2M} \times \frac{1}{2\pi c^n}.$$

Taking dimensions:

$$[\ell_{\mathrm{p}}^2] = \left[\frac{c^3 \lambda_{\mathrm{m}} r_{\mathrm{s}}}{c^n}\right] \quad \Rightarrow \quad L^2 = [c]^{3-n}L^2,$$

which requires $n = 3$.

S10 At first sight there is not enough information available to make the required estimate. Any expression for an energy must include quantities that can provide a mass dimension ($[E] = ML^2T^{-2}$), but neither t nor R does this. As, physically, the rate of expansion of the fireball is determined by the inertial resistance of the air it has to displace, the density of that air must be a factor in determining what happens.

A plausible alternative for providing the 'M-factor' is the atmospheric pressure p_0, which has dimensions $(MLT^{-2})/L^2 = ML^{-1}T^{-2}$. However, from p_0, R and t the *only* way to construct an energy (with dimensions ML^2T^{-2}) would be with the combination $p_0 R^3$; this would not include time, and so could make no use of the recorded data.

So, we seek the dimensional connection between E, t, R and ρ, the density of air. Their dimensions are

$$[E] = ML^2T^{-2}, \quad [t] = T, \quad [R] = L, \quad [\rho] = ML^{-3}.$$

By inspection (or by equating dimensional powers),

$$E = \mu \frac{\rho R^5}{t^2} \qquad (*)$$

where μ is a dimensionless constant (which we assume to have a value close to unity).

If this form is correct, the quantity R^5/t^2 should have (approximately) the same value for each observation. The following table tests this:

$t\,(10^{-3}\,\text{s})$	0.24	0.66	1.22	4.61	15.0	53.0
$R\,(\text{m})$	19.9	31.9	41.0	67.3	106.5	175.0
$R^5/t^2\,(10^{13}\,\text{m}^5\,\text{s}^{-2})$	5.41	7.58	7.78	6.50	6.09	5.84

Given that a large power of R is involved, and consequently any small inaccuracies in its measurement are greatly magnified, the approximate constancy of R^5/t^2 is confirmed. We will use the average numerical value of 6.5×10^{13}.

Finally, at sea-level, the density ρ of air is approximately 1.22 kg m^{-3} and so, with the assumption that $\mu \approx 1$, equation $(*)$ gives the energy of the explosion as

$$E \approx \frac{\rho R^5}{t^2} \approx 1.22\,\text{kg}\,\text{m}^{-3} \times 6.5 \times 10^{13}\,\text{m}^5\,\text{s}^{-2} = 8 \times 10^{13}\,\text{J}.$$

This certainly seems a much easier way of estimating the energy than actually trying to measure it directly!

S11   The first question that arises is that of how the dimensions of a derivative are related to those of its components. A derivative is defined as the limiting value of the ratio of two quantities, each of which can have any or no physical dimensions. Whatever the dimensions of that ratio, they will not be changed just because the magnitudes of its numerator and denominator become arbitrarily small; the natures of both remain unchanged. Thus, a derivative in the form df/du or $\partial f/\partial u$ is equivalent, on dimensional grounds, to dividing $[f]$ by $[u]$, and so $[df/du] = [f]/[u]$.

By way of example, the first term on the LHS of the quoted equation has dimensions $[\psi]L^{-2}$, and similarly for the second and third term. The fourth term has dimensions $[LT^{-1}]^{-2} \times [\psi]T^{-2} = [\psi]L^{-2}$, and so is compatible with the first three. For possible validity, the equation's RHS must also have dimensions $[\psi]L^{-2}$.

Now, as almost everyone knows, mc^2 is equal to E and so is an energy! Planck's constant h, as relatively few people know, has the dimensions of energy $\times$ time; the numerical factor $1/2\pi$ in $\hbar$ does not affect this. So, the quoted RHS has the same dimensions as $(\psi \times \text{energy}^2) \div (\text{energy} \times \text{time})^2$, i.e. $[\psi]T^{-2}$. These are not the same as those on the LHS, being different by a factor of L^2T^{-2}.

The mistake cannot be attributed to one in $[\psi]$, which appears uniformly in every term of (∗) and is therefore irrelevant for the present investigation. But, L^2T^{-2} *are* the dimensions of the square of a speed, and the obvious candidate is c^2. The 'speed' of the particle (actually, not well defined in quantum physics) could also have been considered, but a simple copying error is more likely.

All the evidence suggests that $(mc^2)^2$ should really have been written as m^2c^2 – as it should!

S 12   This situation is described by one or more connections between five variables, and, since neither heat nor electricity plays any part, their dimensions will involve only M, L and T. With only three dimensional equalities to satisfy, and five variables with which to do so, the solution is unlikely to be unique.

In simpler cases, the normal procedure is to try to express one variable in terms of the others by an equation containing an unknown numerical constant. But this procedure can be more fundamentally considered as finding a dimensionless combination q of the variables involved, and then setting q equal to a constant, say μ – formally described by $q = \mu$, or $f(q) = 0$ for the function $f(q) = q - \mu$. If only one dimensionless combination is possible, then the solution is unique, except for the value of μ.

However, if more than one such combination, q_1, q_2, ... can be found, the most that can be said is that $f(q_1, q_2, \ldots) = 0$, where f is a function of more than one variable. Here, the number of q_i is limited by how many *independent* combinations are possible; any combination that can be obtained as the result of multiplying or dividing other combinations is not to be included. Our current problem illustrates this situation (see below).

We start by setting out the dimensions of the five variables describing the flow:

$$[d] = L, \quad [R] = L^3 T^{-1}, \quad [\rho] = ML^{-3},$$
$$[\eta] = ML^{-1}T^{-1}, \quad [p] = ML^{-1}T^{-2}.$$

The only two of these that may not be immediately apparent are

$$[\eta] = \frac{[\text{force per unit area}]}{[\text{velocity gradient}]} = \frac{MLT^{-2}L^{-2}}{LT^{-1}L^{-1}} = ML^{-1}T^{-1}$$

and

$$[p] = \frac{[\text{force}]}{[\text{area}]} = \frac{MLT^{-2}}{L^2} = ML^{-1}T^{-2}.$$

The process of forming dimensionless combinations is somewhat arbitrary, but a suitable set of independent q_i should result, whatever starting point is chosen. We begin by combining η and ρ in such a way that M does not appear, and then incorporate R to eliminate T; we do this because it will leave only some power of L, and this can always be eliminated by factoring in an appropriate power of d. As a series of formulas that will result in a dimensionless variable q_1:

$$\left[\frac{\eta}{\rho}\right] = \frac{M}{LT}\frac{L^3}{M} = L^2 T^{-1} \quad \Rightarrow \quad \left[\frac{\eta}{\rho R}\right] = L^{-1} \quad \Rightarrow \quad q_1 = \frac{\eta d}{\rho R}.$$

As the pressure difference p does not appear in q_1, we base our construction of q_2 on it. A possible sequence is represented by the following series of formulas:

$$\left[\frac{\eta}{p}\right] = \frac{M}{LT}\frac{LT^2}{M} = T \quad \Rightarrow \quad \left[\frac{\eta R}{p}\right] = L^3 \quad \Rightarrow \quad q_2 = \frac{\eta R}{p d^3}.$$

Since we have five variables and three relevant dimensions, two is the number of independent dimensionless variables we can expect to find. As you may care to verify, another dimensionless combination, one that does not contain R, is $q_3 = \eta^2/(p\rho d^2)$. However this is *not* a third independent combination, since $q_3 = q_1 \times q_2$; similarly, $q_4 = (R^2\rho)/(pd^4)$ fails the independence test. Why? In summary, the defining equation for the process is $f(q_1, q_2) = 0$, for some (unknown) function f.

Given this paucity of information, connection can only be made between two situations with different values for the variables, if it happens that the values of q_1 and q_2 are the same *in both*. In the specific problem posed, d is the same for both measurements. If we denote the second set of measurements by a prime, then we must have

$$\frac{\eta d}{\rho R} = q_1 = \frac{\eta' d}{\rho' R'} = \frac{(\eta/3)d}{2\rho R'} \quad \Rightarrow \quad R' = R/6,$$

and

$$\frac{\eta R}{p d^3} = q_2 = \frac{\eta' R'}{p' d^3} = \frac{(\eta/3)(R/6)}{p' d^3} \quad \Rightarrow \quad p' = p/18.$$

In words, for the second liquid, with density 2ρ and viscosity $\eta/3$, the flow rate will be reduced by a factor of 6 if the pressure difference is reduced by a factor of 18.

S 13 Although it may seem natural to work from a figure with Q as its uppermost point and OQ at some angle ϕ to 'the vertical', this is not the simplest way to proceed. Since the lamina, with or without the extra weight, will (in reality) always hang with its centre of gravity directly below the suspension point (Q), all we need to do is find the angle the line OQ makes with that joining Q to the centre of gravity G of the modified lamina.

This is most easily done by using a coordinate system that has AD and AB as the x- and y-axes, as in the figure; the relevant lengths and distances do not then involve any sines or cosines of ϕ. It is clearly convenient to take the original square as having a side of length 2 and a mass of $4m$.

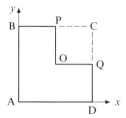

We will find G by formal but straightforward calculation, though some, if not most, aspects of the results can be deduced on symmetry grounds.

Firstly, without the additional particle, the coordinates of the lamina's centre of mass G_1 are

$$x_1 = \frac{2m\frac{1}{2} + m\frac{3}{2}}{3m} = \frac{5}{6}, \qquad y_1 = \frac{2m\frac{1}{2} + m\frac{3}{2}}{3m} = \frac{5}{6}.$$

That the two coordinates would be equal could have been anticipated, since G *must* lie on the line AO.

The angle θ_1 the line OQ makes with the vertical when the lamina is suspended is the same angle as $\angle$ OQG, that is

$$\theta_1 = \tan^{-1}\left(\frac{1 - y_1}{2 - x_1}\right) = \tan^{-1}\frac{(1/6)}{(7/6)} = \tan^{-1}\frac{1}{7} = 8.1°.$$

With the additional mass at the mid point of OB, the mass loss of the (missing) square OPCQ has been compensated for in both magnitude and distance above the line $y = 1$. So we can immediately conclude that the y-coordinate of the new centre of gravity G_2 will be 1. Since both O and Q also have $y = 1$, OQ will be vertical, whatever the value of the x-coordinate of G_2. As a formal calculation:

$$x_2 = \frac{2m\frac{1}{2} + m\frac{3}{2} + m\frac{1}{2}}{4m} = \frac{3}{4}, \qquad y_2 = \frac{2m\frac{1}{2} + m\frac{3}{2} + m\frac{3}{2}}{4m} = 1.$$

As previously, OQ's angle to the vertical is

$$\theta_2 = \tan^{-1}\left(\frac{1 - y_2}{2 - x_2}\right) = \tan^{-1}\frac{(0)}{(5/4)} = \tan^{-1} 0 = 0°.$$

S 14 We measure all vertical heights from the upper surface of the beaker's base. Clearly, the criterion for maximum stability is that the centre of gravity G of the whole system, beaker plus liquid, should be as low as possible.

When the beaker has been filled to a depth x the liquid's centre of gravity is at a height $x/2$, and G is at a height y given by

$$y = \frac{mh + \frac{1}{2}\lambda x^2}{m + \lambda x} \qquad \text{where } \lambda = \pi r^2 \rho. \qquad (*)$$

We need to find the value of x that minimizes y. The mechanical way to do this is using differential calculus and that is done in solution (b) below. But a much more elegant line of reasoning is available, as follows.

Solution (a). When liquid is initially added to the beaker, it begins to lower the overall centre of gravity, but after sufficient liquid to make G lie in its surface has been added, anything further necessarily makes G start to rise. In other words, the system is most stable when the overall centre of gravity lies in the liquid surface – that is when $y = x = x_s$. As an equation:

$$x_s = \frac{mh + \frac{1}{2}\lambda x_s^2}{m + \lambda x_s} \quad \Rightarrow \quad \tfrac{1}{2}\lambda x_s^2 + mx_s - mh = 0,$$

and hence the required (necessarily positive) value of x_s is

$$x_s = \lambda^{-1}(\sqrt{m^2 + 2\lambda mh} - m).$$

The height of G then has the *same* value.

Solution (b). The 'automated' way to minimize y in equation $(*)$ is using differential calculus, as follows:

$$\frac{dy}{dx} = \frac{(m + \lambda x)\lambda x - (mh + \frac{1}{2}\lambda x^2)\lambda}{(m + \lambda x)^2} = 0,$$

i.e.

$$\tfrac{1}{2}\lambda x_s^2 + mx_s - mh = 0 \quad \Rightarrow \quad x_s = \lambda^{-1}(\sqrt{m^2 + 2\lambda mh} - m).$$

To answer the second part of the problem, we now substitute $x = x_s$ into $(*)$.

$$\begin{aligned}
y_s &= \frac{mh + \tfrac{1}{2}\lambda\lambda^{-2}(m^2 + 2\lambda mh - 2m\sqrt{m^2 + 2\lambda mh} + m^2)}{\sqrt{m^2 + 2\lambda mh}} \\
&= \frac{\lambda mh + m^2 + \lambda mh - m\sqrt{m^2 + 2\lambda mh}}{\lambda\sqrt{m^2 + 2\lambda mh}} \\
&= \lambda^{-1}(\sqrt{m^2 + 2\lambda mh} - m).
\end{aligned}$$

As already demonstrated in solution (a), this shows that $y_s = x_s$, i.e. for maximum stability, the overall centre of gravity must lie in the liquid surface.

S 15 In order to analyse this situation, we are going to need to introduce a number of additional parameters, both fixed and variable; we anticipate that some or all of the fixed ones will contribute to the (undeclared) constant of proportionality in the quoted relationship. Let the distance between the anchor points be ℓ, the stone have mass M, and the angles made by the strings with the horizontal be θ_1 and θ_2 (as shown in the figure).

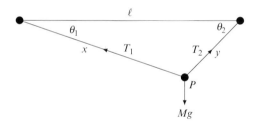

At P, resolving the forces both vertically and horizontally:

$$T_1 \cos\theta_1 = T_2 \cos\theta_2, \tag{1}$$
$$T_1 \sin\theta_1 + T_2 \sin\theta_2 = Mg. \tag{2}$$

Whilst, from geometry,

$$x \sin\theta_1 = y \sin\theta_2, \tag{3}$$
$$x \cos\theta_1 + y \cos\theta_2 = \ell. \tag{4}$$

From (1) and (4)

$$x \cos\theta_1 + y\,\frac{T_1}{T_2}\cos\theta_1 = \ell \quad \Rightarrow \quad \cos\theta_1 = \frac{T_2\,\ell}{T_2 x + T_1 y}.$$

From (2) and (3)

$$T_1 \sin\theta_1 + T_2\frac{x}{y}\sin\theta_1 = Mg \quad\Rightarrow\quad \sin\theta_1 = \frac{Mg\,y}{T_2x + T_1y}.$$

Similarly,

$$\cos\theta_2 = \frac{T_1\,\ell}{T_2x + T_1y} \quad\text{and}\quad \sin\theta_2 = \frac{Mg\,x}{T_2x + T_1y}.$$

Now, $\cos^2\theta_1 + \sin^2\theta_1 = 1 = \cos^2\theta_2 + \sin^2\theta_2$. Hence, omitting a common factor of $(T_2x + T_1y)^{-2}$, we have

$$T_2^2\ell^2 + (Mg)^2y^2 = T_1^2\ell^2 + (Mg)^2x^2,$$

which can be rearranged as

$$T_1^2 - T_2^2 = \left(\frac{Mg}{\ell}\right)^2 (y^2 - x^2), \qquad \text{i.e.} \qquad T_1^2 - T_2^2 \propto y^2 - x^2.$$

S 16 (*a*) The condition for the stability of any sub-pile of successive books (including the cases of both the whole pile and a single book, i.e. the uppermost one), is that there should be some support directly below its centre of gravity. If this condition is satisfied for *every* sub-pile, then the whole pile will not topple; if it is not satisfied for *any* of the sub-piles, at least a part of the pile will fall. We now apply this test to the given situation, with all values implicitly measured in mm.

It is obvious that the fourth (top) book, which overhangs the third by 115, i.e. less than $\frac{1}{2} \times 240 = 120$, passes the test.

For the other sub-piles it is better to have a more generally applicable formulation, and it is convenient to work in terms of the horizontal distance the far side of each of the books is to the right of the table's edge. These are 30, 70, 130 and 245. For the sub-pile consisting of the top two books, its centre of gravity is situated at

$$\tfrac{1}{2}[(245 - 120) + (130 - 120)] = \tfrac{1}{2} \times 135, \text{ i.e. } < 70 \quad\Rightarrow\quad \text{test passed.}$$

For the next sub-pile of the top three books, we similarly have

$$\tfrac{1}{3}[(245 - 120) + (130 - 120) + (70 - 120)]$$
$$= \tfrac{1}{3} \times 85, \text{ i.e. } < 30 \quad\Rightarrow\quad \text{test passed.}$$

Continuing in this way, but slightly more compactly, for the whole pile:

$$\tfrac{1}{4}[(245 + 130 + 70 + 30) - (4 \times 120)] = \tfrac{1}{4} \times -5, \text{ i.e. } < 0 \quad\Rightarrow\quad \text{test passed.}$$

Thus for each and every sub-pile, its centre of gravity is sited immediately above another book or the table; the pile *does not* topple – even though no part of the uppermost book is above the table!

(*b*) We have already made use of the quantities defined as x_n, and the formulation used in the final case (the whole pile) indicates how a general criterion might be expressed. For any particular sub-pile, say one consisting of r books, stability requires that

$$\frac{1}{r}\left[(x_N + x_{N-1} + \cdots + x_{N-r+1}) - \left(r \times \tfrac{1}{2}\ell\right)\right] < x_{N-r}.$$

This condition has to be satisfied by every sub-pile; if it fails for *any* one, the pile topples.

Thus the general criterion can be stated as follows: for *every* individual r in the range 1 to N, we must have

$$\sum_{i=N+1-r}^{N} x_i < r\left(x_{N-r} + \tfrac{1}{2}\ell\right).$$

S 17 If the towel is to hang asymmetrically over the rail, the tensions at the upper ends of the two hanging parts must be different, and the first job is to find how these tensions are connected quantitatively.

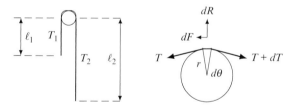

We consider first the forces that act radially on an infinitesimal length of towel that subtends an angle $d\theta$ at the centre of the towel rail, as shown in the right-hand part of the figure. The net effect of the tensions at its two ends is to produce a force directed towards the centre of the rail, and given in magnitude by

$$(T + T + dT) \sin\left(\tfrac{1}{2}d\theta\right) \simeq T\,d\theta \quad \text{as } d\theta \to 0.$$

The normal reaction of the rail on the towel dR is equal and opposite to this, and generates a maximal frictional force $dF = \mu\,dR$ to oppose relative motion between the towel and the rail. Next, equating the balanced tangential forces gives

$$T+dT=T+dF=T+\mu\,dR=T+\mu T\,d\theta \quad\Rightarrow\quad dT=\mu T\,d\theta.$$

To find the cumulative effect of all of these infinitesimal increments in the towel's tension, we integrate this result using elementary calculus:

$$\int \frac{dT}{T}=\int \mu\,d\theta \quad\Rightarrow\quad \ln\frac{T}{T_0}=\mu(\theta-\theta_0),$$

where T_0 is the tension at angular position θ_0 – in this problem, where the shorter length of towel first touches the rail.

Applying the above result to the towel, the tensions T_1 and T_2 shown in the figure are related by

$$T_2=T_1 e^{\mu\pi}.$$

The tensions themselves are equal to the weights (per unit width) of the parts of the towel they support, and so, more explicitly,

$$\ell_2 \rho g = \ell_1 \rho g\, e^{\mu\pi}, \tag{$*$}$$

with ρ being the towel's (irrelevant) surface density.

If x is the maximum allowed length difference, then

$$\ell_1 = \tfrac{1}{2}(L-\pi r-x) \quad\text{and}\quad \ell_2=\tfrac{1}{2}(L-\pi r+x).$$

When these are substituted into $(*)$, the resulting equation can be re-arranged as

$$\frac{x}{L-\pi r}=\frac{e^{\mu\pi}-1}{e^{\mu\pi}+1}.$$

Hence x.

S 18 For a satisfactorily suspended mobile, each of its rods should be horizontal (rather than vertical!) and no rod should have the potential to foul another rod or any string hanging from it. For a rod to be horizontal, it must be suspended at a point about which its loads have net zero moment. The units of the coordinates are irrelevant and will be omitted. The word 'weights' should probably read 'masses'.

- Rod A has only models hung from it, ($3m$ at 2) and (m at -6), and it is clear that it should be suspended at 0. [String 9 $\checkmark$]
- Rod A is hung from rod D, which also carries a load (m at 10). Consequently D should be suspended at x, where

$$(0-x)(3m+m)+(10-x)m=0 \quad\Rightarrow\quad x=2. \qquad \text{[String 2 }\checkmark\text{]}$$

- Rod D is hung from rod C; but C also carries both (4*m* at 10) and rod B. We therefore need to establish B's situation.
- Rod B carries only (2*m* at −2) and (3*m* at −12). It should therefore be suspended at *y*, where

$$(-2 - y)2m + (-12 - y)3m = 0 \quad \Rightarrow \quad y = -8. \qquad [\text{String 7} \checkmark]$$

- Rod C is now known to support D (5*m* at 2), B (5*m* at −8) and (4*m* at 10). It should be suspended at *z*, where *z* is such that

$$(2 - z)5m + (-8 - z)5m + (10 - z)4m = 0. \qquad (*)$$

This final equation implies that $z = \frac{10}{14}$; this is *not* in accord with the entry in the table for string 6, the one that suspends the whole mobile from the hook H.

Given that rods A, B and D are all correctly suspended for hanging horizontally, the most likely error is in the position of, or load on, string 8. For *z* to be equal to 0 in (∗), we must have

$$10m - 40m + p = 0,$$

where *p* is the product 'coordinate × load' for string 8. Since *p* must equal 30*m*, if a single transcription error has been made, it is either that the coordinate of string 8 should be 7.5, or that its load should be 3*m* – the latter seems much more likely!

The assembled mobile should probably look something like this:

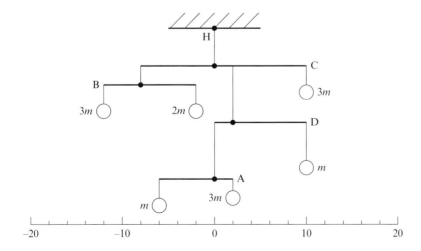

S 19   We need concern ourselves only with the moment balance about the base of the mast, as the ball joint there provides any needed horizontal and vertical reactions. Since the wind is horizontal and the force it exerts uniform throughout the height of the mast, the moment it produces is straightforwardly

$$M_{\text{wind}} = \int_0^{\ell} y \times f \, dy = \tfrac{1}{2} f \ell^2.$$

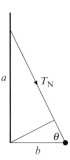

The figure, a section in a plane containing one of the cables, say T_N, shows that its moment about the base is

$$M_N = T_N b \sin \theta = \frac{T_N b a}{\sqrt{a^2 + b^2}}.$$

(a) With a wind from due North, only T_N will be non-zero, and we must have

$$\frac{T_N b a}{\sqrt{a^2 + b^2}} = \tfrac{1}{2} f \ell^2. \tag{$*$}$$

(b) With a NE wind, both T_N and T_E will contribute to the restraining moment, but each, making an angle of $45°$ with the wind, has its effect reduced by a factor of $\sqrt{2}$. Nevertheless, the tension in each will be less than that in (a) above. This is only to be expected, as it is intuitively obvious that the maximum tension must arise when a single cable has to provide all of the restraint.

The total length of cable required is clearly $4\sqrt{a^2 + b^2}$ and this has to be minimised subject to $T_N < T_0$, i.e.

$$\frac{\sqrt{a^2 + b^2}}{ab} < \frac{2T_0}{f\ell^2}.$$

The LHS of this inequality, having dimensions L^{-1}, varies inversely with the overall scale of a and b, and the only optimizing choice available is the ratio of the two. The fraction $\sqrt{a^2 + b^2}/ab$ becomes very large if either of a and b is chosen to be very small, and from its symmetry it is clear that it minimizes if $a = b$. To keep the overall scale, and hence the required length of cable, to a minimum, we can then set a to satisfy the *equality*

$$\frac{\sqrt{a^2 + a^2}}{aa} = \frac{2T_0}{f\ell^2} \quad \Rightarrow \quad a = \frac{f\ell^2}{\sqrt{2}\,T_0} = b.$$

S 20 If there were no friction, it would not be possible to balance the rod, loaded in the same way, at two different fulcrum points. That it is possible, is due to the frictional torque provided by the pivot, and in all three cases the size and direction of that torque are initially unknown; there must be a limiting value L to its magnitude, though it is not necessarily reached in any of the scenarios.

If Q and R are the (signed) resistive torques for the first two situations, then from the equilibrium conditions we have

$$Mx = m(a - x) + Q,$$
$$M\left(x + \tfrac{1}{4}a\right) = m\left(\tfrac{3}{4}a - x\right) + R,$$

from which it follows that

$$ma + Q = (M + m)x = \left(\tfrac{3}{4}m - \tfrac{1}{4}M\right)a + R.$$

This can be rearranged as

$$(M + m)a = 4(R - Q). \tag{$*$}$$

Denoting $\tfrac{1}{4}(M + m)a$ by λ, we now consider the possible ranges of values for Q and R. From the physics involved, we expect them to be of opposite signs, and, whether or not this is so in reality, the minimum values of $|Q|$ and $|R|$ needed to satisfy $(*)$ will be lowest if this is the case; this situation, i.e. $|Q| = \tfrac{1}{2}\lambda = |R|$, is the only safe assumption if a *guarantee* is to be given. Consequently, the *maximum* value of y for which equilibrium is *assured* is $y = \tfrac{1}{2}\lambda \div (M + m) = \tfrac{1}{8}a$, though it might be the case that larger values can be tolerated.

S 21 Although we know that the floor is rough, we are going to need an actual value for the coefficient of friction between the floor and the ladder. It can be deduced from the conditions under which the ladder just starts to slip under its own weight. This situation is shown in figure (i).

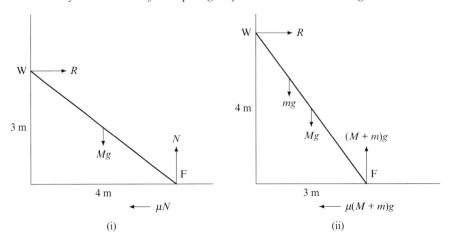

(i)   (ii)

By Pythagoras, the ladder touches the wall at a point W that is 3 m above the floor, and, since the wall is smooth, the reaction R of the wall on the ladder is horizontal. At F, the foot of the ladder, the normal reaction N, which must be equal to Mg, generates a (maximal) frictional force of μN, where μ is the coefficient of friction. The value of μ is found straightforwardly by balancing moments about W, and using the fact that $N = Mg$:

$$Mg \times 2 + \mu Mg \times 3 = Mg \times 4 \quad \Rightarrow \quad \mu = \tfrac{2}{3}.$$

With the man and his load (of combined mass m) on the ladder, as shown in figure (ii), we consider moments about F. It is clear that as the man and his load go further up the ladder the moment of their combined weight mg about F increases. Consequently, so does the opposing moment supplied by the reaction R, becoming maximal when the man is just about to step onto the wall. At this point, taking moments about F quantitatively gives

$$R \times 4 = Mg \times \tfrac{3}{2} + mg \times 3.$$

The ladder will slip (i.e. the combined centre of gravity of the man, his load, and the ladder will move to the right) if $R > \mu(M + m)g$, i.e. if

$$\tfrac{1}{4}\left(\tfrac{3}{2}M + 3m\right)g > \tfrac{2}{3}(M + m)g,$$
$$m\left(\tfrac{3}{4} - \tfrac{2}{3}\right) > M\left(\tfrac{2}{3} - \tfrac{3}{8}\right),$$
$$m > \tfrac{7}{2}M.$$

Since the man himself weighs $3Mg$ the maximum load he can carry is only $\tfrac{1}{2}M$. Health & Safety officers would, of course, be appalled if they knew of this situation – not even $\tfrac{1}{2}M$ would be allowed!

S 22

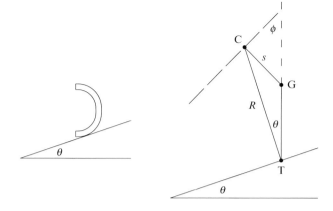

When the bowl is placed on the inclined plane, it will roll a little until its centre of mass G lies vertically above T, its point of contact with the plane. This will determine the angle ϕ which the plane containing its rim makes with the vertical. In the figure on the right, this plane, which also contains C, the centre of curvature of both surfaces of the hemispherical shell, is indicated by a dashed line.

The first task is to locate the position of G on the bowl's axis of symmetry. If the bowl were a solid uniform hemisphere, the distance CG would be equal to $\frac{3}{8}R$, where R is the radius of that hemisphere (here, 10.0 cm).[1] The given bowl is not solid, however; but it can be viewed as a solid hemisphere containing another concentric hemisphere of negative density. If the wall thickness is t, then this second half-sphere has radius $R - t$ (here, 8.0 cm).

The common centre of mass of the two superposed hemispheres (i.e. of the actual bowl), lies on the line CG, and is a distance s from C given by

$$\tfrac{2}{3}\pi\rho\left(R^3 - (R - t)^3\right) s = \tfrac{2}{3}\pi\rho R^3 \left(\tfrac{3}{8}R\right) - \tfrac{2}{3}\pi\rho(R - t)^3 \left[\tfrac{3}{8}(R - t)\right],$$

from which it follows that[2]

$$s = \frac{3}{8}\frac{R^4 - (R - t)^4}{R^3 - (R - t)^3} = 4.537 \quad \text{for } R = 10.0 \text{ and } t = 2.0.$$

Now, recalling that CG is perpendicular to the dashed 'rim line' and that CT is perpendicular to the (tangential) sloping plane, we can apply the sine rule to the triangle CGT:

$$\frac{s}{\sin\theta} = \frac{R}{\sin\left(\tfrac{1}{2}\pi + \phi\right)} = \frac{R}{\cos\phi}.$$

---

[1] This result can be verified from a formula book, or proved as in the *Note* following this solution.
[2] As expected, the result lies between the standard values of $\frac{3}{8}R = 3.75$ for a solid hemisphere, and $\frac{1}{2}R = 5.0$ for a thin hemispherical shell.

It then follows that

$$\cos \phi = \frac{R}{S} \sin \theta = 2.20 \sin \theta.$$

Now, purely numerically, this equation offers the possibility of yielding $\cos \phi > 1$ for sufficiently large θ. This cannot correspond to a physical situation. The limiting physical case occurs when $\cos \phi = 1$, $\phi = 0$, and the rim plane is vertical (this is the situation indicated, but not stated, in the left-hand figure). More descriptively, the limit of stability occurs when the horizontal distance that C is from the vertical through T exceeds the length CG, and the centre of mass can no longer position itself vertically above T; the maximum horizontal component of the vector $\overrightarrow{CG}$ is reached when the rim plane is vertical and the axis of symmetry of the bowl is horizontal. This consideration sets the limit on the slope of the plane as

$$\theta_{\text{max}} = \sin^{-1}\left(\frac{S}{R}\right) = 27.0°.$$

Note. Taking the axis of symmetry of the solid hemisphere as the x-axis, and the equation of its surface as $x^2 + y^2 = R^2$, the mass of an elementary disc of thickness dx is $\pi y^2 \rho \, dx$, where ρ is its density. Then from the definition of the centre of mass, its position $(\bar{x}, 0)$ is given by

$$\tfrac{2}{3}\pi R^3 \rho \bar{x} = \int_0^R \pi y^2 \rho x \, dx = \int_0^R \pi (R^2 - x^2) \rho x \, dx = \tfrac{1}{4}\pi \rho R^4,$$

i.e. $\bar{x} = \tfrac{3}{8}R$.

S 23   The figure shows the forces acting on a small length of the rope.

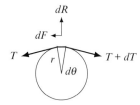

In solution S 17 appearing on page 111 it is shown, for a virtually identical situation, that the cumulative effect of the infinitesimal increases in tension

throughout the rope is represented by

$$\ln \frac{T}{T_0} = \mu(\theta - \theta_0),$$

where T_0 is the tension at angular position θ_0 – in this problem, where the end of the rope carrying the counterweight first touches the cylinder.

Substituting the physical values $T = 100g\,\mathrm{N}$, $T_0 = 1g\,\mathrm{N}$ and $\mu = 0.05$ gives

$$\theta - \theta_0 = \frac{1}{\mu}\ln 100 = 92.1\,\mathrm{rad} = 14.7\,\mathrm{turns}.$$

In practical terms, this requires $15\frac{1}{2}$ turns.

It is interesting to note that the number of turns required does not depend upon the radius r of the cylinder. A larger radius of curvature involves more rope ($d\ell = r\,d\theta$), but, for any given length $d\ell$ of rope, it reduces the component of the tension that acts radially ($\sin\frac{1}{2}d\theta \propto d\ell/r$); the two effects cancel each other. The number of turns required does not even depend upon r being the same at all angular positions. This raises the possibility that the number of turns needed is independent of the size and (convex) cross-sectional form of the cylinder, provided that the rope is sufficiently flexible to adapt to any local radius of curvature, however small.

S 24   To analyse the stability of the floating baton, we are clearly going to need the density of the oil, and this is not given. However, the aimless drifting of the baton in the oil immediately indicates that the oil and the baton have the same densities, i.e. the oil has density ρ'.

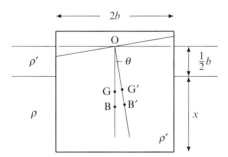

The figure shows a cross-section of the baton floating in the oil-capped pond; its length (perpendicular to the plane of the figure) will not need to be considered. The points marked G and B are the centres of gravity of, respectively, the baton, and the displaced fluid, the latter also being the centre of buoyancy. The points G′ and B′ are the positions of the same two points when the baton's upper surface is tilted through a small angle θ.

If, in equilibrium, x is the depth of wood below the oil layer, then, because the upthrust of the displaced fluids must equal the baton's weight, we have for each unit of its length (ignoring a factor of $2bg$ in all terms)

$$x\rho + \frac{1}{2}b\rho' = 2b\rho' \quad \Rightarrow \quad x = \frac{3\rho'}{2\rho}b.$$

In order to calculate the moments generated when the baton is tilted, we need to evaluate the lengths OG and OB, shown in the figure. Since G is the centre of gravity of the wooden baton,

$$\text{OG} = b - \left(2b - \frac{b}{2} - x\right) = x - \frac{b}{2} = \left(\frac{3\rho'}{2\rho} - \frac{1}{2}\right)b.$$

With B being the centre of gravity of *all* the displaced fluid, OB is more difficult to find. Taking moments about O in the form 'mass $\times$ distance to the centre of gravity' for both fluids (and again ignoring a common factor of $2bg$), we have

$$\text{OB} = \frac{x\rho\left(\frac{1}{2}b + \frac{1}{2}x\right) + \frac{1}{2}b\rho'\frac{1}{4}b}{x\rho + \frac{1}{2}b\rho'}$$

$$= \frac{\frac{3}{2}b\rho'\left(\frac{1}{2}b + \frac{3}{4}\frac{\rho'}{\rho}b\right) + \frac{1}{8}b^2\rho'}{\frac{3}{2}\rho'b + \frac{1}{2}b\rho'}$$

$$= \left(\frac{7}{16} + \frac{9}{16}\frac{\rho'}{\rho}\right)b.$$

As we will see in the next paragraph, we will need the difference between these two lengths, namely

$$\text{length GB} = \left(\frac{7}{16} + \frac{9}{16}\frac{\rho'}{\rho} - \frac{3}{2}\frac{\rho'}{\rho} + \frac{1}{2}\right)b = \frac{15}{16}\left(\frac{\rho - \rho'}{\rho}\right)b.$$

Ignoring for the moment the effects at the surface of the oil, when the baton is tilted, the equal and opposite forces due to the weight of the baton and the buoyancy upthrust (each of magnitude $4b^2\rho'g$), will produce a net capsizing couple C_A (anticlockwise in the figure) of

$$\text{baton's weight} \times (\theta \times \text{length GB}) = \frac{15\,(\rho - \rho')}{4}\frac{}{\rho}b^3\rho'g\theta.$$

This will be a capsizing couple because OB > OG.

The restoring couple, opposing the capsizing one, is the result of an additional mass of $\frac{1}{2}b^2\theta\rho'$ being added on one side of the centre line, and the weight of an equal mass being lost, through missing buoyancy, on the other. The distance from O to the centre of mass, in each case, is $\frac{2}{3}b$. Thus the restoring couple C_R has magnitude

$$2 \times \frac{1}{2}b^2\theta\rho'g \times \frac{2}{3}b = \frac{2}{3}b^3\rho'g\theta.$$

Comparing the two couples, we see that the baton will be stable against small disturbances (i.e. $C_R > C_A$) if

$$\frac{2}{3} > \frac{15\,(\rho - \rho')}{4} \frac{}{\rho},$$

which simplifies to $45\rho' > 37\rho$.

S 25 The required sketches are shown in the figure by the labelled thicker lines.

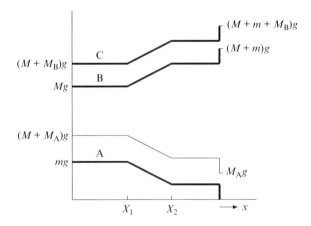

The first deviation from the horizontal at $x = X_1$ marks the point at which the cylinder starts to enter the water, and the sloping section reflects the increasing upthrust as it is gradually submerged. The sketches become horizontal again, at $x = X_2$, when it is fully submerged, with the value registered on balance A being the difference between the cylinder's weight and that of the water it displaces. The vertical parts of the graphs, both upwards and downwards, are all equal in magnitude, and give the (sudden) changes that occur when that part of the cylinder's weight that is not buoyancy-supported is transferred to the base of the bowl.

For the purposes of sketching the readings on the three balances, only the mass of spring balance B is required, those of A and C being irrelevant. However, the force provided by the person or mechanism that holds up balance A, and controls the lowering of the cylinder into the water, does depend on M_A; though not asked for, its variation with x is shown by the thinner line in the figure.

S 26   The situation when the stirrer is just about to slip into the glass is shown in the figure.

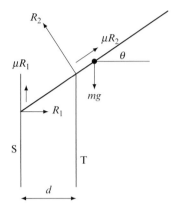

The stirrer's length $2a$ is not marked, so as to not further clutter an already crowded diagram, in which the five forces acting are indicated by single-headed arrows. As the end inside the tumbler is on the point of slipping downwards, the frictional force μR_1 due to surface S acts in an upwards direction, as does the vertical component of the frictional force generated at the top of side T.

Resolving the overall force balance for the static system, both vertically and horizontally:

$$\mu R_1 + R_2 \cos\theta + \mu R_2 \sin\theta = mg, \tag{1}$$
$$R_1 - R_2 \sin\theta + \mu R_2 \cos\theta = 0. \tag{2}$$

Taking moments about the contact point on surface S:

$$R_2 \times \frac{d}{\cos\theta} = mga \times \cos\theta. \tag{3}$$

[Taking moments about the lip of side T would have been equally valid, and would have produced the same final result.]

Now, writing $\mu = \tan\alpha$ and multiplying each of equations (1) and (2) through by $\cos\alpha$, transforms them into

$$(1^*) \quad R_1 \sin\alpha + R_2(\cos\theta\cos\alpha + \sin\alpha\sin\theta) = mg\cos\alpha,$$

and

$$(2^*) \quad R_1 \cos\alpha - R_2(\cos\alpha\sin\theta - \sin\alpha\cos\theta) = 0.$$

More compactly,

$$R_1 \sin\alpha + R_2 \cos(\theta - \alpha) = mg\cos\alpha \quad \text{and} \quad R_1 \cos\alpha = R_2 \sin(\theta - \alpha).$$

Eliminating R_1 and R_2 from these two equations and equation (3), gives the (fearsome-looking) equation

$$\frac{mga}{d}\cos^2\theta\left[\frac{\sin\alpha\sin(\theta-\alpha)+\cos\alpha\cos(\theta-\alpha)}{\cos\alpha}\right]=mg\cos\alpha.$$

But this simplifies to

$$a\cos^2\theta\cos(\theta-\alpha-\alpha)=d\cos^2\alpha,$$

i.e. the relationship given in the question.

It may seem surprising that, with its centre of mass beyond the lip of the tumbler, the stirrer can fall *into* it. But try it out for yourself, using a real uniform rod and a suitable-sized cylinder. To see clearly what happens, the materials employed must be such that there is noticeable friction between them. The equation derived above can be used to give a rough guide as to suitable relative sizes for the rod and cylinder.

S 27 Since the tension required in the ribbon is determined by both the weight of a chocolate segment and the location of its centre of mass, the first task is to find the latter.

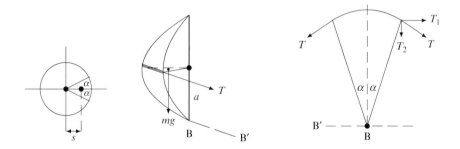

The left-hand figure above shows a plan view of the orange including one of the segments (technically a wedge) of angle 2α, with its centre of mass indicated by a back circle, a distance s from the orange's axis. In spherical polars, $dV = r^2\sin\theta\,dr\,d\theta\,d\phi$ and $x=r\sin\theta\cos\phi$; and so, if $V=(4\pi a^3/3)\times(2\alpha/2\pi)$ is the volume of the segment, we have

$$Vs = \int_{-\alpha}^{\alpha}\int_0^\pi\int_0^a r^3\sin^2\theta\cos\phi\,dr\,d\theta\,d\phi$$

$$= \frac{a^4}{4}\frac{\pi}{2}2\sin\alpha,$$

$$s = \frac{\pi a^4}{4}\frac{3\sin\alpha}{4\alpha a^3} = \frac{3\pi}{16}\frac{\sin\alpha}{\alpha}a.$$

The central figure shows the forces acting on the segment, its weight and a component of the tension in the ribbon. There is also the reaction from the table, but as we will be taking moments about the line BB' this will not come into the calculation.

As shown in the right-hand figure, the tension at one end of the particular piece of ribbon in contact with the segment can be resolved into components parallel to and perpendicular to BB'; the parallel component T_1 produces no moment about BB', but the perpendicular component T_2 produces a couple of magnitude $T\sin\alpha \times a$, as does the other end of the piece of ribbon. For equilibrium, we must have

$$2Ta\sin\alpha = mg \times s = mg\,\frac{3\pi}{16}\,\frac{\sin\alpha}{\alpha}\,a.$$

Chocolate oranges have several equal segments, and so we set $m = M/N$ and $N = 2\pi/2\alpha$ giving

$$T = \frac{Mg}{N}\,\frac{3\pi}{32}\,\frac{1}{\pi/N} = \frac{3Mg}{32}.$$

We note, in passing, that the calculated tension is independent of N (>1) and does not require all segments to have equal angles. It is also independent of a, except through the mass of the orange.

S 28 Before reading the solution given below, try to give a qualitative explanation of how a free ping-pong ball could remain submerged. All experience indicates that, on release, it might be expected to rise to the water surface – and to do so very rapidly!

The figure shows the ball sitting in the hole and it is clear that the angle $\alpha = \sin^{-1}(\sqrt{3}/2) = \pi/3 = 60°$. Thus, our calculations will be for a part-sphere which, in spherical polar coordinates, lies in the range $0 \le \theta \le \pi - \alpha$, with $\sin(\pi - \alpha) = \sin\alpha = \sqrt{3}/2$ and $\cos(\pi - \alpha) = -\cos\alpha = -1/2$.

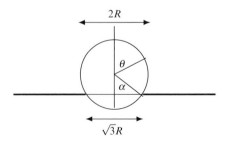

To find the minimum water depth h for which the ping-pong ball remains in place, we need to find the downward component of the force, resulting from hydrostatic pressure, on the part of the ball in the water. It is assumed that the atmospheric pressures above the water and below the hole are equal.

With a general vertical height measured by the polar angle θ (see the figure), and letting $h = \lambda R$, the hydrostatic pressure at level θ is $p = \rho g R(\lambda - \cos\alpha - \cos\theta)$, where ρ is the density of water. The area of the circular strip of the ball's surface, at that level, that subtends an angle $d\theta$ at the centre of the ball, is $2\pi R \sin\theta\, R\, d\theta$; its normal makes an angle θ with the vertical. Consequently, the force this strip contributes is

$$dF = \text{pressure} \times \text{area} \times \text{inclination factor}$$
$$= \rho g R(\lambda - \cos\alpha - \cos\theta)2\pi R \sin\theta\, R\, d\theta\, \cos\theta.$$

Some of this force, the part corresponding to depths with $\theta > \pi/2$, actually acts upward, but the net downward force F due to hydrostatic pressure is given by

$$\frac{F}{2\pi R^3 \rho g} = \int_0^{\pi-\alpha}\left(\lambda - \frac{1}{2} - \cos\theta\right)\sin\theta\cos\theta\, d\theta$$

$$= \left(\lambda - \frac{1}{2}\right)\int_0^{\pi-\alpha}\sin\theta\cos\theta\, d\theta - \int_0^{\pi-\alpha}\sin\theta\cos^2\theta\, d\theta$$

$$= \left(\lambda - \frac{1}{2}\right)\left[\frac{1}{2}\sin^2\theta\right]_0^{\pi-\alpha} + \left[\frac{1}{3}\cos^3\theta\right]_0^{\pi-\alpha}$$

$$= \left(\lambda - \frac{1}{2}\right)\frac{1}{2}\frac{3}{4} + \frac{1}{3}\left(-\frac{1}{8} - 1\right)$$

$$= \frac{3}{8}\lambda - \frac{3}{16} - \frac{9}{24} = \frac{3}{8}\lambda - \frac{9}{16}.$$

When there is little water in the bucket, this is a negative quantity, formally representing a 'net upthrust' on the ball, which therefore has to be held down. However, when λ is sufficiently large for F to be zero, the ball has balancing upwards and downward forces acting on it – and there is no need to hold it in position! The pressure causing the upthrust, acting on the lower part of the ball, is greater, but affects a smaller surface area; that responsible for a downward force, affecting the upper part of the ball, is smaller, but acts over a larger area.

From the final line of the calculation above, we see that $F = 0$ when $\lambda = \frac{3}{2}$, i.e., for this particular pairing of ball and hole radius, when the water is just level with the top of the ball.

Even though the ball is already completely covered, adding further water will result in an increased downward force. This is because, although the 'average pressure' (when weighted by a factor $\sin\theta\cos\theta$) will be increased by the same amount for both the upward and downward forces, the upper area experiencing hydrostatic pressure is greater than the lower one. It follows that with a smaller hole, a sufficiently greater depth of water would still be able to hold the ball in place. But with no hole at all, it would never sit unconstrained on the bottom of the bucket, however deep the water; this corresponds to $\alpha = 0$, equal upper and lower areas, and Archimedes Principle.

The intermediate case $(0 < \text{hole diameter} < \sqrt{3}R)$ can be qualitatively demonstrated by drilling/cutting out a hole about 2.5 cm in diameter from the bottom of disposable plastic container (e.g. an ice-cream tub), placing it in a sink whilst supporting it on blocks (so that the underside of the container is open to the atmosphere), holding the ping-pong ball (diameter 3.73 cm) in position, and then fully filling the container with water. The ball will remain in position, but, as, in practice, water does leak out at the perimeter of the hole, the water level in the container will fall at a moderate rate, and at some point, when the depth is about $4\frac{1}{2}$ cm, the ball will suddenly rise to the surface.

S 29   Clearly, the tension in the right-hand chain section is equal to F; let us denote the left-hand chain and its tension by T_1, and the central tensioned chain section by T_2. Since everything is in equilibrium, there can be no net couple acting on the free pulley (otherwise it would turn), and so we must have that $T_1 = T_2$, and therefore that each of them is equal to $\frac{1}{2}W$.

Now, from the the non-rotation of the compound pulley, we also have, by taking moments about its axle, that

$$Fb + T_2 a = T_1 b.$$

With $T_1 = T_2 = \frac{1}{2}W$, it immediately follows that

$$F = \left(\frac{b-a}{2b} \right) W.$$

Now if F moves downwards at speed v, the rate at which it does work is Fv. Since there is no slipping, and hence no frictional loss, the load must gain potential energy at the same rate. So, it must move upwards, and do so at a speed u such that $Wu = Fv$, i.e.

$$u = \left(\frac{b-a}{2b} \right) v.$$

S 30   The figure illustrates two stages of the process, when the stone is initially released, and just after the collision between the two stones has taken place.

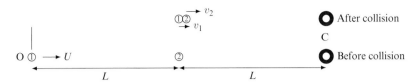

Because its end situation is simple, we start with 'stone 2'. It just reaches C after travelling a distance L against a frictional force of μmg, where m is the mass of a stone. It must therefore have started with kinetic energy μmgL, i.e., in the notation of the upper part of the figure,

$$\tfrac{1}{2}mv_2^2 = \mu mgL \quad \Rightarrow \quad v_2 = \sqrt{2\mu gL}.$$

Now if u is the relative velocity of approach of the two stones just before the collision, from energy conservation for stone 1's travel we have

$$\tfrac{1}{2}mU^2 - \mu mgL = \tfrac{1}{2}mu^2,$$

whilst from momentum conservation at the collision,

$$mu = m(v_1 + v_2) \quad \text{with} \quad v_2 - v_1 = eu.$$

Eliminating v_1 gives $2v_2 = u(1 + e)$, and so

$$\begin{aligned}
U^2 &= u^2 + 2\mu gL \\
&= \frac{4v_2^2}{(1+e)^2} + 2\mu gL \\
&= 2\mu gL \left(\frac{4 + 1 + 2e + e^2}{(1+e)^2} \right), \\
U &= \frac{\sqrt{2\mu gL}}{1+e}(5 + 2e + e^2)^{1/2}.
\end{aligned}$$

We note, in passing, that this gives the expected result, $U = 2\sqrt{\mu gL}$ if $e = 1$; then, the second stone simply 'takes over' from the first, which must start with enough kinetic energy to travel $2L$ against a frictional force of μmg.

S 31 What happens in this scenario depends upon the interplay between the gravitational, kinetic, and frictional-loss energies. Each has a different dependence on x, but they are not independent, being linked through energy conservation and, more physically, through the inextensible string.

When the blocks have each moved a distance x, the work already done against friction will be $2\mu Mgx$, whilst the gravitational energy lost will be

mgy. By conservation of energy, the kinetic energy T of the system will therefore be

$$T = mgy - 2\mu Mgx.$$

The distances x and y are not independent. The extra length of string between the tables is now $2x$, and each sloping string has length $a + x$. Consequently,

$$(a + x)^2 = a^2 + y^2 \quad \Rightarrow \quad y^2 = x^2 + 2ax.$$

The motion will stop when $T = 0$, provided that this happens before $x = a$; if it doesn't, the motion will certainly continue.

Setting $T = 0$, we have $my = 2\mu Mx$ and hence, successively,

$$m^2 y^2 = 4\mu^2 M^2 x^2 \quad \Rightarrow \quad m^2(2ax + x^2) = 4\mu^2 M^2 x^2 \quad \Rightarrow \quad x = \frac{2am^2}{4\mu^2 M^2 - m^2}.$$

Here the solution $x = 0$, corresponding to the starting position, has been ignored. For the blocks to remain on the tables, we must have $x < a$, i.e.

$$a(4\mu^2 M^2 - m^2) > 2am^2 \quad \Rightarrow \quad 4\mu^2 M^2 > 3m^2 \quad \Rightarrow \quad \mu M > \frac{\sqrt{3}}{2}m.$$

This is the required condition.

For the blocks to stop just as they reach the edges of the tables, the inequality must become an equality, i.e. $4\mu^2 M^2 = 3m^2$. The maximal value of T occurs when $\dot{T} = 0$, i.e. when $mg\dot{y} = 2\mu Mg\dot{x}$. Since $y^2 = x^2 + 2ax$,

$$2y\dot{y} = 2x\dot{x} + 2a\dot{x} \quad \Rightarrow \quad (a + x)\dot{x} = y\dot{y},$$

and substituting in $m\dot{y} = 2\mu M\dot{x}$ for either $\dot{x}$ or $\dot{y}$ gives the condition for maximum kinetic energy as

$$\begin{aligned}
m(a + x) &= 2\mu My, \\
m^2(a + x)^2 &= 4\mu^2 M^2(2ax + x^2), \\
&= 3m^2(2ax + x^2) \\
0 &= 2x^2 + 4ax - a^2, \\
x &= \tfrac{1}{2}(\sqrt{6} - 2)a = 0.22a.
\end{aligned}$$

Beyond $x = 0.22a$, the kinetic energy starts to decrease, as friction dissipates energy at a greater rate than the falling particle provides it.

S 32   A golf umbrella is about 1.2 m in diameter, and the density ρ of air about 1.22 kg m^{-3}. If the cross-sectional area of the open umbrella is A and the wind speed is V, then the mass of air stopped by the umbrella in unit time is $\rho A V$ and the momentum destroyed is $\rho A V^2$. Hence, by Newton, the force required to restrain it is

$$F \approx \rho A V^2 = 1.22\,\pi(0.6)^2\,(15)^2 = 312\,\text{N}.$$

The author has never picked up a (canvas) golf bag containing more than three clubs, but airlines allow golf sets of up to about 23 kg to be checked in.[3] In addition, Health & Safety think that 25 kg is the maximum that anybody should lift unaided – and golf is said to be a healthy sport. So, a maximum force for lifting golf bags is $\approx 25 \times 9.81 = 245$ N. The umbrella wins!

S 33 In theory this question is simply a matter of calculating a projectile's motion under gravity, deducing its launch conditions from its subsequent behaviour. However, the algebra involved is somewhat complicated, and to obtain the final (presumed small) net clearance, significant rounding errors in much larger distances must be avoided. The figure shows the relevant parameters.

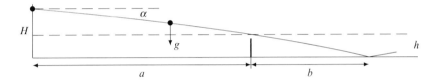

The ball's flight under gravity is an uninterrupted one, as is the first part of it, taken by itself. For the first part, the time of flight t is $a/v\cos\alpha$, and so

$$H - h = v\sin\alpha\,\frac{a}{v\cos\alpha} + \frac{g}{2}\,\frac{a^2}{v^2\cos^2\alpha}$$

$$= a\tan\alpha + \frac{g}{2v^2}\,\frac{a^2}{\cos^2\alpha}. \qquad (*)$$

Similarly, for the whole flight, with $t = (a+b)/v\cos\alpha$,

$$H = (a+b)\tan\alpha + \frac{g}{2v^2}\,\frac{(a+b)^2}{\cos^2\alpha}, \qquad (**)$$

from which it follows that

$$v^2 = \frac{g(a+b)^2(1+\tan^2\alpha)}{2[H - (a+b)\tan\alpha]}, \quad \text{using } \frac{1}{\cos^2\alpha} = \sec^2\alpha = 1 + \tan^2\alpha.$$

---

[3] On one occasion, the author's flight was delayed for nearly an hour because of the excessive number of golf bags that had been checked in, by others. According to the PA announcement, there would be a delay, and a change of boarding gate, because a 'stronger' plane needed to be brought into service!

Next, subtracting (∗) multiplied by $(a+b)^2$ from (∗∗) multiplied by a^2, and then rearranging, gives

$$[H - (a+b)\tan\alpha]a^2 = [H - h - a\tan\alpha](a+b)^2,$$

$$\tan\alpha\left[a(a+b)^2 - a^2(a+b)\right] = H\left[(a+b)^2 - a^2\right] - h(a+b)^2,$$

$$\tan\alpha\left[a(a+b)(a+b-a)\right] = H\left[2ab + b^2\right] - h(a+b)^2,$$

$$\tan\alpha = H\frac{(2a+b)}{a(a+b)} - h\frac{(a+b)}{ab}.$$

For reference, we start by listing the given values, and some derived from them, with all lengths in metres:

$$a = 11.89, \quad b = 6.40, \quad H = 2.60, \quad n = 0.914, \quad d = 0.067, \quad v = 50\,\text{s}^{-1}$$

$$a + b = 18.29, \quad 2a + b = 30.18, \quad g(a+b)^2 = 3281.7\,\text{s}^{-2}.$$

Substituting into the equation for $\tan\alpha$ gives

$$\tan\alpha = 0.1388\,H - 0.2404\,h.$$

If $\tan^2\alpha \ll 1$, then the equation for v^2 can be rearranged as

$$H - (a+b)\tan\alpha \approx \frac{g(a+b)^2}{2v^2} = \frac{3281.7}{2 \times 2500} = 0.6563,$$

and

$$\tan\alpha = \frac{2.60 - 0.6563}{18.29} = 0.1063.$$

Finally,

$$h = \frac{ab}{(a+b)}[0.1388\,H - 0.1063] = 1.0590.$$

As the net is 0.914 m high and the ball's radius is 0.0335 m, the net is cleared by $1.0590 - 0.914 - 0.0335 = 0.111$ m, i.e. by 11 cm.

S 34 The figure shows the layout of the 'five day' clock.

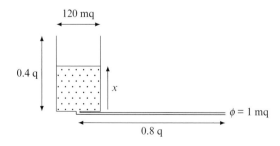

With x as the height in cubits of the water in the cylinder, the pressure drop across the capillary tube is $\rho g x = 125 \times 19.6\,x$, and so the volume flow rate, measured in cubic cubits per second, is

$$\frac{dV}{dt} = -\frac{\pi}{8}\frac{(5 \times 10^{-4})^4}{5 \times 10^{-4}}\frac{125 \times 19.6}{0.8}\,x = -1.503 \times 10^{-7}\,x.$$

Since the cylinder is of radius 60 mq, the rate of fall of the water level in it is

$$\frac{dx}{dt} = -\frac{1}{\pi(0.06)^2}\frac{dV}{dt} = -1.329 \times 10^{-5}\,x\,\mathrm{q\,s^{-1}}.$$

A standard integration gives

$$x(t) = x(0)\exp(-\lambda t) \quad \text{with } \lambda = 1.329 \times 10^{-5}\,\mathrm{s^{-1}}.$$

Refilling is needed when $x(t) = 0.04$ q. As $x(0) = 0.4$ q, this happens at a time

$$t = \frac{\ln(0.4/0.04)}{\lambda} = 1.73 \times 10^5\,\mathrm{s} = 48\,\mathrm{h}, \quad \text{i.e. 2 days.}$$

We can conclude that the clocks' purchasers were quite upset, but had the (small) consolation that the refilling could always take place at the same time of day, albeit on alternate days.

S 35 The normal equation of motion describes the position and velocity of the particle as functions of time. However, in this problem the times for the rise and fall of the particle will clearly be different, and obtaining information from the rise to use in an analysis of the fall will be difficult. What *is* common to the two parts is the distance over which each takes place, i.e. the height to which the particle rises; denote this by X, as shown in the figure.

The figure also shows the notation for the other quantities involved: the ascent is measured upwards by x, with initial and final speeds U and 0 respectively; the descent is measured downwards by y with corresponding speeds of

0 and D. From what we have noted earlier, it follows that we should write the equation of motion in terms of velocity v and either x or y, as appropriate.

For the *upward* motion, Newton's law gives

$$m\frac{dv}{dt} = -mg - kv^2, \quad \text{where } v = \frac{dx}{dt}.$$

To change this into an equation connecting v and x, we re-write dv/dt in the form

$$\frac{dv}{dt} = \frac{dv}{dx}\frac{dx}{dt} = v\frac{dv}{dx},$$

and obtain

$$mv\frac{dv}{dx} = -mg - kv^2.$$

This can be integrated, and the appropriate limits included, as follows:

$$\frac{mv\,dv}{mg + kv^2} = -dx \quad \Rightarrow \quad \frac{m}{2k}\left[\ln(mg + kv^2)\right]_U^0 = [-x]_0^X,$$

i.e.

$$X = \frac{m}{2k}\ln\frac{mg + kU^2}{mg}. \tag{$*$}$$

For the *downward* motion, a similar procedure (with v now equal to dy/dt) gives

$$mv\frac{dv}{dy} = mg - kv^2,$$

which, on integration, leads to

$$-\frac{m}{2k}\left[\ln(mg - kv^2)\right]_0^D = [y]_0^X \quad \Rightarrow \quad X = \frac{m}{2k}\ln\frac{mg}{mg - kD^2}.$$

Equating these two expressions for X:

$$\frac{m}{2k}\ln\frac{mg}{mg - kD^2} = \frac{m}{2k}\ln\frac{mg + kU^2}{mg},$$
$$(mg - kD^2)(mg + kU^2) = m^2g^2,$$
$$D^2(mg + kU^2) = mgU^2,$$

giving as the return speed

$$D = \sqrt{\frac{mg}{mg + kU^2}}\,U.$$

Note Equation (∗) may look 'not quite right', since if $k \to 0$ the ratio $m/2k \to \infty$. However, this is not so, as for small k,

$$X = \frac{m}{2k} \ln \frac{mg + kU^2}{mg} = \frac{m}{2k} \ln \left(1 + \frac{kU^2}{mg} \right)$$

$$= \frac{m}{2k} \left[\frac{kU^2}{mg} - \frac{1}{2} \left(\frac{kU^2}{mg} \right)^2 + \cdots \right] \sim \frac{U^2}{2g} + \cdots,$$

and the expected formula for X is recovered.

S 36 We have been given very little information, and so any constraints on the height reached by the second ball must come from the physics involved. Let upper case variables refer to the ball that is drawn aside, and lower case ones to the one that is hit. Clearly we are seeking the conditions that allow the velocity v of the latter, immediately after the collision, to be maximal; it is intuitively obvious that this will require that the collision is elastic, with no loss of total kinetic energy.

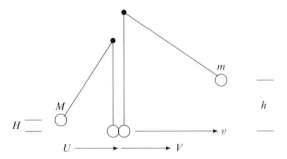

The situation is shown in the figure with the first ball, of mass M, having speeds U and V before and after the impact, respectively. As the ultimate possible height h for the second ball is entirely determined by v, all that is required is that the length of the second string is greater than $\frac{1}{2}h$; in all other respects, the string lengths are irrelevant.

Whether or not the collision is elastic, momentum conservation requires that

$$MU = MV + mv,$$

where m is the mass of the struck ball. The (desired) elastic interaction can be expressed as a formula by [4]

$$\tfrac{1}{2}MU^2 = \tfrac{1}{2}MV^2 + \tfrac{1}{2}mv^2 \quad \text{or by} \quad v - V = U - 0.$$

---

[4] The second of these is in the form 'the final relative velocity is equal to e times the initial relative velocity', where e is the coefficient of restitution, and equal to unity for an elastic collision.

If we write the ratio m/M as λ (>0) and use the momentum conservation equation (above), the first of these 'elastic' conditions becomes

$$U^2 = (U - \lambda v)^2 + \lambda v^2 \quad \Rightarrow \quad v = \frac{2U}{\lambda + 1} \text{ provided } \lambda v \neq 0.$$

The alternative expression of perfect elasticity, taken with that for momentum conservation, becomes

$$U = (v - U) + \lambda v, \quad \text{again leading to } v = \frac{2U}{\lambda + 1}.$$

As might have been anticipated, v will be large when $m \ll M$, and the above result confirms this. The upper limit for physically obtainable values of v is $2U$, approached as $\lambda \to 0$. We now relate these speeds to the heights H and h:

$$\begin{aligned} \tfrac{1}{2}MU^2 &= MgH, \\ \tfrac{1}{2}mv^2 &= mgh, \\ v < 2U \quad \Rightarrow \quad v^2 < 4U^2 \quad &\Rightarrow \quad h < 4H. \end{aligned}$$

At first sight, it may seem that if m were made small enough, the initial speed v that the struck ball is given by the collision could be arbitrarily high. And this would be so, if *energy conservation* were the only constraint. However, as we have seen, *momentum conservation* puts a stop to that!

S 37 In many dynamics problems, balls and other projectiles are treated as 'point objects', but, as any snooker player will tell you, taking account of the finite size of the balls and the orientation of the 'line of centres' of each pair of balls, is absolutely crucial. For our analysis, let the positions of the centres of the balls be denoted by C for the cue ball, and by R_1 and R_2 for the two reds, as in the left-hand figure.

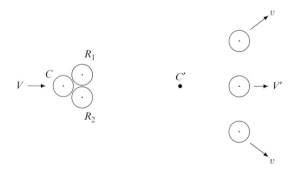

At the moment of collision, $R_1 C R_2$ form an equilateral triangle of side $2a$, and both CR_1 and CR_2 make angles of $30°$ with the cue ball's path. The impulses received by the red balls are in these directions, as is illustrated in the right-hand figure, which also marks the point C', the position of C at impact. This figure 'captures' the moment that the three balls are aligned.

Using the symmetry of the situation, after the collision we need consider only what happens to C and R_1. Let R_1 (and R_2) have velocity v in the directions of their respective lines of centre with C. Then, from the conservation of momentum, we have

$$mV = mV' + 2mv \cos 30°,$$

where V' is the velocity of C after the impact.

Along the line CR_1 we have that the relative velocities before and after the collision are related by

$$v - V' \cos 30° = e(V \cos 30° - 0).$$

Re-arranging these two equations and inserting numerical values gives

$$V' + \sqrt{3}\, v = V,$$
$$-\sqrt{3}\, V' + 2v = \sqrt{3}\, eV,$$

from which

$$v = \frac{\sqrt{3}(1+e)}{5} V \quad \text{and} \quad V' = \frac{2 - 3e}{5} V.$$

Our first conclusion is that, since C did travel forwards, i.e. $V' > 0$, we must have $e < \frac{2}{3}$.

[Not good news for snooker balls!]

As the three balls come into line after time t, and the red balls start $2a \sin 60°$ to the right of the cue ball, the axial component of v must be less than V':

$$v \cos 30° < V',$$
$$\frac{\sqrt{3}(1+e)}{5} \frac{\sqrt{3}}{2} < \frac{2 - 3e}{5},$$
$$3 + 3e < 4 - 6e,$$
$$e < \frac{1}{9}.$$

[Even worse news for snooker balls!]

As the alignment was observed, e must be less than $\frac{1}{9}$, and the time t is determined by

$$v \cos 30° \, t + 2a \sin 60° = V' t \quad \Rightarrow \quad t = \frac{2a \sin 60°}{V' - v \cos 30°}.$$

Now,

$$\frac{V' - v\cos 30°}{V} = \frac{2 - 3e}{5} - \frac{\sqrt{3}(1+e)}{5}\frac{\sqrt{3}}{2} = \frac{1}{10}(1 - 9e),$$

and so

$$t = \frac{10\sqrt{3}a}{V(1 - 9e)}.$$

Needless to say, m does not come into it – on dimensional grounds, if nothing else.

S 38   In an obvious and standard notation, the flight of the ball is given by

$$x = tV\cos\theta \quad \text{and} \quad y = tV\sin\theta - \tfrac{1}{2}gt^2.$$

[Here, and in what follows, we have written the t-dependence of a term in an equation in such a way as to try to avoid, for example, expressions such as $\cos\theta t$ which might be wrongly taken as meaning $\cos(\theta t)$.]

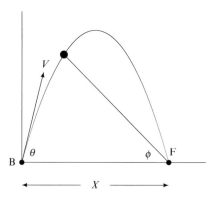

The time of the ball's flight is clearly twice the time it takes a vertical speed of $V\sin\theta$ to be reduced to zero by gravity, i.e. $2V\sin\theta/g$. The corresponding horizontal range X is therefore

$$X = \frac{2V\sin\theta}{g}V\cos\theta = \frac{V^2}{g}\sin 2\theta,$$

and it is at this point that the fielder F is standing (see the figure). The angle of elevation ϕ of his line of sight, as he watches the flight of the ball, is given by

$$\tan\phi = \frac{tV\sin\theta - \tfrac{1}{2}gt^2}{X - tV\cos\theta} = \frac{tV\sin\theta - \tfrac{1}{2}gt^2}{(V^2/g)\sin 2\theta - tV\cos\theta}.$$

To determine the rate of change of $\tan\phi$, whilst avoiding very complicated algebraic expressions, we think of $\tan\phi(t)$ as $f(t)/g(t)$, which has derivative $(gf' - fg')/g^2 \equiv p(t)/q(t)$. With this prescription,

$$
\begin{aligned}
p(t) &= \left(\frac{V^2}{g}\sin 2\theta - tV\cos\theta\right)(V\sin\theta - gt) - (tV\sin\theta - \tfrac{1}{2}gt^2)(-V\cos\theta) \\
&= \frac{1}{2}V\cos\theta\left[\frac{4V^2}{g}\sin^2\theta - t4V\sin\theta + t^2 g\right],
\end{aligned}
$$

$$
\begin{aligned}
q(t) &= \frac{V^4}{g^2}\sin^2 2\theta - t\frac{2V^3}{g}\sin 2\theta\cos\theta + t^2 V^2\cos^2\theta \\
&= \frac{V^2\cos^2\theta}{g}\left[\frac{4V^2}{g}\sin^2\theta - t4V\sin\theta + t^2 g\right].
\end{aligned}
$$

It can be seen that $p(t)$ and $q(t)$ have identical (but complicated) time dependences, and so

$$
\frac{d(\tan\phi)}{dt} = \frac{p(t)}{q(t)} = \frac{g}{2V\cos\theta} = \text{constant.}
$$

S 39 In this situation the mass, as well as the velocity, of the raindrop is continuously changing, and the added mass itself has initial momentum that has to be taken into account.

The equation of motion can be obtained directly by balancing the force mg, with the rates of change of momentum of the drop and the vapour it absorbs; or, alternatively, by considering the momentum changes that occur in time dt:

$$
m\,dv + v\,dm = mg\,dt - U\,dm.
$$

Either method results in

$$
\frac{d(mv)}{dt} = mg - U\frac{dm}{dt}. \tag{$*$}
$$

The rate at which the mass changes is given by

$$
k \times 4\pi r^2 = \frac{dm}{dt} = \frac{4\pi}{3}\rho\frac{d(r^3)}{dt} = 4\pi\rho r^2\frac{dr}{dt} \quad\Rightarrow\quad \frac{dr}{dt} = \frac{k}{\rho},
$$

where ρ is the density of (rain)water.

To obtain an equation for v, (remembering that m changes with t) we recast $(*)$ as

$$
\frac{d}{dt}(mv + Um) = mg = \frac{4\pi}{3}\rho r^3 g,
$$

and then integrate with respect to t from $t=0$ to a general t. The left-hand side is straightforward, but on the right-hand side we need to write dt as $dr \times (dt/dr)$ and use the fact that dt/dr is, as shown above, constant, and equal to ρ/k.

$$mv + Um - 0 - Um_0 = \frac{4\pi}{3}\rho g \int_0^t r^3\, dt = \frac{4\pi}{3}\rho g \frac{\rho}{k} \int_0^r r^3\, dr$$

$$= \frac{4\pi}{3}\rho g \frac{\rho}{k}\frac{1}{4}\left(r^4 - r_0^4\right).$$

Since at all times $m = (4\pi/3)\rho r^3$ we can divide through by $(4\pi\rho/3)$ and obtain

$$r^3 v + U(r^3 - r_0^3) = \frac{\rho g}{4k}(r^4 - r_0^4),$$

$$v = \frac{\rho g}{4k}\left(r - \frac{r_0^4}{r^3}\right) - U\left(1 - \frac{r_0^3}{r^3}\right).$$

Since r grows linearly with time, $r(t) = r_0 + kt/\rho$, eventually $r \gg r_0$ and

$$v \approx \frac{\rho g}{4k}r - U,$$

meaning that the raindrop's acceleration a is given by

$$a = \frac{dv}{dt} \approx \frac{\rho g}{4k}\dot{r} = \frac{\rho g}{4k}\frac{k}{\rho} = \frac{g}{4} = \text{constant}.$$

S 40   At any particular moment, there will two forces contributing to the scale reading: (i) the weight of the chain already on the pan, and (ii) the force created to destroy the momentum of the chain that is still falling. We note that, as the chain is in free fall, with all parts of it subject to the same acceleration g, there are no internal stresses within it.

At time t, when a length x is already on the scale pan, and the top of the chain has fallen by the same amount, the downward speed of the moving part of the chain is everywhere $v = gt$. In a small time dt, a further length $v\,dt = gt\,dt$ is stopped by the pan, resulting in a momentum change $dp = -\rho(gt\,dt)v = -\rho(gt\,dt)gt$ and the corresponding force on the scales is

$$F(t) = -\frac{dp}{dt} = \rho g^2 t^2.$$

This has to be added to the weight of the chain already on the pan $W(t) = \rho g x$. However, since the chain is in free fall under gravity, this will be all the chain that was originally less than $\frac{1}{2}gt^2$ above the scale pan, i.e $x = \frac{1}{2}gt^2$. Thus the scale reading is

$$S(t) = F(t) + W(t) = \frac{3}{2}\rho g^2 t^2.$$

This time dependence will continue until $t = t_0$ where $\frac{1}{2}gt_0^2 = \ell$, after which $F(t) = 0$ and $W(t) = \rho g \ell = Mg$, where M is the mass of the complete chain. At $t = t_0$, $S(t)$ is at its maximum with a reading of $3\rho g \ell = 3Mg$.

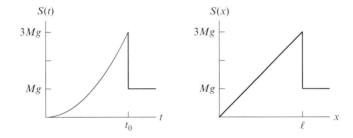

The quadratic time dependence of $S(t)$ is shown in the left-hand figure above. Since x also varies quadratically with t, the corresponding variation of $S(x)$ with x is linear, as shown in the right-hand figure.

S41   As always, momentum conservation determines what happens! But it should be borne in mind that momentum can change as a result of a mass change, just as much as a velocity change.

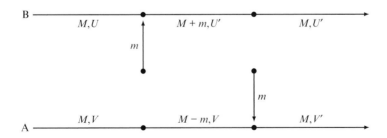

The figure shows schematically the masses and speeds of the two carriages at the three stages of the double transfer. When the mailbag is thrown as described, there is no change in the *speed* of the delivering carriage. However, it does change its momentum parallel to the tracks, since the bag carries some of it away and transfers it to the other carriage.

At the first transfer, A does not change its speed, but, on receiving the bag, B does. Its new speed U' is given by conservation of momentum as follows:

$$(M+m)U' = MU + mV \quad \Rightarrow \quad U' = \frac{MU+mV}{M+m} = U + \frac{m}{M+m}(V-U).$$

Similarly, at the second transfer

$$MV' = (M-m)V + mU',$$

$$V' = \frac{1}{M}\left[(M-m)V + m\frac{MU+mV}{M+m}\right]$$

$$= \frac{MV+mU}{M+m} = V + \frac{m}{M+m}(U-V).$$

Since each carriage has total mass M both before and after the double transfer, the kinetic energy change ΔE is

$$\Delta E = \frac{M}{2}(V^2 - V'^2 + U^2 - U'^2)$$

$$= \frac{M}{2}\left[-\frac{2mV(U-V)}{M+m} - \frac{m^2(U-V)^2}{(M+m)^2} - \cdots\right.$$

$$\left.\cdots - \frac{2mU(V-U)}{M+m} - \frac{m^2(V-U)^2}{(M+m)^2}\right]$$

$$= \frac{M}{2}\left[\frac{2m(V-U)^2}{M+m} - \frac{2m^2(U-V)^2}{(M+m)^2}\right]$$

$$= \frac{Mm(V-U)^2}{M+m}\left(1 - \frac{m}{(M+m)}\right) = \frac{M^2 m}{(M+m)^2}(V-U)^2.$$

This lost mechanical energy, non-zero if U and V are not equal, but independent of which is the greater, is dissipated in bringing the mailbag up to, or down to, the new speed of the receiving carriage; it could, for example, appear as frictional heat loss on a carriage floor, or as sound and heat generated in a catching mechanism. Similar considerations apply to any kinetic energy the bag may have due to motion perpendicular to the tracks, but that does not come into the present calculation.

S 42   When there is freedom to do so, we should always choose the simplest and most convenient coordinate system available; here we choose the origin of y so that the otter's entry point has coordinates $(0,0)$ and it leaves the water at $(a,0)$. Since the stream velocity varies with x, we can only consider what happens in an arbitrarily small time, or over an arbitrarily small distance. This will lead to a differential equation, which we then hope to solve.

Before we can consider the otter's path, we need the full expression for the stream's velocity profile $v(x)$. Since it is quadratic in x and takes zero value at $x = 0$ and $x = a$, it must be of the form $v(x) = Ax(a - x)$, where A has to be chosen so as to make $v(\frac{1}{2}a) = \frac{3}{4}u$, i.e. $A = 3u/a^2$. Thus,

$$v(x) = \frac{3u}{a^2}x(a - x).$$

Now we consider the positional changes occurring in a small time interval dt: $dx = u \sin\theta \, dt$; $dy = [v(x) - u \cos\theta] \, dt$. From these we can construct an expression for dy/dx, and hence for $y(x)$ with the constant of integration chosen ($=0$) so as to make $y(0) = 0$:

$$\frac{dy}{dx} = \frac{3ux(a - x)}{ua^2 \sin\theta} - \frac{u \cos\theta}{u \sin\theta},$$

$$y(x) = \frac{3}{a^2 \sin\theta}\left(\frac{ax^2}{2} - \frac{x^3}{3}\right) - \frac{\cos\theta}{\sin\theta}x.$$

As the otter emerges from the water opposite its starting point, $y(a) = 0$, meaning that

$$\frac{3a}{2} - a - a\cos\theta = 0 \quad \Rightarrow \quad \cos\theta = \frac{1}{2} \quad \Rightarrow \quad \theta = \frac{\pi}{3}.$$

Inserting $\cos\theta = 1/2$ and $\sin\theta = \sqrt{3}/2$ and simplifying, gives the equation for the 'GPS-path' as

$$y(x) = \sqrt{3}a\left[\left(\frac{x}{a}\right)^2 - \frac{2}{3}\left(\frac{x}{a}\right)^3 - \frac{1}{3}\left(\frac{x}{a}\right)\right].$$

The extrema of y occur when $y'(x) = 0$, i.e. when

$$2\left(\frac{x}{a}\right) - 2\left(\frac{x}{a}\right)^2 - \frac{1}{3} = 0 \quad \Rightarrow \quad x = \frac{a}{2}\left(1 \pm \sqrt{\frac{1}{3}}\right).$$

These values, $0.79a$ and $0.21a$ are symmetrically placed about the midstream point, as shown in the figure below.

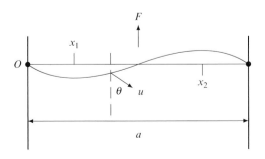

S 43   Because of the 'difficult' boundaries, and the unusual boundary conditions, this is a problem that can only be solved by considering the individual steps one at a time, using the output of each event as the input for the next one. It is hard to anticipate what will happen.

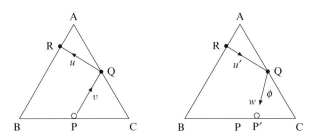

The left-hand figure above is a continuation of the situation shown in the question. If u is the speed of the ball after its impact at Q, and we denote the angle $\angle$AQR by θ, then, since PQ is parallel to AB and therefore $\angle$PQC $= \pi/3 = 60°$, we have from the information about the parallel and perpendicular velocity components, that

$$u \sin \theta = \tfrac{1}{3} v \sin(\pi/3),$$
$$u \cos \theta = v \cos(\pi/3),$$

leading to

$$\tan \theta = \frac{1}{3} \tan(\pi/3) = \frac{\sqrt{3}}{3} = \frac{1}{\sqrt{3}} \quad \Rightarrow \quad \theta = \frac{\pi}{6} = 30°$$

and

$$u^2 = \left(\frac{1}{9} \frac{3}{4} + \frac{1}{4} \right) v^2 = \frac{1}{3} v^2.$$

Now, since $\theta = 30°$ and $\angle$QAR $= 60°$, angle $\angle$QRA must be a right angle. It follows that the billiard ball will be returned back along the path RQ, but with reduced speed $u' = \tfrac{1}{3} u$.

The ball will bounce at Q for a second time, and leave it with speed w, and in a direction making an angle ϕ with QC, as shown in the right-hand figure. For this rebound, the velocity components are related by

$$w \cos \phi = u' \cos \theta,$$
$$w \sin \phi = \tfrac{1}{3} u' \sin \theta,$$

this time leading to

$$\tan \phi = \frac{1}{3} \tan \theta = \frac{1}{3\sqrt{3}}$$

and

$$w^2 = u'^2 \left(\frac{3}{4} + \frac{1}{9}\frac{1}{4} \right) = \frac{u^2}{9} \left(\frac{7}{9} \right) = \frac{v^2}{27} \left(\frac{7}{9} \right).$$

From this $w = 0.17v$.

If x is the distance from C of the impact point P$'$ and 2ℓ is the length of the tables cushions, then, for the triangle P$'$QC, the sine rule gives

$$\frac{x}{\sin \phi} = \frac{\ell}{\sin \left(\pi - \frac{1}{3}\pi - \phi \right)}.$$

Thus

$$\begin{aligned}
\frac{x}{\ell} &= \frac{\sin \phi}{\sin \frac{2}{3}\pi \cos \phi - \sin \phi \cos \frac{2}{3}\pi} \\
&= \frac{1}{\sin \frac{2}{3}\pi \cot \phi - \cos \frac{2}{3}\pi} \\
&= \frac{1}{(\sqrt{3}/2)\, 3\sqrt{3} + (1/2)} = \frac{1}{5}.
\end{aligned}$$

The point of impact P$'$ is therefore a distance of four-tenths of a cushion length away from P, towards the corner C.

S 44 Since, both horizontally and vertically, the object's acceleration depends on it current velocity, we can expect to have to solve differential equations to obtain its velocity components. To translate these velocity components into 'distances travelled' will require further integrations over a time interval T which itself is determined by the result of one of those integrations. This may be tricky!

Let us denote the object's (downward) velocity by v, and the horizontal one (relative to the river's flow) by u. Since the resistive force $\mathbf{F}$ *directly* opposes the object's motion, and is directly proportional to the object's speed relative to the water, we can treat the force as two separate components, one horizontal and one vertical, i.e. $F_x = -\beta u$ and $F_y = -\beta v$.

Horizontally, the relevant equation is

$$m\frac{du}{dt} = -\beta u \quad \text{with } u(0) = -u_0 \quad \Rightarrow \quad u(t) = -u_0 e^{-\beta t/m}.$$

Relative to the riverbank,

$$U(t) = u_0 \left(1 - e^{-\beta t/m} \right).$$

To find how far (X) this carries the object downstream, we need to determine T, the time it takes for it to sink to the bottom. Then, we will have

$$X = \int_0^T u_0 \left(1 - e^{-\beta t/m}\right) dt. \qquad (*)$$

To find T, we use that, vertically, $v(0) = 0$ and the governing equation is

$$m\frac{dv}{dt} = \frac{\rho - \rho_0}{\rho}mg - \beta v.$$

Denoting $g(\rho - \rho_0)/\rho$ by λ, gives

$$\frac{dv}{m\lambda - \beta v} = \frac{dt}{m},$$

$$-\frac{1}{\beta}\ln\left(\frac{m\lambda - \beta v}{m\lambda}\right) = \frac{t}{m},$$

$$v(t) = \frac{m\lambda}{\beta}\left(1 - e^{-\beta t/m}\right).$$

Clearly $v_0 = m\lambda/\beta$ is the terminal downward speed, attained when the object's 'weight less upthrust' ($m\lambda$) is equal to the resistive force (βv).

It now follows that

$$Y = \int_0^T v(t)\, dt = \int_0^T \frac{m\lambda}{\beta}\left(1 - e^{-\beta t/m}\right) dt. \qquad (**)$$

Without even evaluating the integrals in $(*)$ and $(**)$, we can say that

$$\frac{X}{u_0} = \frac{\beta\, Y}{m\lambda} = \frac{\beta\, Y\rho}{mg(\rho - \rho_0)}. \qquad \text{Hence } X.$$

It is not surprising, at least in retrospect, that the integrals giving X and Y have the same (exponential) form; both horizontally and vertically, the speeds of the blunt object approach limiting values (u_0 and v_0 respectively) as a result of it being retarded by a force proportional to the current velocity component in that direction. One consequence of this is that, for the question as posed, we do not even have to find an expression for T!

S 45 We note that when we come to determine the rocket's speed at the end of the burn, we will need to express that speed as a function of its mass, rather than of time or distance, since the end of the burn is determined by a mass condition.

The rocket is subjected to two forces, one due to the expelled fuel, the other to gravity. The first of these can be calculated using the conservation of momentum, as illustrated in the figure. This shows the mass m and speed v

of the rocket at two times separated by dt. The expelled fuel has speed $v - u$ relative to our earthbound reference frame.

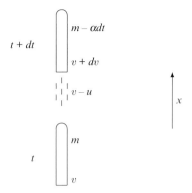

We have

$$(m - \alpha\, dt)(v + dv) + \alpha\, dt(v - u) = mv,$$

from which it follows that, to first order in infinitesimals, $m\, dv - \alpha u\, dt = 0$ and that the first acceleration is $dv/dt = (\alpha u)/m$. The second one is clearly g, acting downwards, and the full equation governing the motion is

$$m\frac{dv}{dt} = \alpha u - mg. \qquad (*)$$

Since the second part of the flight is under the influence of gravity alone, we will need the speed of the rocket at the end of the burn, as this will be the launch speed for the second stage. The end of the burn is determined by the condition $m = \frac{1}{2}m_0$ and so we convert $(*)$ into an equation connecting v and m by replacing α by $-dm/dt$. Then

$$\frac{dv}{dt} = -\frac{u}{m}\frac{dm}{dt} - g,$$

which, with $v(0) = 0$ and $m(0) = m_0$, integrates to

$$v(t) = -u\ln\frac{m}{m_0} - gt.$$

The first stage will end at time $T = \frac{1}{2}m_0/\alpha$ with

$$v(T) = -u\ln\tfrac{1}{2} - gT = u\ln 2 - \frac{m_0 g}{2\alpha}.$$

After the burn the rocket case is in free flight under gravity and will rise through a further distance $h_2 = v^2(T)/2g$, before falling back to earth.

We also need to find the distance h_1 travelled during the burn, and for this we need to express the m in $(*)$ as an explicit function of time, $m = m_0 - \alpha t$ for $0 \le t \le T$, and obtain

$$\frac{dv}{dt} = \frac{\alpha u}{m_0 - \alpha t} - g,$$

$$\frac{dx}{dt} = v(t) = \frac{\alpha u}{(-\alpha)} \ln\left(\frac{m_0 - \alpha t}{m_0}\right) - gt = -u \ln\left(1 - \frac{\alpha t}{m_0}\right) - gt.$$

The contribution of the final term to $h_1 \equiv x(T)$ is clearly $-\frac{1}{2}gT^2 = -gm_0^2/8\alpha^2$.

The first term has to be integrated by parts.[5] Omitting for the moment the factor $-u$,

$$\int_0^T \ln\left(1 - \frac{\alpha t}{m_0}\right) dt = \left[t \ln\left(1 - \frac{\alpha t}{m_0}\right)\right]_0^T - \int_0^T \frac{t}{1 - (\alpha t)/m_0}\left(\frac{-\alpha}{m_0}\right) dt$$

$$= \frac{m_0}{2\alpha} \ln\left(\frac{1}{2}\right) + \int_0^T \frac{\alpha t}{m_0 - \alpha t} dt$$

$$= -\frac{m_0}{2\alpha} \ln 2 - \int_0^T \left(\frac{m_0 - \alpha t}{m_0 - \alpha t} - \frac{m_0}{m_0 - \alpha t}\right) dt$$

$$= -\frac{m_0}{2\alpha} \ln 2 - T + \frac{m_0}{(-\alpha)} \left[\ln(m_0 - \alpha t)\right]_0^T$$

$$= -\frac{m_0}{2\alpha} \ln 2 - T - \frac{m_0}{\alpha} \ln\left(\frac{\frac{1}{2}m_0}{m_0}\right) = \frac{m_0}{2\alpha}(\ln 2 - 1).$$

Collecting together the various contributions, and reinstating the factor $-u$ in h_1, the total height h the rocket attains is

$$h = h_1 + h_2 = \frac{m_0 u}{2\alpha}(1 - \ln 2) - \frac{g}{2}\left(\frac{m_0}{2\alpha}\right)^2 + \frac{1}{2g}\left(u \ln 2 - \frac{m_0 g}{2\alpha}\right)^2$$

$$= \frac{m_0 u}{2\alpha}(1 - \ln 2) + \frac{u^2(\ln 2)^2}{2g} - \frac{m_0 u \ln 2}{2\alpha}$$

$$= \frac{u^2(\ln 2)^2}{2g} + \frac{m_0 u}{\alpha}\left(\frac{1}{2} - \ln 2\right).$$

---

[5] Or, if known, the standard indefinite integral $\int \ln x \, dx = x \ln x - x$ can be applied, with $x = 1 - (\alpha t)/m_0$ and $dx = -(\alpha/m_0) \, dt$.

Note 1. Since $\ln 2 > \frac{1}{2}$ the second term will make a negative contribution to h, but, on physical grounds, the total h cannot be negative. To ensure that the expression for h is positive, we must have

$$\frac{u^2(\ln 2)^2}{2g} > \frac{m_0 u}{\alpha}\left(\ln 2 - \frac{1}{2}\right),$$

$$\frac{(\ln 2)^2}{2} > \frac{m_0 g}{\alpha u}\left(\ln 2 - \frac{1}{2}\right),$$

$$0.240 > 0.193 \frac{m_0 g}{\alpha u}.$$

At this point, the fact that the rocket took off *immediately* (see the question) becomes very relevant. It would only do so if the thrust from burning fuel αu were greater than its initial weight $m_0 g$. Thus the fraction on the RHS of the above inequality is less than unity, and the condition necessary for a positive value of h is fulfilled.

Note 2. Although we will not pursue it here, you may like to consider, in qualitative terms, what would happen when the rocket is ignited if $\frac{1}{2}m_0 g < \alpha u < m_0 g$.

S 46 The ball does not go on bouncing for ever, because each time it contacts the concrete floor it loses some of its energy (which is ultimately transformed into heat shared between the ball and the floor). A measure of this loss for any two materials is e, the coefficient of restitution; it is defined as the ratio of the relative speed of the colliding objects just after they meet to that just before they do so, and lies in the range $0 \le e \le 1$, with the two extremes corresponding to adhesion and an ideally elastic collision, respectively.

The energy loss, by itself, does not guarantee a finite bouncing time, since there will still be an 'infinite' number of bounces. However, with the lower and lower energy of each bounce, the height reached and the time taken will progressively reduce. The question becomes 'will the running *sum* of the bounce times approach a limit, or will it grow without limit?'; the wording of the problem clearly implies the former.

As the ball's total energy, and hence the height it can reach, during the nth bounce is proportional to the square of its speed v_n at the start of that bounce, and $v_{n+1} = ev_n$, we need to determine the appropriate value for e. Since the ball is dropped at time $t = 0$ from a height h, its speed just before it first hits the concrete is $v_0 = \sqrt{2gh}$, and immediately afterwards, at the start of the first bounce, it is $v_1 = e\sqrt{2gh}$.

The only other information we have is the *total* time for all bouncing to cease, and so we need to find an expression for this in terms of e. The time t_n required for the nth bounce is twice the time it takes gravity to reduce an upward velocity of v_n to zero, i.e. $t_n = 2v_n/g$. For the initial drop, using "$s = ut + \frac{1}{2}at^2$", the time taken is $t_0 = \sqrt{2h/g}$. Thus the total time for the drop and all bounces is [6]

$$
\begin{aligned}
T &= \sqrt{\frac{2h}{g}} + \frac{2v_1}{g} + \frac{2v_2}{g} + \frac{2v_3}{g} + \cdots \\
&= \sqrt{\frac{2h}{g}} + \frac{2v_0}{g}(e + e^2 + e^3 + \cdots) \\
&= \sqrt{\frac{2h}{g}} + \frac{2\sqrt{2gh}}{g}\frac{e}{1-e} \\
&= \sqrt{\frac{2h}{g}}\left(1 + \frac{2e}{1-e}\right).
\end{aligned}
$$

Substituting the given data:

$$
12.1 = \sqrt{\frac{2 \times 2}{9.81}}\left(1 + \frac{2e}{1-e}\right) \quad \text{yielding } e = 0.900.
$$

We can now return to the question of the height of a bounce, recalling that it is proportional to the square of the take-off speed. If that of the nth bounce is h_n, then

$$
h_1 = e^2h, \quad h_2 = e^4h, \quad h_3 = e^6h, \quad \ldots,
$$

and so $h_5 = e^{10}h = (0.900)^{10} \times 2 = 0.697$ m.

S 47 It will be difficult not to 'notice' the shapes of the curves in the figures below, but if you do not already have values for the labels marked as V, T_i and X_i, $(i = 1, 2, 3)$, try to find them before reading on.

Since $M \gg Nm$, we can assume that any velocity v induced in the trolley will be $\ll v_0$, and that any resulting displacement x will be $\ll L$.

The first of the three figures above shows the position shortly after $t = 0$, but before the first of the bullets has reached the target, which happens at time $t = T_1 = L/v_0$. During this period, the trolley is absorbing the recoil from the gun, and is therefore subject to a force to the right (of the figure); consequently it accelerates in that direction.

By Newton's laws, the force is constant at fmv_0 and the acceleration uniform with magnitude fmv_0/M. Since this phase of the motion lasts for a time interval T_1, at the end of it the trolley's velocity is $V = fmv_0/M \times L/v_0 = fmL/M$, as

---

[6] In the third line below, we have used the result that, for $-1 < x < 1$, the sum of the infinite geometric series $1 + x + x^2 + x^3 + \cdots$ is $1/(1-x)$.

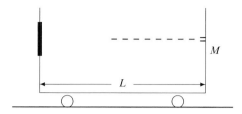

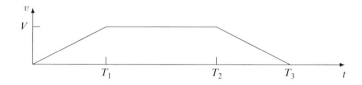

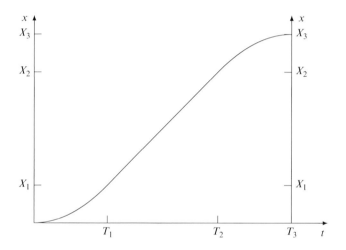

indicated in the second figure. During the same interval, the distance covered by the trolley *to the right* varies quadratically with time, and at T_1 is

$$X_1 = \frac{1}{2} \frac{fmv_0}{M} \frac{L^2}{v_0^2} = \frac{fmL^2}{2Mv_0},$$

as marked on the third figure.

After the first bullet reaches the target, the latter is subject to a recoil force equal but opposite to that experienced by the gun mounting. Consequently there is no net force acting on the trolley, and, correspondingly, no further acceleration. However the trolley will continue to travel to the right with the speed V it has already acquired. This phase will last until the last bullet has just left the gun, i.e. until $t = N/f$, marked as T_2 in both the second and third figures. During this second phase the trolley will travel a further distance

$$\frac{fmL}{M}\left(\frac{N}{f} - \frac{L}{v_0}\right),$$

represented by $X_2 - X_1$ in the third figure.

In the final phase, only the target has a force acting upon it, and that lasts for as long as it takes for the final bullet to travel the length of the trolley. Thus the velocity changes will be the reverse of those that occurred in the first phase, i.e. a deceleration of fmv_0/M and a velocity change of $-fmL/M$, bringing the trolley to a standstill. However, the direction of travel will still be to the right, and a further distance equal in magnitude to X_1 will be covered. The total distance the trolley will have moved to the right is given by

$$X_3 = \frac{fmL^2}{2Mv_0} + \frac{fmL}{M}\left(\frac{N}{f} - \frac{L}{v_0}\right) + \frac{fmL^2}{2Mv_0} = \frac{Nm}{M}L.$$

This very simple result, $M \times X_3 = Nm \times L$, can be understood as saying roughly "since no external forces act on the system as a whole, its centre of mass cannot move – the mass transferred from right to left by the bullets, is compensated for by the trolley moving its mass from left to right". A more rigorous calculation, treating relative masses and velocities without approximations, would result in a more complex result, but the same general principle would still apply.

S 48   This problem is in essence a two-stage version of the situation analysed in S 36 on page 133. If assistance is needed, it would probably help to study that problem first.

The figure shows three stages of the 'magical performance'. On the left is the original situation, and in the centre and on the right, magnified drawings of the situations after the first and second collisions respectively.

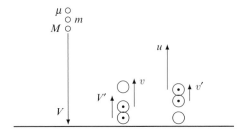

The figure also defines the notation we will use, with M, m and μ as the three masses selected, and V the (common) speed that each attains after falling through a height h, with $V^2 \propto h$. With the conservation of vertical momentum, and the perfect elasticity represented by the final relative velocity of two colliding balls being equal to that before the collision, we have

$$MV - mV = MV' + mv \quad \text{with } v - V' = 2V,$$
$$mv - \mu V = mv' + \mu u \quad \text{with } u - v' = v + V.$$

Eliminating V' and v':

$$MV - mV = M(v - 2V) + mv,$$
$$mv - \mu V = m(u - v - V) + \mu u,$$

which gives

$$v = \frac{MV - mV + 2MV}{M + m} = \frac{3M - m}{M + m} V, \tag{$*$}$$

$$u = \frac{mv - \mu V + mv + mV}{m + \mu} = \frac{2mv - \mu V + mV}{m + \mu}. \tag{$**$}$$

For the topmost mass to rise as high as possible, u has to be maximized, and from $(**)$ this means v has to be maximal. From $(*)$ this means choosing $r = M/m$ so as to maximize $(3r - 1)/(r + 1)$. This function has no turning points for positive values of r and the best that can be done is to make $M \gg m$ with v then tending to $3V$. Using this (physically unattainable) value gives

$$u = \frac{7mV - \mu V}{m + \mu} \quad \text{which tends to } 7V \text{ if } m \gg \mu.$$

Although the theoretically desirable infinite ratios for M/m and m/μ are not realistic, since V^2 corresponds to h, a speed of $7V$ corresponds to a height of $49h$!

Somewhat more realistically, with $\mu = 1$, $m = 10$ and $M = 100$, $v = 2.636V$ and $u = 5.612V$, giving a final height for the ejected ball of $31.5h$. How to make the weighted, yet perfectly elastic, balls appears not to be known to any Physics or Engineering Society – only to members of the Magic Circle.

S 49   We need to find a general prescription for the increase in speed attained by a rocket when its mass is reduced, by burning fuel, to a particular fraction of its original value. The result will then be applied to three different scenarios, the two stages of a double-stage rocket, and the single stage of a simple rocket. Comparisons will then be made.

Let m be the mass of the rocket at a general time, and v be the corresponding speed. When an infinitesimal amount dm of the fuel on board is burned and ejected, the rocket's mass will decrease by dm, and, by conservation of momentum, its speed will increase by dv. Formally $m\,dv + u\,dm = 0$, and pictorially,

in which dm is negative in practice.

As a simple differential equation, momentum conservation can be written as

$$\frac{dv}{dm} = -\frac{u}{m} \quad \Rightarrow \quad \frac{v_f - v_i}{u} = \ln\left(\frac{m_i}{m_f}\right),$$

where the subscripts 'i' and 'f' indicate initial and final quantities. We now apply this result to the two stages of firing.

First stage:

$$v_i = 0,\ m_i = M_1 + M_2,\ m_f = m_1 + M_2 \quad \Rightarrow \quad v_1 = u\ln\frac{M_1 + M_2}{m_1 + M_2}.$$

Second stage:

$$v_i = v_1,\ m_i = M_2,\ m_f = m_2 \quad \Rightarrow \quad v_2 = v_1 + u\ln\frac{M_2}{m_2} = u\ln\frac{M_2(M_1 + M_2)}{m_2(m_1 + M_2)}.$$

Single (combined) rocket:

$$v_i = 0,\ m_i = M_1 + M_2,\ m_f = m_1 + m_2 \quad \Rightarrow \quad v_{1+2} = u\ln\frac{M_1 + M_2}{m_1 + m_2}.$$

The *additional* speed attained by the two-stage design is therefore

$$\Delta v = v_2 - v_{1+2} = u\ln\frac{M_2(M_1 + M_2)(m_1 + m_2)}{m_2(m_1 + M_2)(M_1 + M_2)} = u\ln\frac{M_2(m_1 + m_2)}{m_2(m_1 + M_2)}.$$

If m_1 is very small compared to the other masses, then $v_2 \approx u\ln[(M_1 + M_2)/m_2]$, i.e. the first stage is just an additional (light) fuel tank for the second stage, which behaves as a single-stage rocket carrying a mass $M_1 + M_2 - m_2$ of fuel. The formula for the additional velocity acquired, Δv, approaches the value $u\ln[(M_2 m_2)/(m_2 M_2)] = u\ln 1 = 0$, as might be expected from the previous observation.

S 50 In general terms, it is clear that the lower the downstream wind speed is made (by sail density and design), the greater is the fraction of its momentum that the wind gives up to the mill. However, it is also the case that the wind speed at the mill itself will be lower, and this will reduce the volume flow rate through the mill's sails. The best compromise is needed, and to determine that we need to find how v depends upon α.

The mass of air striking the sails in unit time is that in a cylinder of cross-section A and length v, i.e $Av\rho$, where ρ is the density of air. Its rate of momentum change is therefore $Av\rho(V - \alpha V)$, and this gives the force acting on the sails, and equally the retarding force on the wind.

The rate at which the wind has to do work on the mill is, from the above, equal to $Av\rho(V - \alpha V)v$, and this work must come from the kinetic energy loss in the same volume of wind, i.e. it has to be equal to $\frac{1}{2}(Av\rho)(V^2 - \alpha^2 V^2)$. Equating these two expressions for the loss of power:

$$Av\rho(V - \alpha V)v = \tfrac{1}{2}(Av\rho)(V^2 - \alpha^2 V^2),$$

which can be simplified to give $v = \frac{1}{2}(1 + \alpha)V$.

This result can now be substituted back into the expression for the power lost to the windmill:

$$P = Av^2(1 - \alpha)V = \tfrac{1}{4}A\rho(1 + \alpha)^2(1 - \alpha)V^3.$$

With one factor that increases with increasing α, and another that decreases, this expression for P will have a maximum or minimum at some value of α. Omitting the constant factors, the turning point will occur when the derivative $dP/d\alpha$ is zero, i.e.

$$2(1 + \alpha)(1 - \alpha) - (1 + \alpha)^2 = 0 \quad \Rightarrow \quad \text{when } \alpha = \tfrac{1}{3} \text{ and } v = \tfrac{2}{3}V.$$

By inspection (comparing with the cases $\alpha = 0$ and $\alpha = 1$) this clearly gives a maximum. The corresponding maximal power is

$$P_{max} = A\rho v(V - \alpha V)v = A\rho \frac{2V}{3}\left(\frac{2V}{3}\right)\frac{2V}{3} = \frac{8}{27}A\rho V^3.$$

Since the initial power available is $\frac{1}{2}A\rho V V^2$, the maximum efficiency of a windmill is $(8/27) \div (1/2) = 16/27$ on this model.

S 51 This question is very similar to Q 39, which analyses the motion of a raindrop moving in fog. Apply the same general principles, albeit in two dimensions rather than one.

Consider what is happening to a small portion of the rim of the hole; suppose it subtends an angle $d\theta$ at the hole's centre. The mass of material uniformly distributed along its length $r\,d\theta$ is that of the sector of the film defined by r and $d\theta$, i.e. $dm = \frac{1}{2}w\rho r^2\,d\theta$. In time dt this mass increases its momentum by

$$dp_1 = \tfrac{1}{2}w\rho r^2\,d\theta\,\ddot{r}\,dt.$$

But this is not the only change in momentum, because in that same time interval an additional area of size $r\,d\theta \times \dot{r}\,dt$ has been 'brought up to speed'. This requires a further injection of momentum given by

$$dp_2 = w\rho r\,d\theta\,\dot{r}\,dt \times \dot{r}.$$

The force providing the total increase in momentum is that of the surface tension of the two-sided film, equal in strength to $2\gamma r\,d\theta$, and so

$$2\gamma r\,d\theta = dF = \frac{d(p_1 + p_2)}{dt} = w\rho\left(\tfrac{1}{2}r^2\ddot{r} + r(\dot{r})^2\right)d\theta,$$

i.e.

$$\frac{1}{2}r^2\ddot{r} + r(\dot{r})^2 = \frac{2\gamma}{w\rho}\,r. \qquad (*)$$

To integrate this equation, we must convert the LHS into the derivative of a product that takes the form $f(r,\dot{r}) = r^m(\dot{r})^n$. Either by inspection, or by substituting $f(r,\dot{r})$ and then equating coefficients, we find that $m=4$ and $n=2$. Thus, to make both the LHS and the RHS into the time derivatives of explicit functions, $(*)$ must be multiplied through by $r^2\dot{r}$:

$$\frac{1}{4}\frac{d}{dt}\left(r^4(\dot{r})^2\right) = \frac{1}{2}r^4\ddot{r}\dot{r} + r^3(\dot{r})^3 = \frac{2\gamma}{w\rho}r^3\dot{r} = \frac{1}{4}\frac{2\gamma}{w\rho}\frac{d(r^4)}{dt},$$

implying that

$$r^4(\dot{r})^2 = \frac{2\gamma}{w\rho}\,r^4 + \text{constant}.$$

When r is small $\sim r_0$ (related to whatever punctured the film) the hole grows rapidly ($\dot{r}\sim r^{-2}$), but later, provided the whole film is large enough, $\dot{r}$ becomes essentially constant and the hole's radius grows at a steady rate characterised by the soap films properties, i.e.

$$r(t) \simeq \left(\frac{2\gamma}{w\rho}\right)^{1/2} t + \text{constant}.$$

S52   Although our analysis will necessarily involve velocities, the question provides only lengths r and w, and the dimensionless quantity e; the final answers must be expressed in terms of these three quantities (and perhaps other standard constants).

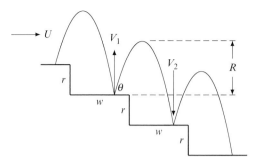

We first note that the bouncing does not affect the horizontal component of the golf ball's velocity; let that be U. However the vertical component is affected by both gravity and the inelastic collisions. As shown in the figure, we denote its value just after a bounce by V_1, and just before the next bounce by V_2.

The height R to which the ball rises above the step on which it last bounced is straightforwardly $R = V_1^2/2g$. The vertical component of the velocity with which it meets the next step is that gained by falling through a distance $R + r$, and given by

$$\tfrac{1}{2}mV_2^2 = mg(R + r).$$

Further, because the bouncing is repetitive, V_2 must be converted to V_1 at each impact, implying that $V_1 = eV_2$. Substituting for R and V_2 in the above equation yields

$$\frac{1}{2}\frac{V_1^2}{e^2} = g\left(\frac{V_1^2}{2g} + r\right),$$

leading to

$$V_1^2 = \frac{2e^2 rg}{1 - e^2}, \quad \text{and hence } R = \frac{V_1^2}{2g} = \frac{e^2}{1 - e^2}\,r.$$

The times taken for the two parts of a full bounce are V_1/g and V_2/g, and so U and w are related by

$$w = U\left(\frac{V_1 + V_2}{g}\right) = \frac{UV_1}{g}\frac{(e + 1)}{e}.$$

The launch angle θ following each bounce is given by

$$\tan\theta = \frac{V_1}{U} = \frac{V_1^2(e + 1)}{egw}$$

$$= \frac{(e + 1)}{egw}\frac{2e^2 rg}{1 - e^2}$$

$$= \frac{2e}{1 - e}\frac{r}{w}.$$

S 53   This question involves relatively little, but somewhat subtle, physics. However, it does require some careful, but theoretically straightforward, mathematics.

It might at first seem that the effects of the wind cancel out – the speed lost when 'heading into the wind' is recovered when 'it becomes a tail wind'. But this is not so. More time is spent covering any particular distance when facing a headwind, than is needed to cover the same distance when assisted by a tail wind; consequently the 'minuses' are effective for longer than the 'pluses', and there is a net overall *addition* to the circuit time.

The second slightly subtle point is that, at any particular position on the circular course, the pilot does not head the plane in the direction of the tangent there – although the plane itself must be made to move along that tangent. He has to direct it at an angle ϕ relative to the tangent, an angle that changes continuously as he goes round the course, and is sometimes positive and sometimes negative.

We will describe the situation using plane polar coordinates (r, θ), but as only the tangent direction and the value of θ are needed, the velocity-vector diagram does not show the whole circle.

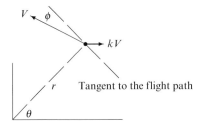

Without losing any generality, we can take the direction of the wind as $\theta = 0$ and a typical position of the plane as (r, θ), where $2\pi r = VT$. The angles θ and ϕ are shown in the figure, and they must be such that there is no net radial velocity for the plane. The net speed in the (positive) direction of the tangent gives the rate at which the circle is being traversed, i.e. is an expression for $r\dot{\theta}$ at that point on the circuit. The new circuit time will be found by integrating $\dot{\theta}^{-1}$ with respect to θ over one complete cycle, as follows.

Resolving the velocity vectors radially and tangentially

$$V \sin \phi - kV \cos \theta = 0,$$
$$V \cos \phi - kV \sin \theta = r\dot{\theta}.$$

Eliminating ϕ between the two equations yields

$$\dot{\theta} = \frac{V}{r} \left(\sqrt{1 - k^2 \cos^2 \theta} - k \sin \theta \right). \tag{$*$}$$

We note in passing that when replacing $\cos\phi$ the positive square root has been used, since, with $k \ll 1$, $\dot\theta$ must be positive.

Since the infinitesimal angle $d\theta$ is covered in the infinitesimal time $d\theta/\dot\theta$, to obtain the new (bigger) circuit time this has to be integrated from θ_0 to $\theta_0 + 2\pi$ for some, but not particular, θ_0.

With $\dot\theta$ as given by $(*)$, the integration cannot be carried out analytically, but since $k \ll 1$ we can make an expansion in powers of k. Retaining only powers of k up to k^2, $\dot\theta$ becomes

$$\dot\theta = \frac{V}{r}\left(1 - \tfrac{1}{2}k^2\cos^2\theta - k\sin\theta + \cdots\right),$$

the first two terms coming from the square root. From this, the expansion of $(\dot\theta)^{-1}$ is[7]

$$(\dot\theta)^{-1} = \frac{r}{V}\left[1 - \left(-\tfrac{1}{2}k^2\cos^2\theta - k\sin\theta\right) + (k^2\sin^2\theta + \cdots) + \cdots\right]. \qquad (**)$$

The final retained explicit term comes from the square of the previous expression in parentheses (), but is the only term in that square that does not contain k^3, or even higher powers of k.

Finally, we are in a position to carry out the integration [of $(**)$]. The unit term contributes 2π to the integral, the sine term contributes nothing, whilst the integrals of $\cos^2\theta$ and $\sin^2\theta$ around a complete cycle are each equal to π. Replacing $(2\pi r)/V$ by T, the new circuit time T' is

$$T' = T + \frac{1}{2}k^2\frac{T}{2} + 0 + k^2\frac{T}{2} = T + \frac{3}{4}k^2 T.$$

In summary, a (relatively modest) steady wind of speed kV increases the circuit time by approximately $75\,k^2\,\%$ – but, perhaps more seriously, also gives the pilot an additional task that needs continuous attention!

S 54 The sequence of events to be expressed quantitatively is that the cube slides down the slope losing gravitational energy, though it recovers some (but not all) of it as kinetic energy. When the cube reaches the bottom of the slope, its front lower edge is brought to a halt, although its centre of gravity continues to move. Whether or not the cube topples is determined by whether or not the kinetic energy it *does* have is sufficient to raise its centre of gravity from its then current position to one $\sqrt{2}b$ directly above the bottom of the slope.

---

[7] Using the binomial expansion $(1 + x)^{-1} = 1 - x + x^2 - \cdots$.

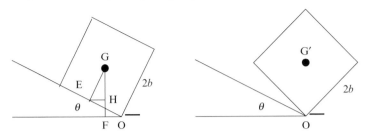

The left-hand figure shows the cube, of mass m, just as it's front edge comes to the rim of the tray, and the right-hand one, in which the cube's centre of gravity is directly above the rim (at O), shows the situation that must subsequently be reached if the cube is to topple over the rim.

The point G′ is the highest point that the centre of gravity attains in a toppling motion. Simple geometry gives the height of G′ as $\frac{1}{2}\sqrt{4b^2 + 4b^2} = \sqrt{2}b$. The height of G above the horizontal through O is a little more difficult to determine, but a similar triangles analysis shows that $\angle EGF$ is equal to θ and hence that FG = FH + HG $= b\sin\theta + b\cos\theta$, since both EO and EG have length b.

The additional energy needed to raise the cube's centre of gravity from height FG to height OG′ must come from the cube's kinetic energy when it reaches O. When it is released from rest with it's front edge a distance s from O (measured along the slope), it is subjected to a gravitational force of $mg\sin\theta$ down the slope and a frictional retardation of $\mu mg\cos\theta$ up it.

Hence it's acceleration down the slope is $g\sin\theta - \mu g\cos\theta$, and using the standard result '$v^2 = u^2 + 2as$' for uniform acceleration gives the square of its velocity when it reaches O as $2(g\sin\theta - \mu g\cos\theta)s$. Therefore, to prevent toppling I need

$$\sqrt{2}mbg > (\sin\theta + \cos\theta)mbg + \tfrac{1}{2}m\,2g(\sin\theta - \mu\cos\theta)s,$$

$$\text{i.e.} \quad s < \frac{\sqrt{2} - \sin\theta - \cos\theta}{\sin\theta - \mu\cos\theta}\,b.$$

Since the cube slides when its net acceleration is just positive, and $\theta = 18°$ at this point, $\mu = \tan 18°$. Substituting this and $\theta = 20°$ gives the result that s must be less than $3.61b$.

S 55   The stated conclusion may be slightly unexpected. The product $T_2(T_2 - T_1)$ could be related to one of the form $(R + x)x$ where x is the extension produced in the elastic chord, but how such a product arises is not clear. However, the principles to be applied are clear – so let's use them.

If the extension of the elastic chord is x, then we have

$$T_1 = \frac{2\pi R}{v} \quad \text{and} \quad T_2 = \frac{2\pi(R+x)}{v} = T_1 + \frac{2\pi x}{v}.$$

Further, x will be such that the tension it induces in the chord will provide the required centripetal force, i.e. $kx = mv^2/(R+x)$.

We re-write this last equation in the form

$$m = k\frac{x}{v}\frac{R+x}{v},$$

and then express the two fractions in terms of T_1 and T_2, as follows

$$m = k\frac{x}{v}\frac{R+x}{v}$$

$$= k\frac{(T_2 - T_1)}{2\pi}\frac{T_2}{2\pi},$$

$$T_2(T_2 - T_1) = \frac{4\pi^2 m}{k},$$

i.e. independent of both R and v.

S 56 (a) The standard equation for an ellipse with its centre C as the origin is

$$\frac{x^2}{a^2} + \frac{y^2}{b^2} = 1.$$

Because of the symmetry of the disc we need to consider only material in the first quadrant $(x > 0, y > 0)$, and we start by computing I_x, the moment of inertia about the x-axis. If the surface density is σ then

$$I_x = 4\sigma \int_0^a dx \int_0^{Y(x)} y^2\, dy, \quad \text{where } Y(x) = b\left(1 - \frac{x^2}{a^2}\right)^{1/2}.$$

The y-integral is elementary and yields

$$I_x = 4\sigma \int_0^a dx \frac{b^3}{3}\left(1 - \frac{x^2}{a^2}\right)^{3/2}.$$

To carry out the x integration, we set $x = a\cos\theta$ with $dx = -a\sin\theta\,d\theta$. Then

$$I_x = \frac{4\sigma b^3}{3}\int_{\pi/2}^0 (-a\sin\theta\,d\theta)\sin^3\theta$$

$$= \frac{4\sigma ab^3}{3}\int_0^{\pi/2}\sin^4\theta\,d\theta$$

$$= \frac{4\sigma ab^3}{3}\frac{3\pi}{16} = \frac{1}{4}Mb^2.$$

Similarly (or by symmetry), $I_y = Ma^2/4$, and so

$$I_z = \tfrac{1}{4}M(a^2 + b^2).$$

Note This same conclusion can be reached by considering the ellipse as a circle of radius b that has been stretched in the x-direction by a factor a/b, and recalling the well known result that, for an axis perpendicular to its plane, the moment of inertia I_z of the unstretched circular disc is $\tfrac{1}{2}Mb^2$. From the perpendicular axis theorem and azimuthal symmetry, it follows that for the circular disc $I_x = I_y = \tfrac{1}{4}Mb^2$. For the ellipse, obtained by making the transformation $x \to x' = (a/b)x$, and $y \to y' = y$, the moment I_x is unaltered, but $I_y \to I'_y = (a/b)^2 I_y$, i.e. it becomes $\tfrac{1}{4}Ma^2$. Applying the theorem again now shows that the required moment of inertia of the ellipse is $I_z = \tfrac{1}{4}M(a^2 + b^2)$.

(b)

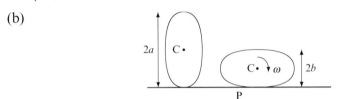

As there is no energy lost through friction, the angular velocity ω of the ellipse, when its minor axis is vertical, can be found directly from energy conservation. Initially, all the energy is gravitational and, measured from the table top, is equal to Mga. At the end, the gravitational energy is Mgb and the total kinetic energy is the sum of the linear kinetic energy of the disc's centre of mass, $\tfrac{1}{2}M(\omega b)^2$, and $\tfrac{1}{2}I_z\omega^2$, the rotational energy about C.[8] Thus,

$$Mga = Mgb + \tfrac{1}{2}M(\omega b)^2 + \tfrac{1}{2}\tfrac{1}{4}M(a^2 + b^2)\omega^2,$$

giving ω as

$$\omega = \sqrt{\frac{8g(a-b)}{a^2 + 5b^2}}.$$

---

[8] The kinetic energy could alternatively be calculated as being purely rotational about P, the point of contact between the disc and the table. However, the relevant moment of inertia is then that for an axis through P, given by the Parallel Axes theorem as $\tfrac{1}{4}M(a^2 + b^2) + Mb^2$, leading to the same result for the total kinetic energy.

Note The quoted result $\int_0^{\pi/2} \sin^4 \theta \, d\theta = 3\pi/16$ can be proved using the double-angle formulae $\cos 2x = 2\cos^2 x - 1 = 1 - 2\sin^2 x$ as follows:

$$\sin^4 \theta = (\sin^2 \theta)^2 = \left[\frac{1}{2}(1 - \cos 2\theta)\right]^2$$

$$= \frac{1}{4}(1 - 2\cos 2\theta + \cos^2 2\theta)$$

$$= \frac{1}{4}\left[1 - 2\cos 2\theta + \frac{1}{2}(1 + \cos 4\theta)\right]$$

$$= \frac{1}{4}\left(\frac{3}{2} - 2\cos 2\theta + \frac{1}{2}\cos 4\theta\right).$$

Hence,

$$\int_0^{\pi/2} \sin^4 \theta \, d\theta = \frac{3}{8}\frac{\pi}{2} - \frac{1}{2}0 + \frac{1}{8}0 = \frac{3\pi}{16}.$$

S 57 There are no outside couples acting about the vertical axis of the turntable and so any angular momentum gained by the dwarves must be equal and opposite to that acquired by the turntable, i.e., in magnitude, $I\omega$ is the same for both. The individual magnitudes of ω_{table} and ω_{dwarves} will be inversely proportional to I_{table} and I_{dwarves} respectively.

Because she is standing on the axis, Snow White contributes nothing to either moment of inertia, and her weight is irrelevant. The fact that they are all bunched together does not affect the contribution of each dwarf to their collective moment of inertia – all that matters is their individual distances from the axis, and these are all the same.

The two moments of inertia are therefore

$$I_{\text{table}} = \tfrac{1}{2} \cdot 150 \cdot (2.5)^2 = 469 \text{ kg m}^2,$$

$$I_{\text{dwarves}} = 7 \cdot \frac{300}{9.81} \cdot (2.5 - 0.5)^2 = 856 \text{ kg m}^2.$$

It follows that the ratio of the two angular speeds is $\omega_{\text{table}} : \omega_{\text{dwarves}} = 0.646 : 0.354$.

So far as covering distance on the table top is concerned, the two angular speeds augment each other. But the GPS only registers movement relative to the ground, and for the miners to cover 1 km relative to the ground requires them to cover $(0.646 + 0.354)/0.354 = 2.82$ times as far on the turntable. So Grumpy does seem to have a point – well, up to a point.

S 58   Since we are not told the skater's initial rotation speed, it may seem that his final speed could be anything. However, we will certainly need values for the moment of inertia of his alter ego, both before and after the arms are lowered. We find them as follows.

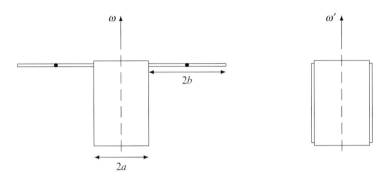

The left-hand figure shows the initial configuration and defines the notation. The contribution of the cylinder to the total moment of inertia I_1 is straight-forwardly $\frac{1}{2}Ma^2$; for the contribution of the 'arms' we need to use the parallel axis theorem. Since the MoI of a thin rod of mass m and length 2ℓ about a perpendicular axis through its centre is $\frac{1}{3}m\ell^2$, that of an 'arm' about the central axis is $\frac{1}{3}mb^2 + m(a+b)^2$. Thus

$$
\begin{aligned}
I_1 &= \frac{1}{2}Ma^2 + 2\left[\frac{1}{3}mb^2 + m(a+b)^2\right] \\
&= \frac{(72-8)(0.2)^2}{2} + 2\left[\frac{1}{3}4(0.35)^2 + 4(0.20+0.35)^2\right] \\
&= 4.027 \text{ kg m}^2.
\end{aligned}
$$

The second MoI (see the right-hand figure) is more straightforward:

$$
I_2 = \tfrac{1}{2}Ma^2 + 2ma^2 = \left(\tfrac{1}{2}64 + 2 \times 4\right)(0.2)^2 = 1.6 \text{ kg m}^2.
$$

If the initial angular speed of the skater was ω, by angular momentum conservation, his final one is

$$
\omega' = \frac{I_1}{I_2}\omega = 2.517\,\omega.
$$

It might seem that this is as far as we can go, but we have not yet used all the information in the question. His kinetic energy of rotational motion has changed, and that energy change must be accounted for: if his final speed is too high, he will have had to use his arm muscles to pull them in (but he makes no effort); if his final speed is too low, his arms would bang into his sides (but they come in gently); for the speed to be neither too high nor too low but 'just right', the gain in rotational energy must just balance the loss in gravitational energy resulting from the lowering of his arms, i.e.

$$\tfrac{1}{2} I_2 \, \omega'^2 - \tfrac{1}{2} I_1 \omega^2 = 2 \, mg \times b,$$

$$\tfrac{1}{2} \, 1.6 \, (2.517 \omega)^2 - \tfrac{1}{2} \, 4.027 \, \omega^2 = 2 \times 4 \times 9.81 \times 0.35,$$

$$3.054 \, \omega^2 = 27.468 \quad \Rightarrow \quad \omega = 3.00 \, \text{rad s}^{-1}.$$

So, the final rotation speed *is* uniquely defined, and given by $2.517 \times 3.00 = 7.55$ rad s^{-1}.

S 59 Most readers will know from experience that door-stops are placed about half way between the centre of the door and its free (unhinged) edge. They will also intuitively feel that, following a door's impact with a door-stop, its hinged edge, if it were not fixed, would rotate about the stop – in one direction if the stop is too close, and in the other direction if it is too far away. These observations need to be made quantitative.

The figure shows the physical layout, with the door's centre of mass C a distance c from the line of the hinges H, and the door-stop P placed a further distance p beyond that.

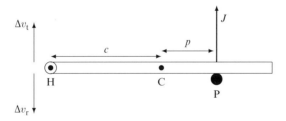

When the door bangs against the door-stop, the impulse J it receives tends to cause both lateral movement of C and rotation about C. The potential speed increment of H due to the translational motion is, as for all parts of the door, $\Delta v_t = J/M$, where M is the mass of the door, and is in the same direction as J. That due to rotation about C is $\Delta v_r = c \Delta \omega$, where $\Delta \omega$ is the potential increment in rotational speed, and given by

$$\Delta \omega = \frac{\text{impulsive couple about C}}{I} = \frac{J \times p}{I}.$$

Here I is the door's moment of inertia *about its centre of mass*. Δv_r must be made to be in the direction opposed to that of J if the impulsive reaction at H is to be kept as small as possible; this is why P is placed at a distance greater than c from H, i.e. beyond the centre of mass.

If there is to be no impulsive reaction at the hinge, we must have $|\Delta v_t| = |\Delta v_r|$, i.e.

$$\frac{J}{M} = \frac{c\,Jp}{I} \quad \Rightarrow \quad p = \frac{I}{Mc}.$$

This gives the 'ideal' distance beyond C for P, which is known as the *centre of percussion*.

For the particular case of a uniform door of width w and density ρ per unit width, $M = \rho w$ and $c = \frac{1}{2}w$, whilst

$$I = \int_{-w/2}^{w/2} x^2 \, \rho \, dx = \rho \left[\frac{x^3}{3} \right]_{-w/2}^{w/2} = \frac{\rho w^3}{12} = \frac{1}{12} M w^2,$$

leading to

$$p = \frac{I}{Mc} = \frac{w}{6},$$

and to the conclusion that the door-stop should be placed $(\frac{1}{2} + \frac{1}{6})w = \frac{2}{3}w$ from the line of the hinges.

S 60   This problem involves gravitational, centripetal and frictional forces simultaneously; nevertheless, applying the basic laws governing each can be expected to produce the required answers.

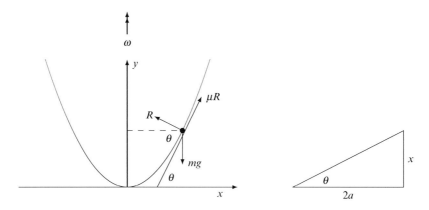

The left-hand figure shows the forces acting on the block when it is on the point of slipping down, and consequently the tangential frictional force, of maximum magnitude μR, has an upward vertical component. Here, R is the normal reaction of the bowl on the block, and is the source of the force needed to maintain the block in its 'orbit'. Since we are only concerned with its vertical height, we can ignore the question of any lateral slippage by the block.

Resolving the force balance vertically and horizontally gives

$$R\cos\theta + \mu R\sin\theta = mg,$$
$$R\sin\theta - \mu R\cos\theta = m\omega_-^2 x,$$

where θ is the angle the tangent (plane) to the rotating bowl makes with the horizontal, as shown in the figure.

Now, since the bowl has the shape $x^2 = 4ay$, elementary differentiation gives $2x = 4a(dy/dx)$ and hence that $\tan\theta = x/(2a)$. The right-hand figure illustrates this, and demonstrates that $\sin\theta : \cos\theta = x : 2a$; this figure is *not* a 'close-up' of any part of the left-hand figure.

Dividing the two force balance equations, and using the above proportionality, yields

$$\omega_-^2 = \frac{g}{x}\frac{\sin\theta - \mu\cos\theta}{\cos\theta + \mu\sin\theta} = \frac{g}{x}\frac{x - 2\mu a}{2a + \mu x}.$$

The corresponding result for ω_+ can be obtained simply by replacing μ by $-\mu$:

$$\omega_+^2 = \frac{g}{x}\frac{x + 2\mu a}{2a - \mu x}.$$

(a) If $\mu \to 0$, then both ω_-^2 and ω_+^2 tend to $\omega_0^2 = g/(2a)$, independently of x in each case. In other words, any value of x, and hence any height y, is possible, but only if ω is precisely equal to ω_0 – something that would not be possible in practice. But then, neither would finding a frictionless block and bowl combination!

(b) For $y = a$, and hence $x = 2a$,

$$\omega_-^2 = \frac{g}{2a}\frac{1 - \mu}{1 + \mu} \quad \text{and} \quad \omega_+^2 = \frac{g}{2a}\frac{1 + \mu}{1 - \mu}.$$

The product $\omega_-^2\omega_+^2$ is equal to $g^2/(4a^2) = \omega_0^4$, showing that ω_0 is the geometric mean of the two limiting rotation speeds, whatever the value of μ.

(c) When $y = a\mu^2$, $x = 2\mu a$ and the expression for ω_- yields zero. Reference to the right-hand figure shows that in this situation $\theta = \tan^{-1}\mu$. This angle gives the slope of a rough inclined plane on which an object just starts to slide under its own weight; the situation is as if the block were placed on the tangential plane to the bowl at $(2\mu a, a\mu^2)$. The value of ω_+, obtained by straightforward substitution, is

$$\omega_+^2 = \frac{g}{2\mu a}\frac{4\mu a}{2a - 2\mu a^2} = \frac{2}{1 - \mu^2}\omega_0^2.$$

(d) If $y = a\mu^{-2}$, and $x = 2a/\mu$, then ω_+ is (formally) infinite and the block cannot be prevented from rising to the rim of the bowl. Theoretically, the minimum angular speed needed is given by

$$\omega_-^2 = \frac{\mu g}{2a} \frac{(2a/\mu) - 2\mu a}{2a + 2a} = \frac{1 - \mu^2}{2} \omega_0^2.$$

But this seems highly irrelevant if the block cannot be retained within the bowl!

S 61   It seems as if Alec is likely to be wrong, because, surely, when the mass goes round in a circle, the string will lengthen further to provide a centripetal force. However, a proper analysis is needed, and the figure shows the arrangement, in which the extension in the string is now the unknown length ℓ_2, the tension in it is T, and its inclination to the vertical is θ. The small solid circles do no more than indicate the length of the unstretched string; the distance from either to P is ℓ_0; the lengths of the string below them are ℓ_1 and ℓ_2 respectively.

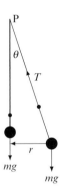

The other relevant but unknown quantities involved are: the mass m, the modulus of the string λ, and the angular velocity ω, which is related to τ by $\tau = 2\pi/\omega$.

The four simple equations that can be obtained from the static and steady dynamical situations are

$$\lambda\ell_1 = mg, \tag{1}$$
$$T\cos\theta = mg, \tag{2}$$
$$T\sin\theta = mr\omega^2 = m(\ell_0 + \ell_2)\sin\theta\,\omega^2, \tag{3}$$
$$\lambda\ell_2 = T. \tag{4}$$

With only four equations, and four unknowns (T, ℓ_2, m and λ) to be eliminated, it might seem that we do not have enough information. But we expect T to be proportional to m and so the two may effectively appear only as the ratio T/m, thus reducing the unknowns to three. So, eliminating λ from (1) and (4) to obtain $T\ell_1 = mg\ell_2$, and then using this equation and (3) to remove ℓ_2, we obtain

$$T\cos\theta = mg,$$
$$T = \left(m\ell_0 + \frac{T\ell_1}{g}\right)\omega^2.$$

Finally, removing m by substitution gives

$$T = \left(\frac{T\cos\theta\,\ell_0}{g} + \frac{T\ell_1}{g}\right)\omega^2,$$

in which, as anticipated, T can be removed by cancellation, and a final re-arrangement gives

$$\omega^2 = \frac{g}{\ell_0\cos\theta + \ell_1} \quad\Rightarrow\quad \tau = 2\pi\sqrt{\frac{\ell_0\cos\theta + \ell_1}{g}}.$$

So the period depends on the angle of the conical pendulum, and Alec was not so smart after all – though he was nearly correct for small-angled conical pendulums. No doubt he was unduly influenced by having timed too many small-angle simple pendulums!

S62 There are two sources from which the mill can extract energy, the kinetic energy of the water entering the buckets and the gravitational energy that the water loses when falling to the lower level. Equally, there are two torques acting upon the mill wheel, one gravitational, the other arising from the momentum change as the water enters the buckets. We need to consider each of the four factors in turn, and then choose the angular speed of the wheel that optimizes its overall efficiency.

With closely spaced buckets, we can treat the water in them as if it formed a uniformly distributed semicircle.

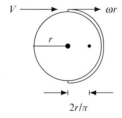

If the angular speed of the wheel is ω, then it takes any particular bucket π/ω to reach the lowest point, and in that time water of mass $M\pi/\omega$ has moved onto the wheel, where M is the mass flow rate of the millrace. This is therefore the total mass of the water semicircle. Its centre of gravity (indicated by a small black circle in the figure) is a distance $2r/\pi$ from the centre of the wheel,[9] and so the torque it produces on the wheel is

$$\tau_1 = \frac{M\pi}{\omega} g \frac{2r}{\pi} = \frac{2Mgr}{\omega}.$$

A second torque, τ_2, arises when the speed of the water from the race is reduced from V to that of the periphery of the wheel, ωr. This impulsive torque is given by

$$\tau_2 = r \times \text{the rate of loss of momentum} = r \times M(V - \omega r).$$

The power generated by the mill, given by the rotation rate × the total torque, $(\tau_1 + \tau_2)$, is therefore

$$P = \omega \left[\frac{2Mgr}{\omega} + Mr(V - \omega r)\right] = 2Mgr + MrV\omega - Mr^2\omega^2.$$

This will be maximized if ω is required to satisfy

$$0 = \frac{dP}{d\omega} = MrV - 2Mr^2\omega \quad \Rightarrow \quad \omega = \frac{V}{2r}.$$

The power then generated is

$$P_{max} = 2Mgr + \tfrac{1}{2}MV^2 - \tfrac{1}{4}MV^2 = 2Mgr + \tfrac{1}{4}MV^2.$$

Producing this power uses, in unit time, $Mg(2r)$ of gravitational potential energy, and $\tfrac{1}{2}MV^2$ of kinetic energy. So the efficiency of the power extraction process is

$$\epsilon = \frac{2Mgr + \tfrac{1}{4}MV^2}{2Mgr + \tfrac{1}{2}MV^2} = \frac{8gr + V^2}{8gr + 2V^2}.$$

Note In plane polar coordinates, the centre of mass of a body that is symmetric about the $\theta = 0$ axis, is given by $\bar{x} = m^{-1} \int x \, dm$. For a uniform semicircular arc of radius a and linear density ρ,

$$\bar{x} = \frac{1}{\pi a\rho} \int_{-\pi/2}^{\pi/2} (\rho a \, d\theta) \, a \cos\theta = \frac{a}{\pi} [\sin\theta]_{-\pi/2}^{\pi/2} = \frac{2a}{\pi}.$$

S 63   This is a frictionless isolated system, and so applying the laws of conservation of energy and angular momentum will answer the questions posed.

---

[9] See the note at the end of this solution, if necessary.

Let the tube length be ℓ, the distance of the particle from the pivot be x, and the angular velocity of the tube be ω_0 initially and ω more generally (see the figure).

As is always the case for an isolated system, angular momentum is conserved, and in the present case, as there is no friction, so is energy. The moment of inertia of the tube is $\frac{1}{3}m\ell^2$ when it rotates about one of its ends[10] and that of the particle is mx^2 when it is at a distance x from the pivot.

Conservation of angular momentum gives

$$\frac{1}{3}m\ell^2\,\omega_0 = \left(\frac{1}{3}m\ell^2 + mx^2\right)\omega \quad \Rightarrow \quad \omega = \frac{\ell^2}{\ell^2 + 3x^2}\,\omega_0, \tag{$*$}$$

whilst energy conservation yields

$$\frac{1}{2}\frac{1}{3}m\ell^2\,\omega_0^2 = \frac{1}{2}\frac{1}{3}m\ell^2\,\omega^2 + \frac{1}{2}m\dot{x}^2 + \frac{1}{2}m(\omega x)^2,$$

which simplifies (a little) to

$$\ell^2\omega_0^2 = \ell^2\omega^2 + 3x^2\omega^2 + 3\dot{x}^2. \tag{$**$}$$

The first equation $(*)$ shows that for $0 < \delta < x < \ell$, for some δ (the particle's initial non-zero distance from the pivot), $\omega_0 > \omega > \frac{1}{4}\omega_0$. It can also be used to re-write $(**)$ as

$$\ell^2\omega_0^2 = \ell^2\omega_0\omega + 3\dot{x}^2,$$
$$\text{i.e. } \dot{x}^2 = \frac{1}{3}\omega_0(\omega_0 - \omega)\ell^2.$$

Since $\delta > 0$ and the particle will start to move when the tube is set in motion, this equation shows that it will speed up as the tube slows down, and emerge from the tube. From $(*)$, $\omega = \frac{1}{4}\omega_0$ when $x = \ell$ and, at that point

$$\dot{x} = \sqrt{\frac{\ell^2}{3}\frac{3\omega_0^2}{4}} = \frac{\omega_0\ell}{2}.$$

With a radial speed of $\frac{1}{2}\omega_0\ell$ and a tangential speed of $\frac{1}{4}\omega_0\ell$, the angle the particle's path makes with the tube axis (see the right-hand figure) is $\phi = \tan^{-1}\frac{1}{2}$.

---

[10] If this is not a familiar result, see the proof in the Note at the end of this solution.

This value can only be approximate, since both the radial speed attained, and the ratio ω/ω_0, are marginally affected by the actual value of δ. Theoretically, if δ were truly zero, the particle would not move, but that would be an unstable situation.

Note For a uniform rod of length ℓ and linear density ρ, the contribution to its moment of inertia about one of its ends by a small length dx situated at a distance x from that end, is $\rho\,dx \times x^2$. And so the moment of inertia of the rod is

$$I = \int_0^\ell \rho x^2\,dx = \tfrac{1}{3}\rho\ell^3 = \tfrac{1}{3}M\ell^2,$$

where M is its mass.

S64 Some care is needed in determining the magnitudes and directions of the various forces that act on the cylinder, in particular the frictional ones. We also need to keep in mind that reactions cannot physically be negative, and formulae that yield such possibilities mathematically should be treated with caution.

The figure below is that given in the question, but with all of the forces acting on the cylinder now included.

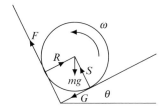

The normal reactions with the planes, S and R, act through the cylinder's centre of mass, whilst the frictional forces $G = \mu S$ and $F = \mu R$, acting tangentially, provide the couple that slows the cylinder's rotation. Straightforward geometry shows that the vertical makes an angle θ with reaction S, and $\tfrac{1}{2}\pi - \theta$ with R.

(a) Resolving the forces involved parallel to each of the (orthogonal) planes gives the equations

$$S + \mu R = mg\cos\theta,$$
$$R - \mu S = mg\sin\theta,$$

which can be solved to give

$$R = \frac{mg(\mu\cos\theta + \sin\theta)}{1 + \mu^2} \quad \text{and} \quad S = \frac{mg(\cos\theta - \mu\sin\theta)}{1 + \mu^2}.$$

Now the retarding couple, provided by friction, is $\mu Ra + \mu Sa$, and since the relevant moment of inertia of the cylinder is $\frac{1}{2}ma^2$,

$$\dot{\omega} = -\frac{\mu(R+S)a}{\frac{1}{2}ma^2},$$

giving the time to come to rest as

$$\tau = \frac{ma\Omega}{2\mu(R+S)} = \frac{a\Omega(1+\mu^2)}{2\mu g[(1+\mu)\cos\theta + (1-\mu)\sin\theta]}.$$

(b) This formula looks 'lopsided', because if we replace θ by $\phi = \frac{1}{2}\pi - \theta$, corresponding to replacing the cylinder and the planes by their mirror images in a vertical plane through the bottom of the groove, the denominator becomes $(1+\mu)\sin\theta + (1-\mu)\cos\theta$, which, for a given θ and μ, produces a different stopping time, despite the apparent symmetry between the two situations.

However, the above is not a valid argument because it has failed to take into account the rotation of the cylinder – this must be reversed for a true mirror image; and if it is, then the frictional forces change their directions in space and their signs in the formulae.

The origin of the difficulty with the calculated result for larger values of θ can be seen in the explicit formula for S. If $\theta > \tan^{-1}(1/\mu)$ then S becomes negative, and although this is mathematically acceptable, physically it is not; without some mechanism for generating an attraction, negative reactions do not occur.

Since $0 < \mu < 1$, the critical value of θ must be greater than $\tan^{-1} 1 = \pi/4$, but less than $\pi/2$. When it is exceeded, the reaction S becomes zero and the frictional force F causes the cylinder to roll away from the groove up the left-hand plane, which by now makes an angle with the horizontal that is less then $\tan^{-1}\mu$. This is not the situation to which the previous calculations apply.

There is no similar difficulty for small values of θ, because in this case the direction of rotation of the cylinder would mean it would have to climb a nearly vertical wall using friction generated by a small normal reaction R, which itself is limited to less than $mg(\mu + \sin\theta)$.

S 65 When a body receives an impulse (a quantum of momentum) it acquires a bodily momentum, and if (as here) the line of action of the impulse does not pass through the centre of mass, it also acquires angular momentum (see solution S 59).

Let O be the centre of the disc and C be the point about which the initial rotation takes place. By symmetry, C must lie on the diameter of the disc that contains the point P at which the impulse is applied, and, intuitively, on the side of O remote from P. Let the distance from O to C be c, and denote the magnitude of the impulse by J. The figure above illustrates this notation.

Firstly, from the conservation of linear momentum, after the impulse is received, O must be moving in a direction parallel to that of J with speed $v = J/M$. If the angular speed of rotation about C is ω, we must also have $v = \omega c$. Together, these observations imply that $J = M\omega c$.

Next, from the conservation of angular momentum (about C), $J(a + c) = I_C \omega$, where I_C is the moment of inertia of the disc about the normal to the disc that passes through C. As (nearly) every schoolboy/girl knows, or at least has in their Formula Booklet, the moment of inertia of such a disc about the normal passing through its centre is $\frac{1}{2}Ma^2$. So, by the parallel axis theorem, $I_C = \frac{1}{2}Ma^2 + Mc^2$.

Combining the three results so as to eliminate J, M and ω, we have

$$M\omega c(a + c) = (\tfrac{1}{2}Ma^2 + Mc^2)\omega \quad \Rightarrow \quad c = \tfrac{1}{2}a.$$

So the disc begins to rotate about the point midway between its centre O and the far end of the diameter through P.

S 66   The analysis of this situation falls into two parts. We first need to know through what angle the rod has fallen before it loses contact with the table, and then we must follow it's flight as a rotating object in free fall.

The figure below shows the situation when the rod is just about to lose contact with the table and has fallen through an angle θ_0 whilst still in contact with it. The rod is also shown as it first touches the floor.

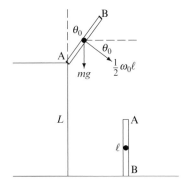

Since the table is very rough, there will be no slippage of the lower end of the rod, whatever the other forces acting on it. The condition for the rod to lose contact with the table edge is that there is no 'action and reaction' between the two, i.e. the centrifugal force acting on the rod becomes just larger than the force on the table due to the rod's weight.

If, at that instant, the angular velocity of the rod about A is w_0, the centrifugal force is $m(\ell/2)w_0^2$, where m is the rod's mass. The component of its weight acting along its length is $mg\cos\theta$ and this must be equal, but opposite to, the centrifugal force, i.e.

$$m\,\frac{\ell}{2}\,w_0^2 = mg\cos\theta_0. \qquad (*)$$

The value of w_0 can be found from the conservation of energy; some of the gravitational potential energy of the rod has been converted to rotational energy. If the rod's moment of inertia about end A is I_A, then

$$\tfrac{1}{2}I_A w_0^2 = mg(1 - \cos\theta_0)\tfrac{1}{2}\ell, \qquad (**)$$

and the downward component of the rod's centre of mass is $\tfrac{1}{2}w_0\ell\sin\theta_0$.

At this point the rod becomes an object rotating with angular speed w_0, and in free fall. If end B hits the floor with BA vertical, then during the fall the rod must rotate through an angle $\pi - \theta_0$ and this takes a time $t = (\pi - \theta_0)/w_0$. This time, therefore, must also be that of the fall, and the corresponding height (using $s = ut + \tfrac{1}{2}at^2$) is

$$s = \frac{w_0\ell}{2}\sin\theta_0\,\frac{(\pi - \theta_0)}{w_0} + \frac{g}{2}\,\frac{(\pi - \theta_0)^2}{w_0^2}.$$

The centre of mass is then $\tfrac{1}{2}\ell$ above the floor, and so

$$s + \tfrac{1}{2}\ell = L + \tfrac{1}{2}\ell\cos\theta_0,$$

where L is the height of the table. Substituting for s and g/w_0^2, followed by re-arrangement and some simplification, gives L as

$$L = \frac{\ell}{2}\left[(\pi - \theta_0)\sin\theta_0 + \frac{(\pi - \theta_0)^2}{2\cos\theta_0} + 1 - \cos\theta_0\right].$$

We now return to $(*)$ and $(**)$ to evaluate θ_0. The moment of inertia of the rod when rotating about A, one of its ends, is $I_A = \tfrac{1}{3}m\ell^2$.[11] Consequently we must have

$$\frac{1}{2}mg\ell(1 - \cos\theta_0) = \frac{1}{2}\frac{1}{3}m\ell^2\,w_0^2 = \frac{m\ell^2}{6}\frac{2g\cos\theta_0}{\ell}.$$

---

[11] If this is not a familiar result, see the proof in solution S 63 on page 168.

It follows that

$$1 - \cos\theta_0 = \tfrac{2}{3}\cos\theta_0 \quad \Rightarrow \quad \cos\theta_0 = \tfrac{3}{5} \quad \text{and} \quad \sin\theta_0 = \tfrac{4}{5}.$$

As an angle, $\theta_0 = 0.9273$ rad $= 53.1°$. Finally,

$$L = \frac{\ell}{2}\left[\frac{4}{5}(2.214) + \frac{5}{6}(2.214)^2 + 1 - \frac{3}{5}\right] = 3.128\,\ell.$$

Theoretically, the rod could have turned through $(2n + 1)\pi$ before hitting the floor, but we have taken the simplest solution, with $n = 0$.

This problem is an analysis of an experiment you <u>can</u>, *and ideally should*, try at home! A normal 30 cm rule (actual length about 33 cm) requires a table top, probably augmented by a few physics books, that together are 1.03 m above the floor, and a towel placed over them to give the assumed non-slip surface. Watch how, and where, it lands!

You should also be able to show theoretically, and *verify experimentally*, that the horizontal distance from the foot of the table of the point of impact is

$$\left[\frac{3}{10}(2.214) + \frac{4}{10}\right]\ell = 1.064\,\ell = 35.1 \text{ cm for the ruler.}$$

S 67 The first task is to decide which properties of the basket need to be known in order to answer the questions posed. For part (a), the 'simple pendulum', we clearly need the position of its centre of gravity; since all other components are 'light', this will essentially determine the effective length of the pendulum. For part (b), the basket's density will play some role, but, given the sort of oscillation involved, it will be in determining the basket's moment of inertia, rather than its total mass.

The 'normal' representation of a parabola with the y-axis as its axis of symmetry and its apex touching the x-axis, is $x^2 = ay$. For the given basket, with its apex at the origin and $(\pm 0.2, 0.2)$ as two points on the parabolic section produced by a plane containing the y-axis, the value of a must be 0.2 m.

For a paraboloid of revolution of height b, the required properties are

$$\bar{y} = \frac{\int y\,dm}{\int dm} = \frac{\int_0^b y\rho\pi x^2\,dy}{\int_0^b \rho\pi x^2\,dy} = \frac{\int_0^b y^2\,dy}{\int_0^b y\,dy} = \frac{(b^3/3)}{(b^2/2)} = \tfrac{2}{3}b,$$

and (recalling that the moment of inertia of a uniform disc of mass m and radius r about its normal axis of symmetry is $\tfrac{1}{2}mr^2$)

$$I = \int_0^b \tfrac{1}{2}x^2\,dm = \int_0^b \tfrac{1}{2}x^2(\rho\pi x^2)\,dy = \int_0^b \tfrac{1}{2}\rho\pi(ay)^2\,dy = \tfrac{1}{6}\pi\rho a^2 b^3.$$

With $a = 0.2$ m, $b = 0.2$ m and $\rho = 1200$ kg m^{-3}, this gives I as 0.201 kg m^2.

(a) From the above, the centre of gravity of the basket, and hence of the pendulum, is $\frac{1}{3}b = 0.0667$ m below its upper surface, whilst the end of the rope is $0.2 \times \tan 60° = 0.3461$ m above it. If the total effective length of the 'simple pendulum' is L, then, from the timed oscillation period, we have

$$3.000 = 2\pi \sqrt{\frac{L}{9.81}} \quad \Rightarrow \quad L = \frac{9}{4\pi^2} 9.81 = 2.236 \text{ m}.$$

It follows that the *actual* rope length is $\ell = 2.236 - 0.346 - 0.067 = 1.823$ m, i.e. 1.82 m, to the nearest centimetre.

(b) If the torsion constant of the rope is k, then the period τ of torsional oscillations is given by

$$\tau = 2\pi \sqrt{\frac{k}{I}} \quad \Rightarrow \quad k = \frac{(5 \times 3)^2}{4\pi^2} 0.201 = 1.146 \text{ N m rad}^{-1}.$$

Therefore the forces needed to produce the initial 20° ($\pi/9$ in radians) twist were each

$$F = \frac{k}{2 \times 0.2} \frac{\pi}{9} = 1.00 \text{ N}.$$

S 68   As only an approximate answer is needed, effects due to air resistance or gravity can be ignored in the investigation of the rapid circulatory motion – though not, of course, when analysing a correctly working ballistic pendulum, where gravity plays an essential role. We will also disregard any tendency of the wire to stretch when under tension.

Initially we will work with symbolic physical quantities (and substitute values later) so that the physics principles remain clear. The analysis will include (i) linear momentum conservation when the bullet is buried in the block, (ii) angular momentum conservation about the beam, and (iii) the tension T in the wire required to maintain the (nearly-)circular motion.

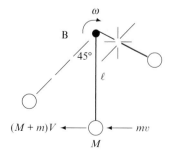

As indicated in the figure, we denote the bullet's mass and initial speed by m and v, the mass of the block by M, and the length and angular velocity of the pendulum by ℓ and ω, respectively.

If V is the speed of the block, with its embedded bullet, immediately after impact – when $\ell = \ell_0$ and $\omega = \omega_0$ – the relevant equations are

(i) $\qquad\qquad (M+m)V = mv,$

(ii) $\qquad\quad (M+m)\omega\ell^2 = (M+m)\omega_0\ell_0^2 = (M+m)V\ell_0,$

(iii) $\qquad\qquad T = (M+m)\omega^2\ell.$

From (ii), $V = \ell^2\omega/\ell_0$, whilst from (iii) $\omega^2 = T/[(M+m)\ell)]$, and so

$$V = \frac{\ell^2}{\ell_0}\sqrt{\frac{T}{(M+m)\ell}}.$$

Now, from the snapping of the wire, $T = 1000$ N when $\ell = 0.50$ m and so

$$V = \frac{0.25}{1.0}\sqrt{\frac{1000}{0.51 \times 0.5}} = 15.6\,\mathrm{m\,s^{-1}}.$$

Then, from (i), the bullet's speed when fired from the rifle was

$$v = \frac{0.51}{0.01}\,15.6 = 798.4 \approx 800\,\mathrm{m\,s^{-1}}.$$

For the ballistic pendulum to have behaved more as expected, i.e. rise by about $45°$ (with no damage to the wire), requires V to satisfy

$$\tfrac{1}{2}(M+m)V^2 = (1 - \cos 45°)(M+m)g\ell_0.$$

In terms of v, and approximating $(M+m)$ by M (since $m \ll M$), this can be rewritten as

$$\left(\frac{m}{M}\right)^2 = \frac{0.586\,g\ell_0}{v^2},$$

leading to $M \approx 3.3$ kg, i.e some six or seven times the mass of the block actually used.

S 69 The physics of the situation is straightforward; the centrifugal force experienced by the object at a radial distance r from the wheel's centre, $mr\omega^2$ in magnitude, is such as to decrease the object's velocity towards the centre. Remembering that r, (a function of time t) is measured radially *outwards*, the equation of motion is

$$m\ddot{r} = mr\omega^2 \quad \text{which has the general solution} \quad r(t) = Ae^{\omega t} + Be^{-\omega t}.$$

The 'boundary conditions' to be satisfied are $r(0) = R$ and $\dot{r}(0) = -V$, i.e.

$$A + B = R \quad \text{and} \quad \omega(A - B) = -V,$$

which have the solution

$$A = \frac{1}{2}\left(R - \frac{V}{\omega}\right) \quad \text{and} \quad B = \frac{1}{2}\left(R + \frac{V}{\omega}\right).$$

If the object arrives at the space-station's centre after a time τ, we must have $r(\tau) = 0$ which, after the substitution of the derived values for A and B, simplifies to

$$(\omega R - V)e^{\omega\tau} + (\omega R + V)e^{-\omega\tau} = 0.$$

This can be re-arranged to give τ as

$$\tau = \frac{1}{2\omega} \ln \frac{V + \omega R}{V - \omega R}. \tag{$*$}$$

For general V this is not an easy function to visualize, but for the special cases indicated:

(i) $V \to \omega R$. Mathematically, the denominator in the fraction in $(*)$ tends to zero and the fraction, as well as its logarithm, tend to infinity. Hence, so does τ. Physically, the object will only reach its destination in a finite time if $V > \omega R$.

(ii) $V \to \infty$. The fraction tends to unity, its logarithm tends to zero, and the object gets there in no time at all! Well, not really, classical mechanics would fail and relativity would step in.

(iii) $V < \omega R$. Formally, τ becomes the logarithm of a negative number; that is imaginary, as is any thought that the object will ever be delivered. See case (i).

(iv) $\omega \to 0$. This seems physically to be absolutely straightforward, $\tau = R/V$; but this is hardly obvious from $(*)$. To see that it is, in fact, correct (or, more accurately, 'not incorrect') we divide both the numerator and denominator of the fraction by V and write the expression for τ in the form

$$\tau = \frac{1}{2\omega} \left[\ln\left(1 + \frac{\omega R}{V}\right) - \ln\left(1 - \frac{\omega R}{V}\right) \right],$$

which can, in turn, be written as[12]

$$\tau = \frac{1}{2\omega}\left\{\left[\frac{\omega R}{V} - \frac{1}{2}\left(\frac{\omega R}{V}\right)^2 + \cdots\right] - \left[-\frac{\omega R}{V} - \frac{1}{2}\left(\frac{\omega R}{V}\right)^2 - \cdots\right]\right\}.$$

As $\omega \to 0$, only the first term from each square bracket survives, and

$$\tau \sim \frac{1}{2\omega}\frac{2\omega R}{V} = \frac{R}{V}.$$

Just as we had expected all along!

S 70 (a) That's for you to know, and for me to remain ignorant of.

(b) We first establish the energy conservation equation for the situation illustrated by the figure in the question, which is reproduced below.

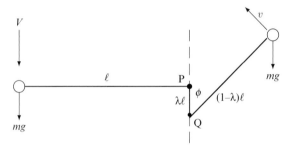

Measuring all gravitational potential energies from the (arbitrary) level of P, we have

$$\tfrac{1}{2}mV^2 = \tfrac{1}{2}mv^2 + mg(1 - \lambda)\ell\cos\phi - mg\lambda\ell,$$

as well as the observation that unless $V > 0$ the mass could never rise above the level of P in the motion that follows release.

Considering, for the moment, only motion in which the string remains taut after its impact with Q, we must have that the centrifugal force acting on the mass exceeds the component of its weight that is directed towards Q. This is clearly only a factor when the mass is above the level of Q, although it is formally a requirement for the whole range $\pi \geq \phi \geq 0$:

$$\frac{mv^2}{(1 - \lambda)\ell} > mg\cos\phi.$$

---

[12] Using the standard expansion

$$\ln(1 + x) = x - \frac{x^2}{2} + \frac{x^3}{3} - \cdots \quad \text{for } -1 < x \leq 1.$$

If we now eliminate v^2 from the equation and the inequality, we obtain the further inequalities

$$V^2 > g\ell(1 - \lambda) \cos \phi + 2g\ell(1 - \lambda) \cos \phi - 2g\ell\lambda,$$

$$\frac{V^2}{g\ell} > 3(1 - \lambda) \cos \phi - 2\lambda.$$

The most critical point of the motion, defining the value of V needed to actually complete one half circle about Q, comes when $\phi = 0$ and $\cos \phi = 1$. Consequently the minimum value of V needed is given by

$$V^2 = g\ell(3 - 5\lambda). \tag{*}$$

This means that for $\lambda > \frac{3}{5}$, no initial impulse is needed, and the kinetic energy generated as the mass falls from its initial position is sufficient for the string to wrap itself at least once around Q.

If $\lambda < \frac{3}{5}$ the string goes slack when $\phi = \cos^{-1}[2\lambda/(3 - 3\lambda)]$, the tension in the string becomes zero and the mass becomes, for a time, a free flight projectile. The analysis of what happens next is too complicated to pursue here, but involves no principles beyond those required for the solutions of other problems in this book. Some observations are included in part (d) below.

For the more general case, when V can be non-zero and the mass can rise above the level of P, wrapping around P (and necessarily at least once around Q) becomes a possibility. Wrapping around P can only occur if $\lambda < \frac{1}{2}$ – otherwise there is not enough string – and so, for $\frac{1}{2} < \lambda < \frac{3}{5}$ there can only be wrapping around Q, however large the additional impulse mV. The minimum needed varies from zero for $\lambda = \frac{3}{5}$, to $m\sqrt{g\ell/2}$ for $\lambda = \frac{1}{2}$.

For $\lambda < \frac{1}{2}$ and a suitable additional impulse, the string could make equal numbers of half circles about both Q and P, with additional whole circles about one of them. For example, if $0.25 < \lambda < 0.33$ then the final position of the string could first encircle both Q and P, then wrap itself around Q a number of times. Or, if $\lambda = 0.21$ and a suitable impulse is provided, the string might encircle P and Q, taken as a unit, twice. The variations are almost endless, and any one needs a significant amount of analysis.

(c) Either way, keep this to yourself.

(d) The table below sets out the observations made by the author (at home, not in a laboratory) using common items found in his workshop.[13] Because of a lack of room and vertical clearance, to say nothing of Health & Safety

---

[13] Though they did include a bench and vice once used by one of Maxwell's assistants, and you may have difficulty in getting hold of one of those!

regulations, measurements with values of λ less than 0.2 were not attempted. For similar reasons, measurements with an additional (uncalibrated) impulse, denoted by '+V' in the table, were limited to a few examples. The effective length ℓ of the pendulum, measured by timing unimpeded small swings, was 0.875 m.

Where 'a.p.' is used in the observations column, it stands for 'asymmetric pendulum', a description of an oscillatory motion (not SHM) in which when the pendulum swings to the left it does so about P and has length ℓ, and when it swings to the right, Q is the pivot and its length is $(1 - \lambda)\ell$. If the maximum angles the string makes with the vertical are θ_L and θ_R, energy conservation (neglecting air resistance) gives

$$mg\ell(1 - \cos\theta_L) = mg(1 - \lambda)\ell(1 - \cos\theta_R).$$

It follows that $\theta_L < \theta_R$, and that the mass moves (sometimes much) more rapidly during the right-hand part of the cycle than during the swing to the left.

The term 'slack' indicates that the string lost tension, and 'fall' that gravity alone took over. The horizontal lines indicate where significant values of λ mentioned above occur.

λ	What happened
0.23	almost immediate a.p.
0.23+V	2-turn wrap of P+Q
0.29	a.p. after erratic 1st swing
0.29+V	1.5-turn wrap of P+Q, and 2 turns round Q
0.35	a.p. after erratic 2nd swing on the LH side
0.41	rises $\sim 20°$ above Q level, slack, fall
0.46	slack, fall, no wrap
0.50	————————
0.53	slack but usually completes one wrap of Q
0.59	one wrap of Q, but on the verge of slack
0.60	————————
0.64	one or two wraps of Q
0.69	two or three wraps of Q
0.76	three or four wraps of Q
0.81	wraps Q completely, all string used
0.94	wraps Q completely, all string used

(e) Over to you!

S 71 The figure defines the notation we use, even though some quantities will or should not appear in the final answer.

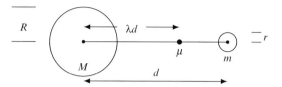

The Moon's gravity 'takes over' when

$$\frac{GM\mu}{\lambda^2 d^2} = \frac{Gm\mu}{(1-\lambda)^2 d^2}.$$ (*)

We can express GM and Gm in terms of the respective gravitational accelerations through

$$GM = g_{\rm E} R^2 \quad \text{and} \quad Gm = g_{\rm M} r^2.$$

Further,

$$\frac{r}{R} = \frac{2r}{d}\frac{d}{2R} = \frac{9.05 \times 10^{-3}}{33.5 \times 10^{-3}}.$$

Making these substitutions into (*) and rearranging yields

$$\frac{(1-\lambda)^2}{\lambda^2} = \frac{g_{\rm M}}{g_{\rm E}}\frac{r^2}{R^2} = \frac{1.62}{9.81}\frac{(9.05)^2}{(33.5)^2} = 1.205 \times 10^{-2}.$$

This means that

$$\frac{1-\lambda}{\lambda} = 0.1097 \quad \text{and that } \lambda = 0.901,$$

showing that the astronauts were more than 90% of the way to the Moon, before its gravity became dominant.

S 72 Since the spacecraft is unpowered and the Moon airless, retro-rockets and parachutes cannot be used to cushion the craft's impact with the Moon. The unavoidable impact speed $v_{\rm min}$ is therefore that acquired from 'falling' onto the moon's surface from a situation of near-zero velocity.

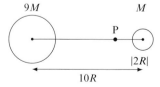

The fall starts from the point P (see the figure) at which the gravitational forces due to the moon and planet are equal (and opposite). The velocity there cannot actually be zero, or the spacecraft might fall back towards the planet, but if this point of minimum velocity is a distance r from the planet's centre, then

$$\frac{9MG}{r^2} = \frac{MG}{(10R-r)^2} \quad \Rightarrow \quad 3(10R-r) = r \quad \Rightarrow \quad r = 7.5R.$$

The launch velocity should be adjusted so that the craft's speed at P is as low as possible.

Between P and the landing point, the spacecraft (of mass m) gathers speed, but loses the equivalent gravitational potential energy E_g in the process. Recalling that the potential energy is always negative, we have

$$\frac{1}{2}mv^2 = E_g(\text{at P}) - E_g(\text{on moon's surface})$$
$$= -\left(\frac{9mMG}{7.5R} + \frac{mMG}{2.5R}\right) + \left(\frac{9mMG}{9R} + \frac{mMG}{R}\right),$$
$$\frac{v^2R}{2MG} = -\frac{12}{7.5} + 2,$$
$$v_{\min} = \sqrt{\frac{4MG}{5R}} \quad \text{is the minimum impact speed.}$$

S 73 We need to use gravity to establish the distance from Earth of a geostationary satellite, and then geography and geometry to answer the question as posed.

A geostationary satellite's period of rotation about C, the centre of the Earth, must be exactly one day. If the radius of its orbit around C is R, then, in an obvious notation, we must have

$$mR\omega^2 = \frac{GMm}{R^2},$$

where ω is the angular rotation rate of the Earth, and M is its mass. The value of M can be deduced from the fact that

$$g = \frac{GM}{R_E^2}, \quad \text{and then used to show that } R^3\omega^2 = gR_E^2.$$

Reference to an atlas shows that $R_E = 6.36 \times 10^3$ km, and that the Galapagos Islands and Lake Victoria both lie on the Equator, as does any location 35° North of Canberra. Thus the three satellites lie in a plane that contains C. Further, as their longitudes are 90° W, 30° E, and 150° E (since New Ireland is *due* north of Canberra), they are equally spaced around an equatorial circle and form an equilateral triangle of side $2R\cos 30° = \sqrt{3}R$ (see the figure).

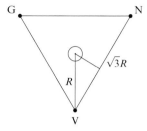

Thus, with $\omega = 2\pi/(24 \times 3600)$ s^{-1}, the time for the signal from V to return is

$$T = \frac{3\sqrt{3}R}{c} = \frac{3\sqrt{3}}{c}\left(\frac{g R_E^2}{\omega^2}\right)^{1/3} = 0.73 \text{ s.}$$

S74 Starting with the given equation,

$$\left(\frac{dR}{dt}\right)^2 = \frac{a^2}{R} + bR^2 - c, \qquad (*)$$

we can obtain the second derivative of R with respect to t by differentiation. An expression for $\ddot{R}(t)$ will give an indication of the curvature of the graph, whilst the original equation might yield information about its slope.

Differentiating $(*)$ with respect to time gives

$$2\left(\frac{dR}{dt}\right)\frac{d^2 R}{dt^2} = -\frac{a^2}{R^2}\frac{dR}{dt} + 2bR\frac{dR}{dt} \quad \Rightarrow \quad \ddot{R} = \frac{d^2 R}{dt^2} = -\frac{a^2}{2R^2} + bR.$$

The other information we have is that dR/dt is positive at at least one point (now!). This means that discarding the solution $dR/dt = 0$, as was done in the above equation, is (probably) justified.

(i) If b is negative, then, since $-a^2/2R^2$ is necessarily negative whatever the sign of a, $\ddot{R}(t)$ is also necessarily negative. In graphical terms, the curvature of the graph is always negative, i.e. its slope becomes more and more negative as time increases. Since there is at least one point (actually a continuous infinity of points) at which the slope is positive, it follows that when the curve is extrapolated, both backwards and forwards, it will cross zero at both ends (see the left-hand figure below). In physical terms, the Universe had zero size at some point in the past (the 'Big Bang'?) and will do so again at some time in the future. These conclusions are independent of the values of a and c.

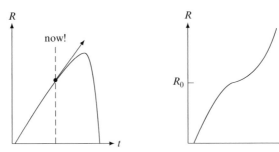

(ii) If $b>0$ and $c<0$, then (∗) shows that dR/dt is never zero, and since it is positive now, it must remain positive throughout. Further, as R increases the negative term in the expression for $\ddot{R}(t)$ becomes smaller in magnitude, whilst the positive term gets larger. Consequently the curvature of the graph is positive, and becomes increasingly so as time goes on; R therefore increases without limit (see the right-hand figure).

When $\ddot{R}(t)=0$, and $R=R_0$, i.e.

$$-\frac{a^2}{2R_0^2}+bR_0=0, \quad \Rightarrow \quad R_0=\left(\frac{a^2}{2b}\right)^{1/3},$$

the graph has zero curvature. For $R<R_0$ the curvature is negative, whilst beyond R_0 it is positive. As we have seen, the slope of the curve can never be zero, and so its general shape is as shown in the right-hand figure, with $R=0$ at some time in the past, and infinite expansion in the future. From the data provided, we cannot say where 'now' is on this graph.

S 75   The definition of the parsec, and hence that of the Megaparsec, uses the average radius of the Earth's orbit (1 AU) as it scale reference; one parsec being the distance of an object whose observed direction changes by two seconds of arc as the viewing platform covers one-half of a solar orbit. If, as in the case of Mars, the orbit radius (R AU) of the viewing platform is greater than 1 AU, then the observed angular deviation $\Delta\phi$ (of the same object) that results from one-half of a platform orbit will be larger. The factor by which it will be larger is the ratio of the two orbit radii, i.e. $R:1$.

The astronomer has forgotten to allow for this factor and his measured value of $\Delta\phi$ should have been reduced by this ratio, or, equivalently, the calculated distance to the galaxy should have been increased by the same factor, i.e. the real distance to Andromeda is $0.51R$ mpc.

It just remains to determine R. Either from Kepler's (third) law, or (using the standard notation) by deducing from

$$m\omega^2 r = \frac{GMm}{r^2} \quad \text{that} \quad \frac{r^3}{T^2} \text{ is the same for all solar planets,}$$

we have

$$\left(\frac{R}{1}\right)^3 = \left(\frac{687}{365}\right)^2 \quad \Rightarrow \quad R = 1.52,$$

and that the correct distance to the Andromeda galaxy is 0.78 Mpc.

S 76 The gravitational effect on any particular part of the cable decreases the further it is from the Earth's centre, whereas the centrifugal effect increases. To sustain the proposed totally fanciful situation requires the latter to just exceed the former for the cable taken as a whole.

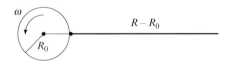

The figure shows the make-believe situation from above the North pole, with ω necessarily being that for the Earth's rotation, i.e. given by $\omega = 2\pi/(24 \times 3600) = 7.27 \times 10^{-5}$ rad s^{-1}.

With R_0 as the Earth's radius, and M as its mass, the 'weight' W of the whole cable is

$$W = \int_{R_0}^{R} \frac{GM\rho}{r^2} \, dr = GM\rho \left(\frac{1}{R_0} - \frac{1}{R}\right),$$

where ρ is the linear density of the cable. The counter-balancing centrifugal force is given by

$$F = \int_{R_0}^{R} \omega^2 r\rho \, dr = \frac{\omega^2 \rho}{2} \left(R^2 - R_0^2\right).$$

The critical value of R for keeping the cable straight is that for which F just exceeds W. In essence, when

$$GM\rho \left(\frac{1}{R_0} - \frac{1}{R}\right) = \frac{\omega^2 \rho}{2} \left(R^2 - R_0^2\right).$$

Now $GM/R_0^2 = g$, and so

$$gR_0^2 \frac{R - R_0}{R_0 R} = \frac{\omega^2}{2} \left(R^2 - R_0^2\right).$$

If $R = \lambda R_0$, then λ satisfies

$$\frac{1}{\lambda(\lambda+1)} = \frac{\omega^2}{2g} R_0 = \frac{(7.27 \times 10^{-5})^2}{2 \times 9.81} \times 6.38 \times 10^6 = 1.719 \times 10^{-3},$$

leading to $\lambda \sim 24$ and a cable 23 times the Earth's radius in length – not far short of half way to the Moon, and all in the realm of science fiction!

S 77   This one-dimensional motion with a position-dependent acceleration is straightforward in principle, but finding the time taken to cover a given distance, rather than the distance covered in a given time, makes it somewhat more demanding mathematically.

The basic equation of motion for the Earth's fall into the Sun is

$$\ddot{r} = -\frac{GM}{r^2},$$

where M is the mass of the Sun, and r is the Earth's distance from it, initially R, the radius of the Earth's (normal) orbit. This equation can be integrated after it has been multiplied through by $\dot{r}$ to yield

$$\frac{1}{2}(\dot{r})^2 = \frac{GM}{r} - \frac{GM}{R},$$

the same result as that obtained by equating the gain in kinetic energy with the loss in gravitational potential energy.

Whichever way it is arrived at, we have

$$\frac{dr}{dt} = -\sqrt{2GM\left(\frac{1}{r} - \frac{1}{R}\right)},$$

and the time T for the fall is

$$T = \int_0^T dt = -\int_R^0 \left[\frac{2GM}{R}\left(\frac{R}{r} - 1\right)\right]^{-1/2} dr.$$

Now, if we set $r/R = x$, with $dr/dt = R\,dx/dt$, the expression for T can be written as

$$T = \sqrt{\frac{R}{2GM}} \int_0^1 \frac{R\,dx}{\sqrt{x^{-1} - 1}} = \frac{\pi}{2}\sqrt{\frac{R^3}{2GM}}.$$

The equation 'keeping the Earth in its normal orbit' is

$$\frac{GM}{R^2} = \omega^2 R \quad \text{and so} \quad \sqrt{\frac{R^3}{GM}} = \frac{1}{\omega}.$$

This gives T as the simple expression

$$T = \frac{\pi}{2\sqrt{2}} \frac{1}{\omega} \approx \frac{\pi}{2\sqrt{2}} \frac{365}{2\pi} = 65\,\text{days}.$$

S 78   When the relativistic correction is added to the Newtonian gravitational force, the force balance equation becomes

$$\frac{GMm}{r^2} + \frac{6GMmv^2}{r^2 c^2} = mr\omega^2.$$

Replacing v by ωr in this equation, and then re-arranging it to express ω explicitly, we have

$$\omega^2 = \frac{GM}{r^3\{1 - [6GM/(rc^2)]\}}.$$

Taking the square-root of both sides, and expanding $(1-x)^{-1/2}$ as $1 - (-\frac{1}{2})x + \cdots$ gives

$$\omega = \sqrt{\frac{GM}{r^3}} \left(1 + \frac{3GM}{rc^2} + \cdots \right).$$

Thus the angular frequency is increased from its Newtonian value by the small fraction

$$\frac{3GM}{rc^2} = \frac{3 \times 6.7 \times 10^{-11} \times 2.0 \times 10^{30}}{5.8 \times 10^{10} \times 9.0 \times 10^{16}} = 7.70 \times 10^{-8}.$$

The Newtonian angular frequency ω_0 has magnitude

$$\omega_0 = \sqrt{\frac{GM}{r^3}} = \sqrt{\frac{6.7 \times 10^{-11} \times 2.0 \times 10^{30}}{(5.8 \times 10^{10})^3}} = 8.29 \times 10^{-7}\,\text{rad s}^{-1}.$$

With each orbit requiring 2π radians, this angular frequency corresponds to 416 orbits per century, with a corresponding phase gain of $7.70 \times 10^{-8} \times 416 \times 2\pi = 2.0 \times 10^{-4}$ radians in that time. This result could have been derived directly, but would have missed out on computing the Mercurial year of just under 88 (Earth) days ($100 \times 356.2 \div 416$).

S 79   For a circular orbit of radius r and angular frequency ω, the force balance requires

$$\frac{k}{r^3} = mr\omega^2. \tag{$*$}$$

The two contributions to the total energy are the kinetic energy,

$$T = \frac{1}{2}m(r\omega)^2,$$

and the potential energy

$$V = -\int_r^\infty \frac{k}{x^3}\,dx = -\frac{1}{2}\frac{k}{r^2}.$$

The total energy, on substituting for ω^2 from $(*)$ is therefore

$$E = \frac{1}{2}m(r\omega)^2 - \frac{1}{2}\frac{k}{r^2} = \frac{1}{2}\frac{k}{r^2} - \frac{1}{2}\frac{k}{r^2} = 0.$$

This rather surprising result is that the total energy of the particle is zero, *whatever the radius of its orbit*.

The angular momentum is

$$L = mvr = mr^2\omega = mr^2\sqrt{\frac{k}{mr^4}} = \sqrt{km} \quad \text{for any } r.$$

In view of the first result, this second one is probably no real surprise, since from the fact that all orbits have the same (zero) energy, any arbitrary small radial perturbation should, in principle, be able to move the particle from any one particular orbit to any other. If this had appeared to involve a change in its angular momentum, there would have been an unexplainable contradiction for motion taking place in a *central* force field. In brief, the particle is in neutral equilibrium in any orbit (including one at ∞)!

S 80   The left-hand figure below shows the satellite orbiting the Earth with angular velocity ω. Before we begin any calculation, we note that ω cannot be less than Ω, because, if it were, it would take more than 24 hours for the observatory to again be positioned directly below the satellite.

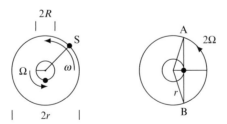

Since in 12 hours the Earth has made half a revolution, and during that time the satellite has *not* passed overhead, the latter must have made exactly 1.5 revolutions in that time; it therefore has an angular velocity of $\omega = 3\Omega$.

To keep the satellite in an orbit of radius r, we must have

$$\frac{GMm}{r^2} = mr\omega^2,$$

whilst on Earth

$$\frac{GM}{R^2} = g,$$

where M is the mass of the Earth, and G and g have their usual meanings. Combining these two equations to eliminate GM gives

$$r^3\omega^2 = gR^2 \quad \Rightarrow \quad \frac{r}{R} = \left(\frac{g}{R\omega^2}\right)^{1/3} = \left(\frac{g}{9R\Omega^2}\right)^{1/3}.$$

We now move to a frame of reference in which the Earth is stationary; in this frame, the satellite has angular velocity 2Ω, as shown in the right-hand figure. The observatory's view of the satellite is limited to the arc that is cut off by the line AB; this subtends an angle at the centre of the orbit of $2\cos^{-1} R/r$. As the time for a complete orbit is $2\pi/2\Omega$, the observable transit time τ is

$$\tau = \frac{2\pi}{2\Omega}\frac{1}{2\pi} 2\cos^{-1}\left(\frac{R}{r}\right)$$

$$= \frac{1}{\Omega}\cos^{-1}\left(\frac{9R\Omega^2}{g}\right)^{1/3}.$$

With $R = 6.37 \times 10^6$ m, $g = 9.81$ m s^{-2}, and $\Omega = 7.27 \times 10^{-5}$ rad s^{-1}, the calculated transit time is $\tau = 4.78$ hours. If the Earth were of very small radius compared to that of the satellite's orbit, we would expect τ to be $\frac{1}{2} \times (24/2) = 6$ hours, or $\frac{1}{2} \times (24/3) = 4$ hours if the Earth did not rotate. The non-negligible size of R produces an horizon that reduces these transit times, and so a time of $4\frac{3}{4}$ hours is consistent with expectations for the (actual) rotating Earth.

S 81   The 'stark' nature of the question could lead to a mass of formulae aimed at finding a connection between a visual measurement and a density. It could also raise the question of whether enough data is available to compute the required answer. However, as we are here concerned with ratios, rather than absolute values, factors that have the same value in two different situations, e.g. $4\pi/3$ or G, will not need to be quantified. This will reduce the complexity of formulae, and the consequent chance of a 'copying error'.

For an estimate, we can assume, in an obvious notation, that $M_S \gg M_E \gg M_M$, and that the Moon–Sun distance is equal to the Earth–Sun distance (which, averaged over a month, it is).

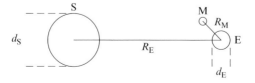

There are two motions to consider, the Sun–Earth orbit, and the Earth–Moon orbit. The figure defines the geometric notation we will use, apart from the angular diameters as measured on the Moon, ϕ_S and ϕ_E, the former also being the angular diameter of the Sun as seen from the Earth. In both cases we have

$$mr\omega^2 = \frac{GMm}{r^2} \quad \Rightarrow \quad GM = r^3\omega^2,$$

where M is the mass of the large central body, and m that of the orbiting one.[14]

Now, with densities ρ_S and ρ_E, we have the ratio relationship

$$\frac{R_E^3\,\omega_E^2}{R_M^3\,\omega_M^2} = \frac{M_S}{M_E} = \frac{d_S^3\,\rho_S}{d_E^3\,\rho_E}.$$

We also have that $\omega_E/\omega_M \approx 28/365$ and that the angular diameters are given by

$$\phi_S = \frac{d_S}{R_E} \quad \text{and} \quad \phi_E = \frac{d_E}{R_M}.$$

Putting these results together, it follows that

$$\frac{\rho_S}{\rho_E} = \frac{d_E^3}{R_M^3}\frac{R_E^3}{d_S^3}\frac{\omega_E^2}{\omega_M^2} = \left(\frac{\phi_E}{\phi_S}\right)^3 \left(\frac{\omega_E^2}{\omega_M^2}\right) = (3.6)^3 \left(\frac{28}{365}\right)^2 = 0.28.$$

This estimate is reasonably close to the actual value of 0.255.

S 82   Clearly, the total energy of the particle must reduce, as it is doing work against the resistive force due to the gas. However, the total energy is made up of two parts, the gravitational energy and the kinetic energy. We need to determine their individual changes in order to answer the questions posed.

Let us denote the particle's kinetic energy by T and its gravitational potential energy by V; its total energy E is then $E = T + V$. As the resistive effects per revolution are small, we can ignore any radial velocities arising as the particle changes orbit.

For the circular orbit

$$R\omega^2 = \frac{GM}{R^2} \quad \text{and} \quad E = T + V = \frac{1}{2}mR^2\omega^2 - \frac{GMm}{R} = -\frac{1}{2}\frac{GMm}{R}, \qquad (*)$$

where M is the mass of the Sun.

---

[14] This is essentially Kepler's third law applied to a circular orbit.

(a) The moment of the resistive force about the Sun is $-kv \times R$, whilst the particle's angular momentum $J = mv \times R$. Since a retarding couple/torque is acting on the system, its angular momentum must change in accordance with

$$\frac{dJ}{dt} = -kvR = -\frac{k}{m}J \quad \Rightarrow \quad J = J_0 e^{-kt/m}.$$

Thus the angular momentum does change – it decreases exponentially, with time constant m/k.

(b) The energy loss per revolution ΔE is equal to resistive force $\times$ distance moved, i.e. $\Delta E = -kv \times 2\pi R = -2\pi kwR^2$. From differentiating ($*$) we also have that,

$$\Delta E = \frac{1}{2} \frac{GMm}{R^2} \Delta R.$$

Hence

$$\frac{\Delta R}{R} = \frac{-2\pi kwR^2}{R} \frac{2R^2}{GMm} = -\frac{4\pi wk}{m} \frac{R^3}{GM} = -\frac{4\pi wk}{mw^2},$$

i.e. a fractional *decrease* in R of $4\pi k/mw$ per revolution.

(c) To obtain the change in v, we express ($*$) in the form (note the sign) $E = -\frac{1}{2}mv^2$ with $\Delta E = -mv\,\Delta v$, then

$$\frac{\Delta v}{v} = -\frac{\Delta E}{mv^2} = \frac{2\pi kR^2w}{mv^2} = \frac{2\pi kR^2w}{mR^2w^2} = \frac{2\pi k}{mw}.$$

This represents an *increase* in v, but this does not contradict the law of conservation of energy. Energy is lost, ultimately in the form of heat, to the interplanetary gas, and so an overall loss by the particle must be sustained. This is accounted for by a decrease $-\Delta V$ in V that provides *both* the gain in T and the frictional loss; the latter two are equal in magnitude, and $|\Delta V|$ is twice either of them. As shown above, R decreases and this makes the gravitational potential energy even more negative than it already was.

S 83 We first of all note that for a gravitational circular orbit the kinetic energy T of the orbiting body is equal to one half of the magnitude of its (negative) potential energy V. This can be seen from the three equations

$$T = \frac{1}{2}mv^2, \quad V = -\frac{GMm}{r}, \quad \frac{mv^2}{r} = \frac{GMm}{r^2}.$$

To just escape the solar system, spacecraft A must double its kinetic energy so that it becomes just greater than the magnitude of its original potential energy. To fall straight into the Sun (after the rocket burn), spacecraft B must be brought to rest relative to the Sun.

As shown in the figure, the two rockets will not be released at the same phase of the space station's orbit. Given the senses of rotation of the Earth around the Sun, and the space station around the Earth, craft A will be launched when the station is furthest from the Sun, and craft B when it is nearest. In this way the speed of the space station will be used to both boost the speed of craft A, and reduce that of B; both manoeuvres will aid fuel economy.

Since spacecraft A must double its kinetic energy, its speed needs to become $\sqrt{2}v_E$, where v_E is the Earth's speed in its orbit around the Sun. For craft B, we need simply that $v_E + v_B = 0$. The speed changes needed are $\Delta_A = (\sqrt{2} - 1)v_E - v_S$ and $\Delta_B = -v_E + v_S$. Since $v_E \gg v_S$, the second change is clearly much greater in magnitude, and spacecraft B is the one needing the more powerful rocket. The difference in the impulse needed is

$$m[v_E - v_S] - m[(\sqrt{2} - 1)v_E - v_S] = (2 - \sqrt{2})mv_E = (2 - \sqrt{2})m\sqrt{\frac{GM}{R}}.$$

We note that v_S does not appear in this difference; it was used equally to reduce the necessary values of both v_A and v_B, and so does not effect their difference.

S 84   All three modes of travel are driven by a central gravitational force, though the acceleration is different in each case. As the latter is determined by the amount of matter closer to the Earth's centre than the traveller, we denote the radius and mass of each Earth by R and M, and the mass of all material within a centred sphere of radius r by $M(r)$, with $M = M(R)$.

(a) If ω is the angular frequency of the orbit, then balancing the centrifugal and gravitational forces gives

$$\omega^2 R = \frac{GM}{R^2} = G\frac{4\pi\rho_0}{3}\frac{R^3}{R^2}.$$

Hence

$$\omega^2 = \frac{4\pi G\rho_0}{3} \quad \Rightarrow \quad T_a = \frac{2\pi}{\omega} = \sqrt{\frac{3\pi}{G\rho_0}}.$$

(b) When the traveller is at radius r, all the material at a greater radius produces no net force on the traveller, and the gravitational force is due to an inner sphere of mass $M(r) = (4\pi\rho_0 r^3)/3$ acting as if it were all at $r = 0$. So the acceleration of the traveller is

$$a(r) = -\frac{G}{r^2}\frac{4\pi\rho_0 r^3}{3}.$$

As the acceleration is $\propto -r$, the 'ride' is one of simple harmonic motion of angular frequency ω given by

$$\omega^2 = \frac{4\pi G\rho_0}{3} \quad \Rightarrow \quad T_b = \frac{2\pi}{\omega} = \sqrt{\frac{3\pi}{G\rho_0}}.$$

(c) Here the Earth's density takes the form $\rho(r) = k/r$ and we must first determine k through the equality

$$\int_0^R \frac{k}{u} 4\pi u^2\, du = \frac{4\pi R^3}{3}\rho_0,$$

$$\frac{k}{2} 4\pi R^2 = \frac{4\pi R^3}{3}\rho_0,$$

$$k = \frac{2}{3}R\rho_0 \quad \Rightarrow \quad \rho(r) = \frac{2R\rho_0}{3r}.$$

So, in this case

$$M(r) = \int_0^r 4\pi u^2 \frac{2R\rho_0}{3u}\, du = \frac{4\pi}{3}R\rho_0 r^2,$$

and the acceleration at radius r is

$$a(r) = -G\frac{4\pi}{3}R\rho_0\frac{r^2}{r^2} = G\frac{4\pi}{3}R\rho_0 = A = \text{constant}.$$

With a constant acceleration, the time to the Earth's centre is given by $R = \frac{1}{2}At^2$, and so

$$T_c = 4 \times \sqrt{\frac{2R}{A}} = 4\sqrt{\frac{3}{2\pi G\rho_0}} = \sqrt{\frac{24}{\pi G\rho_0}}.$$

(d) In units of $(G\rho_0)^{-1/2}$ the respective travel times, T_a, T_b and T_c, are in the ratio $\sqrt{3\pi} : \sqrt{3\pi} : \sqrt{24/\pi}$, i.e. $3.07 : 3.07 : 2.76$. Results (b) and (c) show that rearranging the density so that it is greater near the centre of the sphere, reduces the travel time and, since $T_a = T_b$, a journey via the centre of the Earth will be a little quicker, but decidedly less comfortable, than one taking the longer route round.

S 85   This is *not* a Doppler effect, as, to first order, the Earth is not moving towards or away from the Sun. The change in wavelength is the net result of the light having to overcome the gravitational field of the Sun, and of then falling through that of the Earth.

To calculate the size of the effect, we consider the light as made up of photons, with the energy of each equal to $hf = hc/\lambda$ and having a corresponding mass $hf/c^2 = h/\lambda c$. If λ' is the wavelength of the light when it has effectively left the Sun's gravitational field,

$$\frac{hc}{\lambda} - \frac{MG}{R}\frac{h}{\lambda c} = \frac{hc}{\lambda'},$$

where M and R are the mass and radius of the Sun, respectively.

The change in wavelength $\Delta\lambda = \lambda' - \lambda$ will be (very) small, and the previous equation can be written in the form

$$hc\frac{\Delta\lambda}{\lambda^2} = \frac{MG}{R}\frac{h}{\lambda c} \quad \Rightarrow \quad \frac{\Delta\lambda}{\lambda} = \frac{MG}{Rc^2}.$$

We need to estimate M, G and R from 'everyday' data.

For R, we can use the Sun–Earth distance [$D = 3.0 \times 10^8 \times (8 \times 60 + 20) = 1.5 \times 10^{11}$ m], and the angle R subtends, giving

$$R = D \times \frac{1}{2} \times \frac{0.53 \times \pi}{180} = 6.9 \times 10^8 \text{ m}.$$

We can estimate the product MG from the motion of the Earth around the Sun; its angular speed is $2\pi/(365 \times 24 \times 3600) = 1.99 \times 10^{-7}$ rad s^{-1}. Consequently,

$$MG = \omega^2 D^3 = \left(1.99 \times 10^{-7}\right)^2 \times \left(1.5 \times 10^{11}\right)^3 = 1.34 \times 10^{20} \text{ m}^3 \text{ s}^{-2}.$$

And so, finally,

$$\frac{\Delta\lambda}{\lambda} = \frac{MG}{Rc^2} = \frac{1.34 \times 10^{20}}{6.9 \times 10^8 \times 9.0 \times 10^{16}} = 2.1 \times 10^{-6}.$$

This corresponds to an increase in the wavelength of about 10^{-12} m and hardly seems worth bothering about – except for the principle!

The change from falling through the Earth's gravitational field is even smaller; the effect varies as M/R and this ratio is about 3000 times smaller for the Earth than for the Sun.

S 86   A qualitative inspection of the given equation reveals some characteristics of its solution. For $\dot{R}$ to be real, whether positive or negative, the RHS of the equation must be positive, i.e. $R < R_0$. Thus the size of the cloud can never exceed R_0, and $R = R_0$ must mark a point at which either the process is no longer described by the equation, or $\dot{R}$ changes sign from positive to negative. It cannot be the reverse of this, since, as already noted, R must be less than R_0 for the equation to have any physical meaning at all.

Consider a dust particle (of mass m) that lies on the cloud's surface, a distance R from its centre. Although the equation to be proved can be deduced almost immediately from conservation of energy, we will start from its equation of motion.

As the particle lies on the surface, the whole cloud contributes to the gravitational force acting upon it, and since the cloud is spherically symmetric, it does so as if all the mass were situated at its centre. Thus $m\ddot{R} = -GMm/R^2$, which, after multiplication through by $\dot{R}$, can be integrated to give

$$\frac{1}{2}\dot{R}^2 = \frac{GM}{R} - k, \qquad (**)$$

where k is a constant. Clearly, $\dot{R} = 0$ if R has the value $R_0 = GM/k$; with this choice of k the above equation becomes the same as $(*)$ in the question, with R_0 more precisely specified as the radius at which the cloud is neither expanding nor contracting.

As the suggested solution

$$R = a \sin^2 \phi, \quad \text{and} \quad t = b(\phi - \sin \phi \cos \phi),$$

is parametric in ϕ, the evaluation of $\dot{R}$ requires finding both $dR/d\phi$ and $dt/d\phi$:

$$\frac{dR}{d\phi} = 2a \sin \phi \cos \phi,$$

$$\frac{dt}{d\phi} = b(1 + \sin^2 \phi - \cos^2 \phi) = 2b \sin^2 \phi,$$

$$\dot{R} = \frac{2a \sin \phi \cos \phi}{2b \sin^2 \phi} = \frac{a}{b} \frac{\cos \phi}{\sin \phi}.$$

Substitution in $(**)$ gives the equation

$$\frac{1}{2} \frac{a^2}{b^2} \frac{\cos^2 \phi}{\sin^2 \phi} = \frac{GM}{a \sin^2 \phi} - \frac{GM}{R_0},$$

i.e.

$$\frac{1}{2} \frac{a^2}{b^2} \cos^2 \phi = \frac{GM}{a} - \frac{GM}{R_0} \sin^2 \phi.$$

For this to be valid for all ϕ, the terms on the right-hand side imply that $a = R_0$. The equation itself then implies that

$$\frac{1}{2}\frac{a^2}{b^2} = \frac{GM}{a} \quad \Rightarrow \quad b^2 = \frac{a^3}{2GM} = \frac{R_0^3}{2GM}.$$

With $a = R_0$, the cloud radius starts at zero when $\phi = 0$, and all the mass is contained in zero volume. i.e. infinite density, and then grows to reach a maximum of R_0 when $\phi = \frac{1}{2}\pi$ (at time $t = \pi b/2$). After that, R decreases until, at $\phi = \pi$, it is again zero and an infinite density returns. This happens after a time

$$T = \pi b = \pi \sqrt{\frac{R_0^3}{2GM}}.$$

Cosmologists assure us that this is not any time soon!

S 87   The figure below shows the 'more realistic' pendulum, and we must begin by calculating its moment of inertia $I_p(A)$ about the suspension point A.

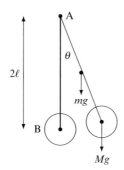

The MoI of the rod about its own centre is $\frac{1}{3}m\ell^2$ and so, by the Parallel Axis theorem, $I_r(A) = \frac{1}{3}m\ell^2 + m\ell^2 = \frac{4}{3}m\ell^2$. Similarly, $I_d(B) = \frac{1}{2}Ma^2$ implies that $I_d(A) = \frac{1}{2}Ma^2 + 4M\ell^2$. Combining these results,

$$I_p(A) = \frac{4}{3}m\ell^2 + \frac{1}{2}Ma^2 + 4M\ell^2.$$

When the pendulum makes an angle θ with the vertical, the restoring couple it generates is

$$C = mg\,\ell\sin\theta + Mg\,2\ell\sin\theta.$$

And so for small swings, in which $\sin\theta \approx \theta$, the equation of motion,

$$I_p(A)\ddot{\theta} = -(m + 2M)g\,\ell\theta,$$

leads to an angular frequency ω given by

$$\omega^2 = \frac{(m + 2M)g\,\ell}{\frac{4}{3}m\ell^2 + \frac{1}{2}Ma^2 + 4M\ell^2},$$

with a corresponding period $T = 2\pi/\omega$.

For this period to match that of the idealized simple pendulum, we require

$$\frac{g}{2\ell} = \omega^2 = \frac{(m + 2M)g\,\ell}{\frac{4}{3}m\ell^2 + \frac{1}{2}Ma^2 + 4M\ell^2}.$$

Substituting $\ell = 3a$, and cross-multiplying, converts this into

$$18a^2(m + 2M) = 12\,ma^2 + 36.5\,Ma^2,$$

$$6m = 0.5M,$$

i.e. the required mass ratio $\lambda = \dfrac{m}{M} = \dfrac{1}{12}$.

S 88 The figure shows the particle in its initial and final positions.

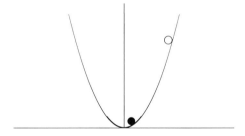

As the bowl has the form $y = bx^2$, the gravitational potential energy of the particle is

$$V(x) = mgbx^2,$$

and its equation of motion, in the absence of friction, would be

$$m\ddot{x} = -\frac{dV}{dx} \quad \Rightarrow \quad \ddot{x} + 2gbx = 0.$$

This is the equation for SHM with angular frequency $\omega = \sqrt{2gb}$, independent of the amplitude of the oscillation.

(a) We can approximate the frictional retardation as an energy loss per cycle of $2 \times 2\ell \times \mu mg = 4\mu mg\ell$. This will not be correct when the motion is confined to the rough region of the bowl, but as a first approximation to the total time for the motion, we can use

$$\text{time} = \frac{\text{initial total energy}}{\text{energy lost per cycle}} \frac{2\pi}{\omega} = \frac{mgbx_0^2}{4\mu mg\ell} \frac{2\pi}{\omega} = \sqrt{\frac{b}{2g}} \frac{\pi x_0^2}{2\mu\ell}.$$

(b) To refine our estimate we divide the motion into two parts, (i) when the amplitude A of the oscillations is greater than ℓ, and (ii) when $A < \ell$.

(i) This part can be treated as in part (a), with the energy loss in this phase equal to $mgb(x_0^2 - \ell^2)$, and hence taking a time

$$T_1 = \frac{mgb(x_0^2 - \ell^2)}{4\mu mg\ell} \frac{2\pi}{\omega} = \sqrt{\frac{b}{2g}} \frac{\pi(x_0^2 - \ell^2)}{2\mu\ell}.$$

(ii) When the amplitude of the oscillation is A (with $0 \le A < \ell$), and the motion is entirely within the rough region, the particle's total energy loss per cycle is $4\mu mgA$. The 'period' of a cycle is not exactly ω (because of the non-linear friction), but we can still use this value as an approximation. This gives the rate of loss of energy as $(4\mu mgA)/(2\pi\omega^{-1})$, i.e.

$$-\frac{4\mu mgA\omega}{2\pi} = -\frac{dE}{dt} = -\frac{d(mgbA^2)}{dt} = -2mgbA \frac{dA}{dt}.$$

Hence $dA/dt = (\mu\sqrt{2gb})/(\pi b)$ and the time taken for the initial amplitude ℓ (for this phase) to fall to zero is

$$T_2 = \frac{\pi\ell\sqrt{b}}{\mu\sqrt{2g}} = \sqrt{\frac{b}{2g}} \frac{\pi\ell}{\mu}.$$

Thus the 'improved estimate' for the total time of the motion is

$$T = T_1 + T_2 = \sqrt{\frac{b}{2g}} \frac{\pi}{\mu} \left(\frac{x_0^2}{2\ell} - \frac{\ell}{2} + \ell \right) = \sqrt{\frac{b}{2g}} \frac{\pi}{2\mu\ell} (x_0^2 + \ell^2).$$

It may seem slightly surprising that the total time needed is greater than that found in part (a), since in phase (b)(ii) there is no time at which the particle is not losing some of its total energy. However, the qualitative explanation is that

- The period of each half oscillation is actually greater than π/ω because of the frictional drag; this we have *not* specifically taken into account.
- The distance over which the frictional force acts during each oscillation is A, and not ℓ, thus reducing the energy lost in one period, and consequently extending the time needed to dissipate the energy, $mgb\ell^2$, that the particle still has. This is the factor that *has* been taken into account.

S 89   As always, we aim to choose the simplest description of the situation that is compatible with all the given data. We have, immediately, that the angular frequency of the motion must be $\omega = 2$ rad s^{-1}. We next note that, as we are dealing only with velocity observations, the position of the centre of vibration is irrelevant; let it be $x = 0$. Similarly, we are dealing with time intervals and so the time origin is irrelevant. The simplest choice of descriptive variables is therefore to take the time of maximum displacement as $t = 0$, i.e.

$$x(t) = A \cos \omega t \quad \text{with} \quad \dot{x}(t) = -\omega A \sin \omega t \quad \text{and} \quad \omega = 2\,\text{rad s}^{-1}.$$

From the first observation at time $t = T$, we have $4 = -2A \sin(2T)$, whilst from the second at $t = T + \frac{1}{2}$,

$$-3 = -2A \sin(2T + 1),$$

$$= -2A[\sin 2T \cos(1) + \cos 2T \sin(1)],$$

$$= -2A \left[\frac{-4}{2A} \cos(1) \pm \sqrt{1 - \left(\frac{-4}{2A}\right)^2} \sin(1) \right],$$

$$3 = -4\cos(1) \pm 2\sqrt{A^2 - 4}\,\sin(1),$$

$$\pm 2\sqrt{A^2 - 4} = \frac{3 + 4\cos(1)}{\sin(1)} = 6.134,$$

giving $A = 3.661$ cm. In the last line above, the positive sign has to be the correct choice, though this is irrelevant in the current context.

The first velocity observation time T satisfies $\sin(2T) = -4/2A$, leading to $2T = 3.7195$ or 5.7053. These two times are both possible: the first, $T = 1.86$ s, corresponds to a negative value of x, but a positive velocity, just after the particle has reversed direction for the first time; the second, $T = 2.85$ s, occurring just before it completes a full cycle. To decide between them, we draw up the following table based on $x(t) = 3.661 \cos 2t$:

T	$x(t)$	$\dot{x}$	$x(t + \frac{1}{2})$	$\dot{x}(t + \frac{1}{2})$
1.860	−3.067	4.000	0.026	7.322
2.853	3.067	4.000	3.340	−3.000

Only the second possibility $T = 2.853$ gives the correct value for the second velocity, and is therefore the required answer. You may find it instructive to graph (by hand or computer) $x(t)$ and $\dot{x}(t)$, and identify all the points listed in the table.

S 90 At first sight we might think there is not enough information available to determine ω, though the quoted answer clearly suggests otherwise. In particular, it does not contain any reference to a spring constant for the string, something that must affect the oscillation period; perhaps it can be obtained from the equilibrium data. For our analysis, let the particle have mass m and the elastic string have spring constant k. Further, denote by θ the angle that each string makes with the vertical.

Then, the extension of each half-string is $\sqrt{a^2 + b^2} - \ell$, and the corresponding tension T is k times this. In equilibrium, the vertical components of the tension in the string balance the particle's weight, i.e.

$$mg = 2T\cos\theta = 2k(\sqrt{a^2 + b^2} - \ell)\,\frac{a}{\sqrt{a^2 + b^2}}.$$

Thus the equilibrium relationship between m and a is

$$\frac{mg}{2k} = a\left(1 - \frac{\ell}{\sqrt{a^2 + b^2}}\right). \qquad (*)$$

To determine the angular frequency of small oscillations about the equilibrium situation, we need to find the relationship between a small change Δa in a and the consequent restoring force ΔF experienced by the particle. But that restoring force will be of the same magnitude (but opposite in direction) as the weight of the small additional particle that would be needed to produce the change Δa.

This relationship *could* be found by writing down equation $(*)$ for mass $m + \Delta m$ and $a + \Delta a$, subtracting the two equations, and then expanding square roots and denominators in Δa power series. But it is probably more easily found by differentiating $(*)$ with respect to a, as follows.

$$\frac{g}{2k}\frac{dm}{da} = a\left[-\left(-\frac{1}{2}\right)\frac{2a\ell}{(a^2 + b^2)^{3/2}}\right] + \left[1 - \frac{\ell}{(a^2 + b^2)^{1/2}}\right]$$

$$= \frac{a^2\ell - a^2\ell - b^2\ell}{(a^2 + b^2)^{3/2}} + 1$$

$$= 1 - \frac{b^2\ell}{c^3} \quad \text{where} \quad c^2 = a^2 + b^2.$$

From this, we have, in turn,

$$g\,dm = 2k\left(1 - \frac{b^2\ell}{c^3}\right)da,$$

$$\Delta F = 2k\,\Delta a\left(1 - \frac{b^2\ell}{c^3}\right),$$

$$\text{acceleration} = \frac{2k}{m}\,\Delta a\left(1 - \frac{b^2\ell}{c^3}\right),$$

$$\omega^2 = \frac{2k}{m}\left(1 - \frac{b^2\ell}{c^3}\right).$$

But, from $(*)$, $2k/m = (g/a)(1 - \ell/c)^{-1}$, and so, finally,

$$\omega^2 = \frac{g}{a}\left(\frac{1 - (b^2\ell/c^3)}{1 - \ell/c}\right).$$

As we knew/anticipated all along, m and k do not appear explicitly in ω – though, of course, their ratio is hidden in a, which does appear.

S 91   Several of the suggested factors can be rejected immediately: the disc is wooden and so B_V can play no part; the moment of inertia of the uniform disc depends only on its mass and radius and so ρ and σ can also be disregarded. Further, the disc is supported by the tensions in the threads at their actual length, whether or not they have been stretched; hence the Young modulus is irrelevant.

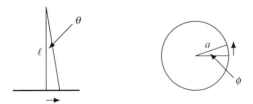

The figure shows (most of) the quantities that could be relevant; on the left is a side view of the disc and a typical thread, when the former has been rotated through the (small) angle ϕ shown in the plan view on the right. The essential connection between the two figures is that the lengths indicated by the short arrows are equal, both being the same small portion of the disc's circumference. Since all angles are small we may write

$$\ell\theta = a\phi \quad \text{as well as} \quad nT = Mg.$$

The restoring torque due to one of the threads is $T\sin\theta \times a \approx Ta\theta \approx Ta\,(a\phi/\ell)$ and so the equation of motion for the disc is

$$I\ddot{\phi} = -n\frac{a^2}{\ell}T\phi,$$

$$\frac{1}{2}Ma^2\ddot{\phi} = -\frac{a^2}{\ell}Mg\,\phi,$$

$$\ddot{\phi} + \frac{2g}{\ell}\phi = 0.$$

It follows that the angular frequency of the torsional oscillations is $\sqrt{2g/\ell}$ and that

$$\tau = 2\pi\sqrt{\frac{\ell}{2g}}.$$

So, only those who predicted ℓ and g – and no other – were right. Were you?

S 92   As the string is elastic, once the system is set in motion (no matter how) it will undergo SHM with an angular frequency $\omega = \sqrt{k/(M+m)}$. However, the point about which it will oscillate will not be the original position of the pan L, but the new equilibrium point P, which is a distance $d = mg/k$ below that position. The figure shows three situations, before the putty is released, at some later time, and when the pan has reached its lowest level.

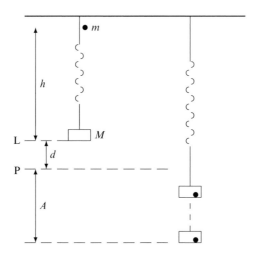

The situation immediately after the putty has landed is one 'snap-shot' of the motion, and as the pan will necessarily be moving (and therefore *not* at an extremum of an oscillation), but still at its original height, it is clear that when the pan does come to reverse its direction at the highest point of the movement, that point will be *above* L. How far above, will be determined by the amplitude A of the oscillation.

In falling through a distance h, the putty acquires a speed $\sqrt{2gh}$, and so when the 'snap-shot is taken', the (downward) velocity v of the pan plus putty is given, as a result of momentum conservation, by

$$(M+m)v = m\sqrt{2gh}.$$

Its displacement, x say, from the new equilibrium position is d, involving a stored elastic energy (actually gravitational potential energy relative to P) of $\frac{1}{2}kd^2$. When the loaded pan is at its lowest position and (temporarily) stationary, $x = A$ and all the energy is elastic. Therefore, as the total energy is constant,

$$\tfrac{1}{2}kA^2 = \tfrac{1}{2}kd^2 + \tfrac{1}{2}(M+m)v^2,$$

$$A^2 = \left(\frac{mg}{k}\right)^2 + \frac{(M+m)}{k}\frac{m^2\,2gh}{(M+m)^2},$$

$$A = \frac{mg}{k}\left(1 + \frac{2hk}{(M+m)g}\right)^{1/2}.$$

The maximum subsequent height above L for the loaded pan is therefore

$$A - d = \frac{mg}{k}\left[\left(1 + \frac{2hk}{(M+m)g}\right)^{1/2} - 1\right].$$

S 93 Since $a \ll \ell$ we can ignore vertical velocities and work directly in terms of horizontal distances. However, the angular frequency of the swings will always be $\omega = \sqrt{g/\ell}$, and the total energy E of the bob during a free swing can be expressed as $\frac{1}{2}m\omega^2 A^2$, where A is the amplitude of the swing. We need to apply these ideas several times, sometimes to swings that don't actually take place.

For the initial motion, with P as the x-origin,

$$x = 2a\cos\omega t, \qquad \dot{x} = -2\omega a \sin\omega t, \qquad E = \tfrac{1}{2}m(2a\omega)^2 = 2m\omega^2 a^2.$$

The bob will hit the wall when $t = T_1$, where

$$-a = 2a\cos\omega T_1 \quad \Rightarrow \quad T_1 = \frac{1}{\omega}\cos^{-1}\left(-\frac{1}{2}\right) = \frac{2\pi}{3\omega}.$$

At this point $\sin\omega t = \sqrt{3}/2$, and the bob's *kinetic* energy is

$$K = \frac{1}{2}m(-2\omega a)^2 \sin^2\omega t = \frac{3}{2}m\omega^2 a^2.$$

One half of this is lost as a result of the impact with the wall, and the ensuing motion has a total energy of

$$E' = \left(2 - \frac{3}{4}\right)m\omega^2 a^2.$$

This new total energy carries the bob to a stationary state at R, and so must be equal to that of a motion that starts from rest at R. If the distance of R from P is b, then b is the amplitude of such a motion, and its *total* energy is $\frac{1}{2}m\omega^2 b^2$. Hence

$$\frac{1}{2}m\omega^2 b^2 = E' = \left(2 - \frac{3}{4}\right)m\omega^2 a^2 \quad \Rightarrow \quad b = \frac{\sqrt{10}}{2}a.$$

The time taken for the bob to rebound to R, must be the same as that taken to reach the wall starting from rest at R. The equation for that motion takes the form $z = b\cos(\omega t')$, and the time T_2 to hit the wall is given by

$$-a = b\cos(\omega T_2) \quad \Rightarrow \quad T_2 = \frac{1}{\omega}\cos^{-1}\left(\frac{-a}{b}\right) = \frac{1}{\omega}\cos^{-1}\left(\frac{-2}{\sqrt{10}}\right).$$

Thus the bob stops at R after a total time T given by

$$T = T_1 + T_2 = \sqrt{\frac{\ell}{g}}\left[\frac{2\pi}{3} + \cos^{-1}\left(\frac{-2}{\sqrt{10}}\right)\right] = 4.35\sqrt{\frac{\ell}{g}}.$$

S 94   We start by noting that if the displacements are small (compared with a), then we can ignore changes in the tensions in the springs for oscillations that are transverse to their lengths. Mathematically, for a *transverse* displacement of z the new length of the spring is $\sqrt{a^2 + z^2}$ and the corresponding change in tension

$$\delta T = k\left(\sqrt{a^2 + z^2} - a\right) = k\left[a\left(1 + \frac{1}{2}\frac{z^2}{a^2} - \cdots\right) - a\right] = k\left(\frac{1}{2}\frac{z^2}{a^2} - \cdots\right),$$

i.e. of second-order in z/a and therefore ignorable, by comparison with the first-order terms that give rise to the SHM.

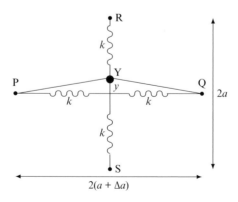

Using an intuitive '*x-y*' notation, we can say immediately that the vibrations in the *x*-direction are not affected by the *y*-springs attached at R and S (see the figure), as they are not tensioned when in equilibrium. The *x*-equation of motion is therefore

$$m\ddot{x} = -kx - kx \text{ leading to an angular frequency } \omega_1^2 = \frac{2k}{m}.$$

Because they *are* tensioned in equilibrium, the *x*-springs *do* affect the *y*-angular frequency ω_2. The figure shows the mass oscillating in the *y*-direction, and at a distance *y* from its equilibrium position. The *y*-springs together provide a restoring force of $2ky$, but, in addition, the *x*-springs contribute $2T\sin\theta$, where θ is the angle $\angle YPQ$ and $T = k\Delta a$ is the tension in each of the *x*-springs. Now, to first order, $\sin\theta = y/a$, and so the *y*-equation of motion is

$$m\ddot{y} = -2ky - 2k\Delta a\frac{y}{a} \quad \Rightarrow \quad \omega_2^2 = \frac{2k}{m}\left(1 + \frac{\Delta a}{a}\right).$$

The defining characteristic of the initial displacement direction is that at all points on it, $x = y$. If the starting point is (b, b) then the two coordinates have later time dependencies

$$x = b\cos\omega_1 t \quad \text{and} \quad y = b\cos\omega_2 t = b\cos(\omega_1 t + \Delta\omega t),$$

where $\Delta\omega = \omega_2 - \omega_1$. It is clear that the original situation will be recovered whenever $\Delta\omega t = 2\pi n$, and for the first time when $n = 1$. More formally, we require

$$\cos\omega_1 t = \cos(\omega_1 t + \Delta\omega t) = \cos\omega_1 t\cos\Delta\omega t - \sin\omega_1 t\sin\Delta\omega t.$$

Now, $\cos\omega_1 t$ and $\sin\omega_1 t$ can never both be zero at the same time, and so we require both $\sin\Delta\omega t = 0$ and $(1 - \cos\Delta\omega t) = 0$. Both are first satisfied at a time τ given by $\tau = 2\pi/\Delta\omega$, with

$$\Delta\omega = \left(\frac{2k}{m}\right)^{1/2}\left[\left(1 + \frac{\Delta a}{a}\right)^{1/2} - 1\right] = \left(\frac{2k}{m}\right)^{1/2}\left(\frac{1}{2}\frac{\Delta a}{a} + \cdots\right).$$

Finally, we deduce that

$$\tau \approx \left(\frac{m}{2k}\right)^{1/2}\frac{4\pi a}{\Delta a}.$$

S 95 The minimal sketch below shows the set up, but with the angle θ that the line joining the centre of the solid sphere and the centre of the spherical bowl makes with the vertical greatly exaggerated. The value of *n* shown is about 7.5.

Because both the sphere and the bowl are rough, the sphere will roll on the latter's surface rather than slide across it, and, because it is released from rest, the path it follows lies in the vertical plane containing both its release point and the lowest point of the bowl.

The interplay between gravitational forces, linear dynamics, and rotational dynamics, makes constructing the equations of motion for the system quite complicated. But we know that they would lead to a solution that is consistent with the law of conservation of total energy E.

As there are no frictional losses and air resistance is negligible, we can give explicit expressions for all the relevant energies:

- the gravitational energy (measured from a level a above the bottom of the bowl) is $mg(n-1)a(1-\cos\theta)$;
- the linear kinetic energy is $\frac{1}{2}mv^2 = \frac{1}{2}m[(n-1)a\dot\theta]^2$;
- the rotational kinetic energy is $\frac{1}{2}I\omega^2 = \frac{1}{2}I(v/a)^2$, with $v = (n-1)a\dot\theta$.

Here, m is the mass of the sphere, v its linear speed, ω its rotational speed, and I its moment of inertia about an axis through its centre (a horizontal axis in this case). For the uniform sphere $I = \frac{2}{5}ma^2$.

Thus we have that

$$E = mg(n-1)a(1-\cos\theta) + \frac{1}{2}m(n-1)^2a^2\dot\theta^2 + \frac{1}{5}m(n-1)^2a^2\dot\theta^2 = \text{constant}.$$

Since the sphere is released near the bottom of the bowl, θ starts off small and can never get bigger than its initial value, and we may therefore, for an approximate answer, replace $1-\cos\theta$ by $\theta^2/2$. After some simplification, the approximate energy conservation equation reads

$$\frac{1}{2}g\theta^2 + \frac{7}{10}a(n-1)\dot\theta^2 = \text{constant}. \tag{$*$}$$

The value of this constant is determined by the position of the sphere's release and is unknown. However, if we now differentiate $(*)$ with respect to time we obtain

$$g\theta\dot\theta + \frac{7}{5}a(n-1)\dot\theta\ddot\theta = 0,$$

i.e.

$$\ddot\theta + \frac{5}{7}\frac{g}{(n-1)a}\theta = 0.$$

This is the equation for simple harmonic motion of angular frequency $\omega_0 = \sqrt{5g/7(n-1)a}$, and so, at the same level of approximation as $1 - \cos\theta \approx \theta^2/2$, the time for the sphere to return to its release point is

$$T = \frac{2\pi}{\omega_0} = 2\pi\sqrt{\frac{7(n-1)a}{5g}}.$$

S 96   When the car is travelling at speed V the period of the 'bumps' is λ/V, and the angular velocity with which they compress and extend the spring is $\omega = 2\pi V/\lambda$. If we denote the vertical deviation of the car's centre of mass from it equilibrium position (when stationary or on a smooth level road) by z, then the equation of motion, taking into account both the restoring force due to any non-zero value of z and the 'driving force' due to the road's undulations, is

$$M\ddot{z} = k(-z + y) \quad \text{where } y(t) = A\sin(\omega t) = A\sin\left(\frac{2\pi V t}{\lambda}\right).$$

This can be written in the standard form for a driven undamped harmonic oscillator:

$$M\ddot{z} + kz = kA\sin(\omega t),$$

with a solution of the form $z(t) = a\sin(\omega t)$.

Note The choice of $\sin(\omega t)$, rather than the more general $C\sin\omega t + D\cos\omega t$, for both the driving term and the solution is permissible, since we may choose the time origin arbitrarily, and, because there is no damping term, the phase difference between the driving force and the car's response to it, can only be 0 or π. The two possibilities are catered for by allowing a to take both positive and negative values.

Substituting the given solution form leads directly to

$$-aM\omega^2 + ka = kA \quad \Rightarrow \quad a = \frac{kA}{k - \omega^2 M} \text{ and } v = a\omega = \frac{k\omega A}{k - \omega^2 M}.$$

Setting $|v|^2$ equal to the critical value v_0^2 gives a quadratic equation in ω^2. Squaring both sides of that equation in order to remove the modulus signs, and writing $k/M = \omega_0^2$ (the square of the natural vibration frequency of the spring and mass combination) gives

$$\omega_0^4 - 2\omega_0^2\omega^2 + \omega^4 = \omega^2\omega_0^2\left(\frac{\omega_0 A}{v_0}\right)^2 \equiv \omega^2\omega_0^2\mu^2.$$

This has the standard solution

$$\omega^2 = \frac{\omega_0^2}{2}\left[(2 + \mu^2) \pm \sqrt{4\mu^2 + \mu^4}\right].$$

The two (positive) values of ω lie one on either side of ω_0.

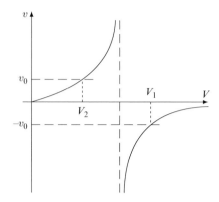

The figure shows the general variation of v with ω, and consequently with V, on an arbitrary scale. As expected, the curve is in two parts, with an 'infinite jump' as ω passes through ω_0 – physically, the response becomes very large, and the phase difference between the road ripples and the car body vibration changes abruptly from 0 to π.

The two values of ω calculated above correspond to the two values of V marked as V_1 and V_2; they are the two road speeds that make $|v|$ equal to v_0. For speeds between these two values, the vibrations exceed what is allowed, and there is more to worry about than the uncomfortable ride!

With the numerical values given: $k = 5000$, $M = 500$, $v_0 = 0.2$, $A = 0.05$ and $\lambda = 10$, the other relevant quantities are $\omega_0^2 = 10$, $\mu^2 = 0.625$ and $\sqrt{4\mu^2 + \mu^4} = 1.700$. Thus the two critical values of ω and their corresponding road speeds are

$$\omega^2 = \tfrac{1}{2}\omega_0^2(2.625 \pm 1.700),$$

$$\omega_1 = 2.151 \quad \Rightarrow \quad V_1 = 3.42 \text{ m s}^{-1},$$

$$\omega_2 = 4.650 \quad \Rightarrow \quad V_2 = 7.40 \text{ m s}^{-1}.$$

The driver of this model car should get out of this speed range, $12.3 - 26.6$ kph ($7.6 - 16.5$ mph), as quickly as possible. But, of course, real cars *do* have plenty of damping built in.

S 97   We first note that, within elastic limits, the period of oscillation for a mass-spring assembly is independent of the amplitude of the oscillation; the value of X will not be relevant to the question posed.

Before the apparatus was broken, the trolley performed SHM with the restoring force per unit displacement of 4λ; on release, 3λ from the compressed RH spring and 1λ from the stretched LH spring; later the roles were reversed, but the total effective spring constant was unchanged. If m was the mass of the trolley, the governing equation was

$$m\ddot{x} + 4\lambda x = 0 \quad \Rightarrow \quad T = \frac{2\pi}{2\omega_0}, \quad \text{where} \quad \omega_0^2 = \frac{\lambda}{m}.$$

After the breakage the oscillation is not pure SHM, and has to be analysed in stages.

(i) $X \geq x \geq 0$. This stage is the same as for the undamaged motion, and so takes a time of $\left(\frac{1}{2}\pi\right)/(2\omega_0)$, i.e. $\frac{1}{4}T$.

(ii) $x < 0$. In this part of the motion, the RH spring plays no part, and a half-cycle of motion governed by the equation $m\ddot{x} + \lambda x = 0$ takes place.[15] This takes a time π/ω_0, i.e. T.

(iii) $0 \leq x \leq X$. This final stage is under the same conditions as stage (i) and takes the same amount of time for its quarter-cycle, i.e. $\frac{1}{4}T$.

Thus the total period of one oscillation is $\frac{3}{2}T$, and the numerical ratio T'/T is 1.5.

S 98 A sketch of the equilibrium set-up is repeated in the left-hand figure below.

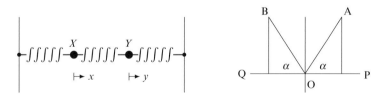

When mass X is at x and Y is at y, the extension of the central spring is $|y - x|$ and the tension in it is k times this length; on X the tension acts to increase x, but on Y its effect is to decrease y. Clearly, the left- and right-hand springs act to return their associated masses to their equilibrium positions. Thus the equations of motion for each mass, under the influence of the two relevant spring tensions are

$$m\ddot{x} = -kx + k(y - x),$$
$$m\ddot{y} = -ky + k(x - y).$$

---

[15] The trolley goes from $x = 0$ to $x = -2X$ and back again.

We separate these coupled differential equations, either by inspection, or by both adding them together and (separately) subtracting them from each other, and obtain a new pair of equations:

$$m\ddot{u} = -ku,$$
$$m\ddot{v} = -3kv,$$

where $u = x + y$ and $v = x - y$. Now, each of these equations represents simple harmonic motion, and their general solutions are of the form

$$u(t) = A\cos\omega t + B\sin\omega t,$$
$$v(t) = C\cos\sqrt{3}\,\omega t + D\sin\sqrt{3}\,\omega t,$$

where $\omega^2 = k/m$.

The initial physical conditions, $x = y = \dot{y} = 0$ and $\dot{x} = V$, translate into $u = v = 0$ and $\dot{u} = \dot{v} = V$. The former imply that $A = C = 0$, whilst the latter show that $V = \dot{u} = \omega B \cos(0)$ and, correspondingly, that $V = \dot{v} = \sqrt{3}\,\omega D$. More explicitly,

$$u(t) = \frac{V}{\omega}\sin\omega t \quad \text{and} \quad v(t) = \frac{V}{\sqrt{3}\,\omega}\sin\sqrt{3}\,\omega t.$$

Re-expressing this in terms of the physical displacements:

$$x = \frac{1}{2}(u+v) = \frac{V}{2\omega}\left(\sin\omega t + \frac{1}{\sqrt{3}}\sin\sqrt{3}\,\omega t\right),$$
$$y = \frac{1}{2}(u-v) = \frac{V}{2\omega}\left(\sin\omega t - \frac{1}{\sqrt{3}}\sin\sqrt{3}\,\omega t\right).$$

From this we see that neither x nor y executes SHM of a well-defined frequency, but each is a superposition of two such motions with different frequencies. The two 'pure' SHMs are known as the *normal modes* of the system; with carefully chosen initial conditions, each mode can be excited on its own, but establishing what those conditions have to be is beyond the scope of the questions in this book.

Rather easier is the task of answering the original question. Displacement x will reach a maximum when $\dot{x} = 0$ – physically when X stops and reverses its direction. Differentiating the above expression for $x(t)$, and setting it equal to zero gives

$$\frac{V\omega}{2\omega}\left(\cos\omega t + \frac{\sqrt{3}}{\sqrt{3}}\cos\sqrt{3}\,\omega t\right) = 0,$$

i.e.

$$\cos\omega t + \cos\sqrt{3}\,\omega t = 0.$$

Since $\sqrt{3} > 1$, when this condition is first satisfied ωt will be in the range $0 < \omega t < \frac{1}{2}\pi$ and $\frac{1}{2}\pi < \sqrt{3}\,\omega t < \pi$, with the two corresponding cosines having equal magnitudes but opposite signs. Now, either from the general geometric identity

$$\cos(\pi - \alpha) = \cos \pi \cos \alpha + \sin \pi \sin \alpha = -\cos \alpha,$$

or by examining the right-hand figure above, in which the condition

$$\cos \angle POB = -\cos \angle BOQ = -\cos \angle POA$$

is satisfied when $\angle POB + \angle BOQ = \angle POQ = \pi$, we conclude that, when x is maximal,

$$\omega t = \pi - \sqrt{3}\,\omega t \quad \Rightarrow \quad t = \frac{\pi}{\omega(1 + \sqrt{3})}.$$

S 99 As indicated in the figure, let us denote the displacements of A, B and C by x, y and z respectively, all nominally to the right of their equilibrium positions, but recognizing that some values could be negative. We start by writing the equation of motion for each atom under the forces produced by the extended or compressed springs that are attached to it.

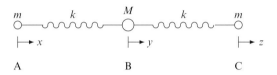

The general equations of motion for the three atoms are thus

$$
\begin{aligned}
m\ddot{x} &= k(y - x), \\
M\ddot{y} &= k(x - y) + k(z - y), \\
m\ddot{z} &= k(y - z),
\end{aligned}
$$

and the constant position of the centre of mass is expressed by $mx + My + mz = 0$.

In mode 1, since B is stationary, as is the molecule's centre of mass, A and C, having equal masses, must move symmetrically but in opposite directions. We therefore need consider only one of them, say A. Its equation of motion is

$$m\ddot{x} = k(0 - x) \quad \Rightarrow \quad \omega_1^2 = \frac{k}{m}.$$

In mode 2, we are told that $x = z$, and from the centre of mass condition can conclude that $x = -My/2m$, and that B's equation of motion is

$$M\ddot{y} = 2k(x - y) = -\frac{k}{m}(My + 2my) \quad \Rightarrow \quad \omega_2^2 = k\frac{(M + 2m)}{mM}.$$

A plot of ω_1^2 against m is simply a rectangular hyperbola with the two axes as asymptotes, and the value of M is irrelevant, except that when $m = M$ the value of ω_1^2 must be $\omega_0^2 = k/M$.

For ω_2^2, some sample values are

m/M	0.5	1.0	1.5	2.0	3.0	5.0
ω_2^2/ω_0^2	4.00	3.00	2.67	2.50	2.33	2.20

As can be seen from the figure below, ω_2^2 tends to the asymptotic value of $2k/M$ for $m \gg M$.

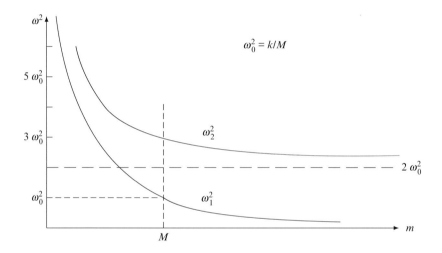

S 100   Although in practice clock pendulums are not simple point masses on the ends of 'light' rods,[16] that model will suffice for our present purposes. In an obvious notation, when the pendulum makes an angle θ with the vertical, it experiences both a gravitational restoring force of $mg \sin \theta$, and, due to the turntable's rotation, a centrifugal force of $m\ell \sin \theta \, \Omega^2$. As the oscillations are small, we can set $\sin \theta \approx \theta$ and obtain as the equation of motion

$$m\ell\ddot{\theta} = m\ell\theta\,\Omega^2 - mg\theta.$$

This represents SHM of angular frequency ω given by

$$\omega^2 = \frac{g}{\ell} - \Omega^2 = \omega_0^2 - \Omega^2.$$

---

[16] See solution S 87 on page 196 which deals with a more realistic model.

Thus, as Ω increases, the frequency of the pendulum's oscillations decreases; and when $\Omega = \omega_0$ the pendulum does not oscillate at all, i.e. Ω_0 has the same value as ω_0. Physically, the situation is that the centrifugal force is just sufficient to prevent gravity from returning the pendulum to its original equilibrium position; any further increase in Ω merely increases the angle θ.

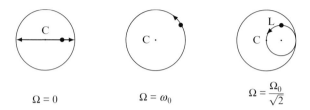

$$\Omega = 0 \qquad \Omega = \omega_0 \qquad \Omega = \frac{\Omega_0}{\sqrt{2}}$$

When $\Omega = 0$ the pendulum behaves in the normal way and oscillates along a straight line, say the x-axis, with angular frequency ω_0, as shown in the left-hand sketch above. If Ω equals Ω_0, which is also the natural frequency ω_0 of the clock pendulum, the inclination of the pendulum remains fixed, and its bob follows a circular path about C (see the second sketch).

Much more difficult to visualize, but still the most straightforward case that has $0 \neq \Omega \neq \omega_0$, occurs when $\Omega = \Omega_0 / \sqrt{2}$. Then $\omega^2 = \omega_0^2 - \frac{1}{2}\omega_0^2 = \frac{1}{2}\omega_0^2$, and both ω and Ω are equal to $\omega_0 / \sqrt{2}$. A qualitative analysis, starting with the bob at $(x, y) = (R, 0)$, or $(r, \theta) = (R, 0)$ in polar coordinates, might proceed as follows:

- r starts to decrease, but at the same time the bob is carried into the region $y > 0$. This continues until a quarter cycle has been completed, by which time $r = 0$, but the oscillation plane has rotated (anticlockwise) by $\pi/2$.
- The 'plus half' of the oscillation plane moves into the second quadrant, but by now the pendulum has swung into the 'minus half' and so the bob is in the fourth quadrant $(x > 0, y < 0)$
- When the pendulum reaches its maximum 'minus value' (half a cycle completed), the clock case has turned through π – with the result that the bob is back where it started, at $(R, 0)$ (in either coordinate frame!)
- As the pendulum – and the turntable – complete the second half of their cycles, the bob simply repeats its closed trajectory L in the table reference frame. Thus the bob moves around L at twice the rotation rate of the turntable.

It is not clear from the above what shape the closed curve L takes, but if we describe the system in polar coordinates (r, θ), the time variation of r is that of a simple pendulum, $r = R \cos \omega t$, whilst that of θ is simply $\theta = \omega t$. Consequently, L and its x- and y-coordinates are:

$$r = R\cos\theta, \quad x = r\cos\theta = R\cos^2\theta, \quad y = r\sin\theta = R\cos\theta\sin\theta.$$

So, x is constrained to be positive in the range $0 < x < 1$, but y ($\sim \sin 2\theta$) will be equally often positive and negative. Clearly L has some symmetry about the point $(\frac{1}{2}R, 0)$ and so we make the substitution $z = x - \frac{1}{2}R$, yielding

$$z = x - \tfrac{1}{2}R = R\cos^2\theta - \tfrac{1}{2}R = \tfrac{1}{2}R(2\cos^2\theta - 1) = \tfrac{1}{2}R\cos 2\theta,$$

$$y = R\cos\theta\sin\theta = \tfrac{1}{2}R\sin 2\theta.$$

It follows that L is a circle of radius $\frac{1}{2}R$ centred on the point $(\frac{1}{2}R, 0)$, as shown in the final sketch. The 2θ-dependence of its coordinates, as opposed to the θ-dependence of the turntable, confirms the qualitative deduction made earlier.

S 101   The period of oscillation is clearly dependent on the force constant k of the spring, but this is not given and will have to be determined from the stated equilibrium conditions.

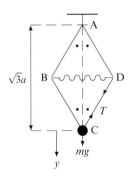

In equilibrium, when the reaction at A must support the weight mg of the mass, and, by symmetry, all the angles CAD, CAB, ACB and ACD are equal (denoted by θ and marked in the figure by black dots), the tensions in the rods have a common value T.

Further, since the spring has equilibrium length a, we have immediately that the value of θ is $\sin^{-1}(\frac{1}{2}a/a) = \pi/6$. Next, resolving forces vertically and horizontally gives

$$2T\cos\theta = mg,$$

$$2T\sin\theta = k(\sqrt{2}a - a).$$

On cross-multiplying to eliminate T:

$$ka(\sqrt{2}-1) \times \frac{\sqrt{3}}{2} = mg \times \frac{1}{2},$$

$$k = \frac{mg}{a} \frac{1}{\sqrt{6}-\sqrt{3}}.$$

To investigate the oscillations, we could define new variables giving the deviations of the various quantities, θ, T, and y, the vertical displacement of the mass, from their equilibrium values; but this would involve the trigonometric functions of the sums of angles, and making small angle approximations. Instead, we stick to our original coordinates, together with y; as we will see, this approach, whilst involving a little more algebra, provides an internal check on possible copying or manipulation errors.

During the motion, when θ, T and y are all varying, we still have the connections

$$y = 2a\cos\theta - \sqrt{3}\,a,$$
$$2T\sin\theta = k(\sqrt{2}\,a - 2a\sin\theta),$$

whilst the equation of motion of the mass is

$$m\ddot{y} = mg - 2T\cos\theta.$$

Substituting into the latter for $\cos\theta$ and $2T$, and then for k, yields, successively,

$$m\ddot{y} = mg - \frac{ka(\sqrt{2}-2\sin\theta)}{\sin\theta}\frac{(y+\sqrt{3}\,a)}{2a}$$

$$= mg - \frac{mg}{(\sqrt{6}-\sqrt{3})}\frac{(\sqrt{2}-2\sin\theta)}{\sin\theta}\frac{(y+\sqrt{3}\,a)}{2a}$$

$$= mg - \frac{mg}{2a(\sqrt{6}-\sqrt{3})}\left[\frac{\sqrt{2}\,y}{\sin\theta} + \frac{\sqrt{6}\,a}{\sin\theta} - 2y - 2\sqrt{3}\,a\right].$$

It is at this point that we impose the condition that the oscillations are small, so that we can approximate $\sin\theta$ by its equilibrium value of $\frac{1}{2}$. When this is done, we have

$$m\ddot{y} = mg - \frac{mg}{2a(\sqrt{6}-\sqrt{3})}\left[2\sqrt{2}\,y + 2\sqrt{6}\,a - 2y - 2\sqrt{3}\,a\right]. \qquad (*)$$

Now, the terms on the right-hand side that do *not* involve y cancel each other – they represent the equilibrium position, without k appearing explicitly – and leave us with

$$m\ddot{y} = -\frac{mg}{2a(\sqrt{6} - \sqrt{3})}(2\sqrt{2}\,y - 2y) = -\frac{mg}{\sqrt{3}\,a}\,y.$$

This is SHM with a period of $2\pi(\sqrt{3}\,a/g)^{1/2}$.

Note 1. It should be noted that if the terms not containing y had not cancelled in (∗), then an error in logic, algebra or transcription, would have been indicated.

Note 2. The period of the *vertical* oscillations of the weight is the same as that for small *horizontal* oscillations of a pendulum of length $\sqrt{3}\,a$, i.e the equilibrium length of AC. The algebra needed to prove it is complicated (and prone to error!), but this connection with the length of AC seems to be a general result for any equilibrium compressed spring length λa with $0 < \lambda < \sqrt{2}$, and not just for $\lambda = 1$.

S 102   When considering a new physics problem, it is very useful to have, if possible, a similar, but well-understood, situation that can be used as a reference or as the starting point for an analysis. For the current problem the 'benchmark potential' is the parabolic one, $V(x) = \frac{1}{2}\alpha x^2$; in this quadratic potential the oscillation frequency is

$$f_0 = \frac{1}{2\pi}\sqrt{\frac{\alpha}{m}},$$

independent of the amplitude of the motion.

If the potential in the neighbourhood of the equilibrium point increases (in both directions) more rapidly than quadratically, then the frequency increases as the amplitude increases, and the system is known as a *hard oscillator*. Conversely, a slower than quadratic increase results in a decrease of frequency with amplitude and a *soft oscillator*.

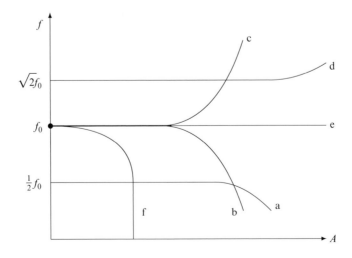

The figure shows in a qualitative way what would be expected for each of the given potential wells. At very low amplitudes the sketch curves appear completely horizontal; in fact, except for curve (e), they should all bend very slightly.

(a) $V(x) = \frac{1}{2}k\frac{x^2}{4 + x^2}$. For small values of x this is a 'benchmark potential', but one with 'spring constant' $k/4$ (rather than k) and a consequent frequency of $f_0/2$. For large amplitudes the potential tends to a constant and so the oscillator is *soft*.

(b) $V(x) = k(1 - \cos x)$. The Taylor expansion of $(1 - \cos x)$ is

$$1 - \left(1 - \frac{x^2}{2!} + \frac{x^4}{4!} - \cdots \right),$$

and so the potential is

$$V(x) \approx \frac{k}{2}\left(x^2 - \frac{x^4}{12} + \cdots \right),$$

resulting in a *soft* oscillator and a falling off of the frequency with amplitude.

(c) $V(x) = k(\cosh x - 1)$. A similar expansion to that in (b) yields

$$V(x) \approx \frac{k}{2}\left(x^2 + \frac{x^4}{12} + \cdots \right),$$

with a *positive* quartic term and a consequent increase in frequency with amplitude, i.e. a *hard* oscillator.

(d) $V(x) = k(\cosh x - \cos x)$. Summing the expansions in (b) and (c) shows that this potential is almost purely parabolic, the first non-quadratic term being $2x^6/6!$ (and even that has a very small numerical coefficient). Thus, the oscillation frequency is essentially independent of A, but it is $\sqrt{2}f_0$, rather than f_0.

(e) $V(x) = \frac{1}{2}k(x^2 - 2ax)$. If this potential is rewritten as $\frac{1}{2}k(x - a)^2 - \frac{1}{2}ka^2$, then it is clear that it is, in fact, a parabolic potential, but one with $x = a$ as the equilibrium point, and a change in the zero of potential. Thus the frequency has no amplitude-dependence, as shown by the solid horizontal line labelled 'e' in the figure.

(f) $V(x) = \frac{1}{2}k(x^2 - x^3)$. This potential is clearly parabolic very close to $x = 0$, but almost equally clearly, it ceases to be an oscillator at all, for amplitudes greater than $\frac{2}{3}$, i.e. where dV/dx becomes zero and the potential has a turning point. Once beyond that point, the particle moves away to infinity, and the frequency is formally zero (see curve f in the figure).

S 103   In this problem we will have to make use of qualitative observations when deciding how to obtain a good approximation to the required answer.

Since the cork takes 2.5 s to complete one cycle, the frequency of the wave-train must be $f=0.4$ Hz, and its velocity must be $v=f\lambda=0.4\lambda$. Substituting this into the given equation for v^2 gives the (cubic) equation

$$f^2\lambda^2 = \frac{g\lambda}{2\pi} + \frac{2\pi\gamma}{\lambda\rho} = 1.56\lambda + \frac{4.4 \times 10^{-4}}{\lambda}.$$

There is no standard method for solving such a cubic equation, but if we examine the terms involved we can simplify it. If the two term on the RHS are to be comparable (and therefore require both to be taken into account), λ must be of the order of

$$\lambda = \sqrt{\frac{4.4 \times 10^{-4}}{1.56}} \approx 1.7 \times 10^{-2}\,\text{m}.$$

But this cannot be the case, since, if it were, the plank would simply have ripples along its edges, and not pitch and toss on the wave-train. Clearly the actual wavelength is of the order of metres, and the gravity-controlled part of the formula for v is totally dominant. So, we have

$$f^2\lambda^2 \approx \frac{g\lambda}{2\pi} \quad \Rightarrow \quad \lambda = \frac{g}{2\pi f^2} = 9.8\,\text{m}.$$

As $9.8 \gg 1.7 \times 10^{-2}$, our neglect of surface tension effects is retrospectively justified.

S 104   The two lens arrangements are shown in the figure below – Paula's on the left and Peter's on the right.

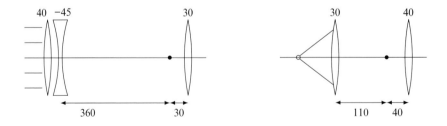

In normal adjustment, an astronomical telescope has a magnification equal to the ratio of the focal lengths of the objective and eyepiece lenses. So the former should be made as large as possible, whilst still forming an image for the eyepiece to view. With the lenses available, the longest effective focal length f' of the objective is achieved by placing the weaker of the converging lenses ($f = 40$ mm) and the diverging lens in as close contact as possible, thus subtracting the magnitudes of their powers and giving

$$\frac{1}{f'} = \frac{1}{40} - \frac{1}{45} \quad \Rightarrow \quad f' = 360 \text{ mm}.$$

This is the distance from the objective at which the image of the Moon is formed, and this must be at the focal point of the eyepiece ($f = 30$ mm) so that Paula can see the final image at infinity. The overall length of the telescope is 390 mm, and three tubes will need to be fastened together. The magnification achieved is $M' = f_{obj}/f_{eye} = 360/30 = 12$. The arrangement is shown in the left-hand figure, with the black dot indicating the location of the intermediate image.

For the microscope, as much optical (focusing) power as possible is needed, and so the diverging lens should not be used. Given the intended objective-eyepiece separation of 150 mm, the converging lenses produce an effective overall focal length f'', where

$$\frac{1}{f''} = \frac{1}{30} + \frac{1}{40} - \frac{150}{30 \times 40} \quad \Rightarrow \quad f'' = -15 \text{ mm}.$$

If Peter's least distance of distinct vision is D, then the magnification provided by using the compound microscope is $M'' = D/f''$. But we are told that $M'' = M'$, and we have shown that $M' = 12$, and so $D = 12 \times 15 = 180$ mm.

Note For the microscope, the two lenses could be interchanged, with the same result for M'', although the seed being viewed would have to be placed a different distance from the objective, at 60 mm rather than 41.3 mm. You may like to calculate, and hence verify, these two distances.

S 105 An integrated time-dependent analysis of the whole situation is mathematically complicated, but the problem can be tackled in stages, with each one a 'relatively small perturbation' on the previous one. The three stages are: the transmitters have exactly the same frequency and are in phase; they have the same frequency but the left-hand one is, say, $\frac{1}{2}\pi$ ahead in phase; the question as posed.

The simple first situation is exactly that of the Young's double-slit experiment, and the detected pattern is a set of fringes centred on the symmetry axis and separated by (in the usual notation)

$$w = \frac{\lambda D}{s} = \frac{30 \times 10^{-3} \times 100}{3.00} = 1 \text{ m}.$$

This situation is as in the figure (shown sideways to save space).

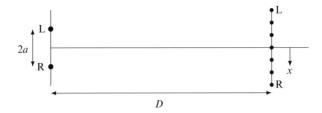

Next we consider what changes if the left transmitter L is $\frac{1}{2}\pi$ ahead of the right one, R. Let us denote sideways displacements at the detection 'screen' by x, counted positive to the right. Then the phase at x of the radiation from L (relative to that for a symmetry axis wave) is

$$\phi_L(x) = \frac{2\pi}{\lambda}\left(\sqrt{(a+x)^2 + D^2} - D\right) \approx \frac{2\pi}{\lambda}\frac{(a+x)^2}{2D},$$

where $D = 100$ m and $a = 1.5$ m. Similarly

$$\phi_R(x) = \frac{2\pi}{\lambda}\left(\sqrt{(a-x)^2 + D^2} - D\right) \approx \frac{2\pi}{\lambda}\frac{(a-x)^2}{2D}.$$

Constructive interference at x requires that $\phi_R(x)$ is $\pi/2$ ahead of $\phi_L(x)$, so as to compensate for the phase lead of the left transmitter:

$$\phi_R(x) = \frac{\pi}{2} + \phi_L(x) \quad \Rightarrow \quad -\frac{4ax\pi}{D\lambda} = \frac{\pi}{2}.$$

Substituting actual values gives

$$x = -\frac{D\lambda}{8a} = -\frac{100 \times 30 \times 10^{-3}}{8 \times 1.5} = -0.25\,\text{m},$$

i.e. the pattern is moved to the *left* by 25 cm.

Finally, in the given situation L's phase gains 2π relative to that of R every second, and consequently the interference pattern moves to the left at a speed of $0.25 \times (2\pi)/(\pi/2) = 1.00$ m s^{-1}.

S 106　(a) The observer receives in a time interval T all the waves that are contained in a length $(v + u)T$. Since each of these is of wavelength $\lambda_s = v/f_s$, the number of them is $(v + u)T/\lambda_s$, and the perceived frequency is

$$f_r = \frac{1}{T}\frac{(v+u)T}{\lambda_s} = \frac{(v+u)}{v}f_s.$$

(b) Let the source be a distance X from the receiver at the start of a time interval T, then, at the end of that interval, the front of the wave train is at $X - vT$, and the back at $X - uT$. When the waves are received they are contained in a train $(X - uT) - (X - vT) = (v - u)T$ long. There are $f_s T$ of them and so their measured wavelength is $\lambda_r = (v - u)T/f_s T$. They pass the stationary receiver with (their actual) speed v and are measured as being at a frequency

$$f_r = \frac{v}{\lambda_r} = \frac{v}{v - u} f_s.$$

(c) With the car travelling towards the sound source S with speed u, the detected sound frequency at the car is given by (a) above as

$$f_1 = \frac{v + u}{v} f_0,$$

where f_0 is the source frequency. After the waves have rebounded from the car, it becomes a moving source of frequency f_1 and S becomes the stationary receiver. From (b) above S records a frequency for the returned signal of

$$f_2 = \frac{v}{v - u} f_1 = \frac{v}{v - u} \frac{v + u}{v} f_0 = \frac{v + u}{v - u} f_0.$$

When these returned waves are mixed with a sample of the original waves, the beats that are produced have a frequency f_b equal to the frequency difference between the two input wave trains, i.e.

$$f_b = f_2 - f_0 = \left(\frac{v + u}{v - u} - 1 \right) f_0 = \frac{2u}{v - u} f_0.$$

With the given data, the car's speed can be calculated:

$$121 = \frac{2u}{350 - u} \times 10^3 \quad \Rightarrow \quad u = \frac{350 \times 121}{2121} = 20.0 \text{ m s}^{-1}.$$

S 107   As we are trying to find a frequency at which both the viola can emit sound, and the piano can resonantly detect it, we have to take into account the physical limits that the structures of each impose on the range of frequencies that can be considered.

We first need to know the possible frequencies for the piano string, and for this we require both its linear density ρ' and the tension T within it. If A is the cross-sectional area of the string, then its linear density is $\rho' = A\rho$ and, as the strain is 1%, its tension is $T = E \times 10^{-2} \times A$. The wave velocity on the string is therefore

$$v = \sqrt{\frac{A\,E\,10^{-2}}{A\,\rho}},$$

and, since $v = f\lambda$ and 1.5 m represents one half of the fundamental wavelength (see the figure), the fundamental frequency of the piano string is

$$f = \frac{1}{\lambda}\sqrt{\frac{A\,E\,10^{-2}}{A\,\rho}} = \frac{1}{3.0}\sqrt{\frac{2.2 \times 10^{11} \times 10^{-2}}{7.7 \times 10^{3}}} = 178 \text{ Hz}.$$

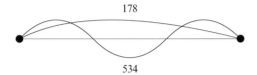

178

534

As the viola is tuned to 440 Hz, shortening its A string will only result in a frequency even further away from 178 Hz; only frequencies above 440 Hz will be possible. The lowest frequency of the piano string that comes in this range is $3 \times 178 = 534$ Hz, with a vibration pattern as shown in the figure. Since the viola string is tuned, and the velocity of transverse waves on it predetermined, we have that the product λf is fixed, leading to

$$2 \times 0.40 \times 440 = 2 \times \ell \times 534,$$

where $\ell = 33.0$ cm is the required length of the string. It therefore has to be shortened by 7.0 cm to give the best chance of exciting a (harmonic) vibration in the piano string.

S 108   As the 'guiding principle' is expressed in terms of the tangent to the path taken by the light ray, we will need to establish the connection between the angle ϕ mentioned in the guide and dy/dx which gives the direction of that tangent in the *x-y* plane. This will lead to a differential equation, which will then have to be solved, either exactly or approximately.

Since the the temperature falls as y gets larger, both the air density and its refractive index increase with height; α is a positive constant. It follows from the statement in the question that as y increases, $\sin\phi$, and hence ϕ, decreases. The path of the light ray that starts at D with $\phi = \frac{1}{2}\pi$, and finally enters the man's eye, must have a positive, and ever increasing, slope, as illustrated in the figure.

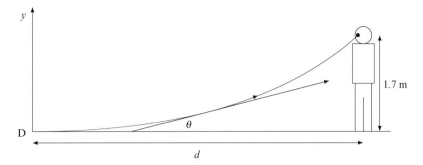

(a) The sin ϕ mentioned in the statement is identical to $\cos\theta$, where $\tan\theta$ is the slope of the tangent to the curve. Consequently,

$$\sin\phi = \cos\theta = \frac{1}{\sec\theta} = \frac{1}{(1+\tan^2\theta)^{1/2}} = \frac{1}{[1+(dy/dx)^2]^{1/2}}.$$

So, Snell's generalized law reads

$$\frac{n_0(1+\alpha y)}{\left[1+\left(\dfrac{dy}{dx}\right)^2\right]^{1/2}} = \frac{n_0(1+0)}{[1+0]^{1/2}} = n_0.$$

The equation satisfied by $y(x)$ is therefore

$$(1+\alpha y)^2 = 1 + \left(\frac{dy}{dx}\right)^2. \qquad (*)$$

(b) The complete solution[17] to $(*)$ requires more advanced mathematics than is usually needed for a physics course, but, if the inequality specified in the question is valid, we can try the simple solution $y = \beta x^2$, giving

$$(1+\alpha\beta x^2)^2 = 1 + (2\beta x)^2,$$
$$2\alpha\beta x^2 + \alpha^2\beta^2 x^4 = 4\beta^2 x^2.$$

If $\alpha^2 x^2 \ll 1$, then the second term on the LHS is negligible compared to the RHS, and we have $\beta = \frac{1}{2}\alpha$. When $y = 1.7$, x, and hence d, is

$$d = x \approx \sqrt{\frac{y}{\beta}} = \sqrt{\frac{2y}{\alpha}} = \sqrt{\frac{2 \times 1.7}{1.5 \times 10^{-6}}} = 1510\,\text{m}.$$

In retrospect $\alpha d \ll 1$ and so the approximation is at least self-consistent.

---

[17] The full solution is $\cosh(\alpha x) = 1 + \alpha y$, which can be checked by substitution if you are familiar with hyperbolic functions and their derivatives.

S 109   The fluctuation in the received signal is due to the phase between the two contributing components changing, itself a result of variations in the distances they have to travel. The surface signal does not change its path length, but that reflected from the Heaviside layer does, if the latter is moving vertically.

The figure shows the paths of the two signals with both horizontal and vertical distances expressed in units of 10^5 m, i.e. the actual height of the Heaviside layer is $h \times 10^5$ m.

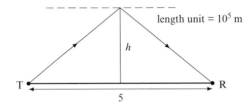

The extra distance travelled by the reflected wave is straightforwardly

$$x = 2\sqrt{h^2 + (2.5)^2} \times 10^5 - 5.00 \times 10^5 \text{ m},$$

and produces a phase lag of

$$\phi = 2\pi \frac{x}{\lambda} = 2\pi \frac{xf}{c} = \frac{2\pi}{3}\left[2\sqrt{h^2 + (2.5)^2} \times 10^5 - 5.00 \times 10^5\right],$$

since $c = 3.00 \times 10^8$ m s^{-1} and $f = 10^8$ Hz. To relate this to the observed effect we need its rate of change, with the only time-varying quantity on the RHS being h. Differentiating with respect to time t, we have

$$\frac{d\phi}{dt} = \frac{4\pi}{3} 10^5 \frac{h}{\sqrt{h^2 + (2.5)^2}} \frac{dh}{dt}.$$

Now, the observed rate of variation of ϕ is 2π in $60/8$ seconds, and at this stage we can set $h = 2$. Thus

$$\frac{dh}{dt} = \frac{3}{4\pi \times 10^5} \frac{\sqrt{4 + 6.25}}{2} \times \frac{2\pi \times 8}{60} = \frac{0.32}{10^5} \text{ m s}^{-1}.$$

As the actual height is $h \times 10^5$, this is a speed of 0.32 m s^{-1}, either up or down.

Note The structure of the mathematics in this question is almost the same as that in solution S 112 appearing on page 228, even though one question seeks a change in frequency, and the other a change in phase.

S 110   Although the physics of this problem is important, its solution is largely concerned with identifying the physical consequences of mathematical relationships.

(a) To examine the behaviour of

$$z = \sqrt{\frac{1+\beta}{1-\beta}} - 1 \qquad\qquad (*)$$

when $v/c = \beta \ll 1$, we expand the square root of the fraction as if it were $(1+\beta)^{1/2}(1-\beta)^{-1/2}$ using the binomial theorem, i.e.

$$z = (1+\beta)^{1/2}(1-\beta)^{-1/2} - 1$$

$$= \left[1 + \frac{1}{2}\beta - \frac{1}{2}\frac{1}{2}\frac{1}{2!}\beta^2 + \cdots\right]\left[1 + \frac{1}{2}\beta + \frac{1}{2}\frac{3}{2}\frac{1}{2!}\beta^2 + \cdots\right] - 1$$

$$= \left[1 + \frac{1}{2}\beta + \frac{1}{2}\beta + \left(\frac{1}{4} - \frac{1}{8} + \frac{3}{8}\right)\beta^2 + \cdots\right] - 1$$

$$= \beta + \frac{1}{2}\beta^2 + \cdots.$$

This expansion answers both questions in this part: for $v \ll c$, i.e. for small values of β, the red shift $z \sim \beta = v/c$; the next term in the expansion is positive, showing that the relativistic shift exceeds the non-relativistic one. This latter result is almost obvious from the presence of '$(1-\beta)$' in the denominator of the argument of the square root; it starts at 1 when $\beta = 0$ and goes to ∞ as $\beta \to 1$; the non-relativistic expression cannot exceed unity. There is no reason to think that plots of the two expressions would cross at some intermediate value. One easily calculated case is $\beta = 0.5$, when the square root is equal to $\sqrt{3} = 1.732$ and $z_{rel} = \sqrt{3} - 1 = 0.732$ which is greater than $z_{nonrel} = \beta = 0.5$.

(b) Since $z_{rel} \geq z_{nonrel}$ for all physical β, we have to determine a positive value for β such that

$$\sqrt{\frac{1+\beta}{1-\beta}} - 1 - \beta = \frac{\beta}{10},$$

i.e.

$$\frac{1+\beta}{1-\beta} = (1 + 1.1\,\beta)^2 = 1 + 2.2\,\beta + (1.1)^2\beta^2.$$

This looks like (and is) a cubic equation, but one (trivial) known solution, $\beta = 0$, when there is no red shift, will make it effectively a quadratic equation. Pedantically cross-multiplying,

$$1 + \beta = 1 + 2.2\,\beta + (1.1)^2\beta^2 - \beta - 2.2\,\beta^2 - (1.1)^2\beta^3,$$

$$0 = 0.2 + \left[(1.1)^2 - 2.2\right]\beta - (1.1)^2\beta^2,$$

$$0 = 1.21\,\beta^2 + 0.99\,\beta - 0.2, \quad \text{(after multiplying through by } -1)$$

$$\beta = \frac{-0.99 \pm 1.3957}{2.42} = -0.986 \text{ or } 0.168.$$

Clearly we need the (positive) value of 0.168, which implies that $v = 5.030 \times 10^7$ m s^{-1}.

To turn this into a galactic distance d we use Hubble's law, $v = H_0 d$, where $H_0 = 65$ km s^{-1} Mpc$^{-1} = 2.11 \times 10^{-18}$ s^{-1}:

$$d = \frac{v}{H_0} = \frac{5.030 \times 10^7}{2.11 \times 10^{-18}} = 2.28 \times 10^{25} \text{ m} = 2.5 \times 10^9 \text{ light-years.}$$

This is a slight overestimate of the distance from Earth that a star or galaxy had to be (when the light was emitted), for there to be a 10% error in the calculated distance if the non-relativistic formula for the red shift is used.

S 111   We label the slits A, B, C and D, as in the figure in the question, and the associated phases (relative to some reference phase) by ϕ_A, ϕ_B, etc.

As we will be using it many times, we first develop a formula for the distance from $(0, y)$ to (L, Y). It is $\sqrt{L^2 + (Y - y)^2}$, and is greater than that from $(0, y)$ to (L, y) by

$$\sqrt{L^2 + (Y - y)^2} - L = L \left(\sqrt{1 + \frac{(Y - y)^2}{L^2}} - 1 \right)$$

$$= L \left(1 + \frac{(Y - y)^2}{2L^2} + \cdots - 1 \right) \approx \frac{(Y - y)^2}{2L}.$$

This additional distance produces a phase lag of

$$\phi = \frac{(Y - y)^2}{2L} \frac{2\pi}{\lambda}$$

for the ray that travels at a slight angle to the x-axis.

(a) For a maximum intensity at $(L, 0)$, we require light from all four sources to be in phase there. The pair B and C are clearly in phase with each other, as are the pair A and D. The lag of the first pair, for which $Y = 0$ and $|y| = a$, is $\pi a^2 / L\lambda$, together with one of π because of the phase inversion; for the outer pair $\phi_A = \phi_D = \pi 4a^2 / L\lambda$. For a common phase for all four at the screen, we must have

$$\frac{\pi a^2}{L\lambda} + \pi + 2\pi n = \frac{\pi 4a^2}{L\lambda}, \quad \text{for some non-negative integer } n,$$

i.e.

$$L = \frac{3a^2}{(2n + 1)\lambda}.$$

We note, in passing, that $L = 3a^2/\lambda$ is the greatest screen distance for which there is an axial maximum.

(b) $L = a^2/\lambda$ corresponds to $n = 1$ in (a), and so $(L, 0)$ is immediately identified as an intensity maximum. For (L, a),

$$\phi_A = \frac{\pi a^2}{L\lambda}, \quad \phi_B = \pi, \quad \phi_C = \frac{\pi 4a^2}{L\lambda} + \pi, \quad \phi_D = \frac{\pi 9a^2}{L\lambda},$$

or, on setting $L = a^2/\lambda$,

$$\phi_A = \pi, \quad \phi_B = \pi, \quad \phi_C = 5\pi, \quad \phi_D = 9\pi.$$

These are all of the form $(2n + 1)\pi$, and so all four contributions are in phase and a maximum intensity is the result. By symmetry, $(L, -a)$ is also a maximum.

For $(L, 2a)$, a similar calculation gives $\phi_A = 0$, $\phi_B = 2\pi$, $\phi_C = 10\pi$, and $\phi_D = 16\pi$. Again all are in phase, being of the form $2n\pi$, and $(L, 2a)$ is a point of maximum intensity. By symmetry, or further calculation, so is $(L, -2a)$. In summary, all five of the points marked on the original figure are intensity maxima.

(c) We have already seen that at $L = 3a^2/\lambda$, all four amplitudes reinforce each other at $(L, 0)$ and the net amplitude is $4S$, say. At $(2a^2/\lambda, 0)$,

$$\phi_A = \phi_D = 2\pi \quad \text{and} \quad \phi_B = \phi_C = 3\pi/2,$$

and so we have two amplitudes of $2S$ superimposed in quadrature (out of phase with each other by $\pi/2$).

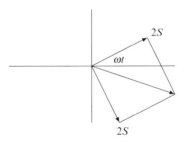

Either from a vector diagram, as in the figure, or more mathematically as follows,

$$
\begin{aligned}
\text{Resultant} &= 2S\cos(\omega t + 2\pi) + 2S\cos\left(\omega t + \tfrac{3}{2}\pi\right) \\
&= 2S\cos\omega t + 2S\cos\omega t \cos\tfrac{3}{2}\pi - 2S\sin\omega t \sin\tfrac{3}{2}\pi \\
&= 2S\cos\omega t + 2S\sin\omega t \\
&= 2\sqrt{2}S\left(\frac{1}{\sqrt{2}}\cos\omega t + \frac{1}{\sqrt{2}}\sin\omega t\right) \\
&= 2\sqrt{2}S\cos\left(\omega t + \tfrac{7}{4}\pi\right),
\end{aligned}
$$

their resultant has magnitude $2\sqrt{2}S$. The ratio of intensities $I(2a^2/\lambda, 0)$: $I(3a^2/\lambda, 0)$ is therefore $(2\sqrt{2})^2 : 4^2 = 1 : 2$.

S112   The figure is that of the question with the addition of the path to and from the $\pm n$th rod, counting the one opposite the man as $n = 0$; in the figure $n = 4$.

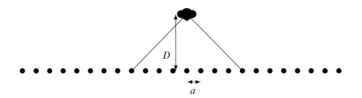

From symmetry, we need only concern ourselves with positive n. Denote the time of the nth echo by t_n with $t_0 = 2D/v$, where v is the velocity of sound in air. From geometry

$$t_n = \frac{2\sqrt{D^2 + n^2 a^2}}{v}, \tag{$*$}$$

and the frequency of echoes experienced at that time is approximately

$$f(t_n) = \frac{1}{t_{n+1} - t_n} \approx \frac{dn}{dt_n} = \left(\frac{dt_n}{dn}\right)^{-1}.$$

In taking this step, we have assumed that time and the perceived frequency can be treated as a continuous variables when n is reasonably large.[18] From $(*)$, it follows that

$$\frac{dt_n}{dn} = \frac{2}{v}\frac{na^2}{\sqrt{D^2 + n^2 a^2}}.$$

To write f as a function of t we need to express both $\sqrt{D^2 + n^2 a^2}$ and na^2 in terms of time:

$$\sqrt{D^2 + n^2 a^2} = \frac{1}{2}vt_n \quad \text{and} \quad (na)^2 = \frac{v^2 t_n^2}{4} - D^2 = \frac{v^2}{4}\left(t_n^2 - t_0^2\right).$$

Substituting these expressions and changing from t_n to a continuous time t yields

$$f(t) = \frac{v}{2a}\frac{t}{\left(t^2 - t_0^2\right)^{1/2}} = f_0\frac{t}{\left(t^2 - t_0^2\right)^{1/2}},$$

---

[18] The author expects this to be truer acoustically than it is mathematically.

where $f_0 = v/2a$ is the ultimate frequency of the echoes, and corresponds to the extra time it takes for the initial bang of the pole to reach the next rod. The value of D becomes irrelevant at this point.

The frequency dependence is therefore that of a falling tone which starts abruptly at t_0, but with no well-defined frequency (for the first two echoes the 'notional frequency' is $v/(2\sqrt{D^2 + a^2} - 2D))$, and falls monotonically to a limiting value of f_0. The figure below is a sketch of this

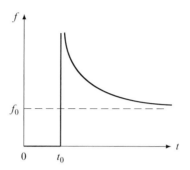

Note The structure of the mathematics in the above question is almost the same as that in the solution to Q 109 appearing on page 224, even though one question seeks a change in frequency, and the other a change in phase.

S 113 The equation governing the diffraction grating is '$d\sin\theta = n\lambda$', but here we are not given the value of d, the line spacing of the grating. Normally this would make establishing the wavelengths of all the lines impossible. However, we are given the wavelength of the green light and this would implicitly give us the line spacing of the grating, but only if we knew the order n of the green spectrum. Perhaps the approximate knowledge of the various wavelengths implicit in the colours can be used to resolve this particular point.

The grating spacing is unknown, but it *is* constant, and so the values of '$n\lambda$' for the three groups must be in the ratio

$$\sin 24.5° : \sin 38.4° : \sin 43.6° = 0.4147 : 0.6211 : 0.6896 = 1 : 1.50 : 1.66.$$

Where there is a colour common to two groups, the final ratios indicate the relative values of n, i.e. the spectral order for that colour, in the two groups. Thus the likely values of n_R in groups (i) and (ii) are 2 and 3, respectively. Similarly, n_B is likely to be 3 in group (i) and 5 in group (iii).

Using the NASA 'rough guide' for the red and blue lines, and the suggested group (i) values for n_R and n_B, we obtain 2×650 to be compared with 3×475, which are quite far from being equal, but are not impossible. When the accurate values for the wavelengths have been established, the match *must* be perfect (within the accuracy of the data).

Since we know the true value of λ_G, we next consider group (ii). With n_R as 3, we need the nearest integer value for n_G that makes

$$3 \times 650 \approx n_G \times 518,$$

clearly $n_G = 4$. With this established, we can revisit and refine previous conclusions:

- For group (ii), $n\lambda = 4 \times 518 = 2072$ and therefore $\lambda_R = 2072 \div 3 = 691$ nm.
- Since $\lambda_V \approx 400$ nm, the integer $n_V \approx 2072 \div 400$, i.e. $n_V = 5$ and $\lambda_V = 2072 \div 5 = 414$ nm.
- From group (i), with $2 \times 691 = 1382 = 3 \times \lambda_B$, we conclude that $\lambda_B = 461$ nm.
- From group (iii), we must have the integer $n_Y \approx (5 \times 461) \div 570$, clearly 4, and from this it follows that $\lambda_Y = 576$ nm.

In summary, for the lines actually observed,

Colour	Violet	Blue	Green	Yellow	Red
Wavelength (nm)	414	461	518	576	691

S 114   The quantities I_i and I_r are the intensities of the incident and reflected ultrasonic waves when they meet a boundary between two materials at normal incidence. Each material is characterized by its impedance Z, which is equal to the product of its density and the speed of sound within it (or, more fundamentally, the ratio of the sound pressure at a point to the local speed of the particles constituting the material).

The stated formula gives the *reflected* sound intensity, and so the transmitted intensity must be

$$I_t = I_i - I_r = \left[1 - \left(\frac{Z_2 - Z_1}{Z_2 + Z_1} \right)^2 \right] I_i$$

$$= \frac{(Z_2 + Z_1)^2 - (Z_2 - Z_1)^2}{(Z_2 + Z_1)^2} I_i$$

$$= \frac{4 Z_1 Z_2}{(Z_2 + Z_1)^2} I_i.$$

After two successive transmissions the intensity will be

$$I_{tt} = \frac{4 Z_1 Z_2}{(Z_2 + Z_1)^2} \frac{4 Z_2 Z_3}{(Z_3 + Z_2)^2} I_i.$$

This is a complicated function of Z_2, but will be maximal when its numerator N and denominator D are related by[19]

$$D \times \text{derivative of } N = N \times \text{derivative of } D$$

$$(Z_2 + Z_1)^2 (Z_3 + Z_2)^2 \, 32 \, Z_1 Z_2 Z_3 = 16 \, Z_1 Z_2^2 Z_3 \left[2(Z_2 + Z_1)(Z_3 + Z_2)^2 + \cdots \right.$$

$$\left. + 2(Z_3 + Z_2)(Z_2 + Z_1)^2 \right],$$

$$2(Z_2 + Z_1)(Z_3 + Z_2) = 2Z_2(Z_3 + 2Z_2 + Z_1),$$

$$Z_2 Z_3 + Z_1 Z_3 + Z_2^2 + Z_1 Z_2 = Z_2 Z_3 + 2Z_2^2 + Z_2 Z_1,$$

$$Z_1 Z_3 = Z_2^2.$$

In words, the two-step transmission is most efficient when the impedance Z_2 is the geometric mean of Z_1 and Z_3 (and therefore lies between them). In this case,

$$I_{\text{tt}} = \frac{16 \, Z_2^4}{(Z_1 Z_2 + Z_2^2 + Z_1 Z_3 + Z_2 Z_3)^2} = \frac{16 \, Z_2^2}{(Z_1 + 2Z_2 + Z_3)^2}.$$

The impedance of each of the five given materials is obtained by multiplying its density by the speed of sound within it, as below

Material	Blood	Fat	Lung	Muscle	Water
Density (kg m^{-3})	1060	920	400	1070	1000
Speed of sound (m s^{-1})	1566	1446	650	1598	1480
Impedance (10^6 kg m^{-2} s^{-1})	1.66	1.33	0.26	1.71	1.48

We are looking for the three values that most closely form a geometric progression, and so can immediately reject that for lung tissue. Of the remaining four possibilities, we need only consider omitting them one at a time, and taking the other three in ascending (or descending) order. The four combinations are (using first letter abbreviations)

Combination	Z_1	Z_2	Z_3	$Z_1 Z_3$	Z_2^2
FBM	1.33	1.66	1.71	2.27	2.75
FWB	1.33	1.48	1.66	2.21	2.19
WBM	1.48	1.66	1.71	2.53	2.76
FWM	1.33	1.48	1.71	2.19	2.75

And the winner is – fat–water–blood! Or, of course, blood–water–fat.

---

[19] The derivative of a function $f(x)$ of the form $f(x) = g(x)/h(x)$ is $f'(x) = (hg' - gh')/h^2$.

S 115   This is a two-slit experiment in which we know the fringe spacing, but have no direct information about either the slit separation or the 'screen distance'. We need to deduce the latter two quantities from the data provided by the lens experiment.

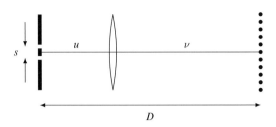

When the lens is in place and sharp images have been formed, the sum of the object and image distances must be equal to the separation between the slits and the microscope, i.e. $u + v = D$, as shown in the figure (though this depicts fringes, rather than a separated slit image, at the microscope).

Between the two sharp-image situations the object and image distances have interchanged, and consequently the magnification of the actual slit separation s has been inverted. So, arbitrarily assigning u and v, we have

$$\frac{v}{u} s = 0.490 \qquad \text{and} \qquad \frac{u}{v} s = 0.250.$$

Multiplying the two equations together gives

$$s^2 = 0.490 \times 0.250 \quad \Rightarrow \quad s = 0.35\,\text{mm}.$$

Dividing the same two equations shows that

$$\frac{v^2}{u^2} = \frac{0.490}{0.250} \quad \Rightarrow \quad v = \frac{7}{5}u \quad \left[u = \frac{7}{5}v \text{ would be equally valid} \right].$$

The lens formula can now be written as

$$\frac{7}{5}\frac{1}{v} + \frac{1}{v} = \frac{1}{f} = \frac{1}{0.210} \quad \Rightarrow \quad v = 0.504\,\text{m and } u = 0.360\,\text{m}.$$

Thus the slit-microscope separation is $D = 0.504 + 0.360 = 0.864\,\text{m}$. With both D and s known, we can use the standard Young's slits formula to calculate the wavelength of the light from the fringe spacing w:

$$\lambda = \frac{s\,w}{D} = \frac{0.35 \times 10^{-3} \times 30.1 \times 10^{-3}}{0.864 \times 22} = 554\,\text{nm}.$$

The light is green!

S 116   For a problem such as this, there is no straightforward algorithm to follow. As in research, arriving at the solution involves keen observation, insight, and a certain amount of luck! However the raw data can be presented in a more transparent form, and the first job is to convert the observed diffraction angles into energy differences between supposed levels.

Line	A	B	C	D	E	F	G
θ (degrees)	8.34	11.16	11.48	18.48	35.49	24–25	10–12
$n\lambda$ (nm)	290	387	398	634	1161	?	?
$E(n=1)\,(10^{-19}\,\mathrm{J})$	6.83	5.11	4.97	3.12	1.71	?	?
$E(n=2)\,(10^{-19}\,\mathrm{J})$	13.66	10.23	9.95	6.24	3.42	?	?

Here we have used the grating formula $d\sin\theta = n\lambda$, with $d = 2 \times 10^{-6}$ m, to obtain values for $n\lambda$, which are then converted to energy differences using $E = hc/\lambda$. The computed values correspond to $n = 1$, but the doubled values for $n = 2$ (i.e. the observed line might be from a second-order spectrum) are also given; this should make recognizing significant numerical relationships easier. (a) Now comes the hard bit! We need to identify additive 'coincidences' between the energy entries in the table, allowing for rounding errors of ± 2 in the final digits. In the $n = 1$ line, one obvious connection is that $E_A = E_B + E_E$, but there are no others. We therefore need to include the $n = 2$ entries in our search. It may be a bit more difficult to spot, but $E_C(2) = E_A(1) + E_D(1)$.

As we now have each of the observed lines involved in an energy relationship, we see whether we can construct an energy level diagram, without having to consider higher values of n. The base energy level is arbitrary, and we take it as zero and denote it by (0).

Level (1), providing the B-line must be at 5.11, and lines A and E must both have the same higher energy level at $5.11 + 1.71 = 6.83$ [level (2)].

As noted earlier $E_C(n=2) = E_A + E_D = 9.95$, and so the C-line represents the second-order spectrum for a transition from a level (3) at 9.95 to level (0); it is marked in the figure below as C_2. This accounts for the established lines.

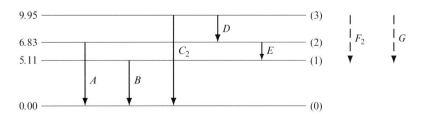

(b) The suspected *F*-line in the range 24°–25° corresponds to a value of E/n of 2.34–2.43 × 10^{-19} J. With the energy levels so far established, no transition can give rise to a first-order diffraction line in the observed angular range.

However, if $n = 2$, the corresponding energy range becomes 4.68–4.86 × 10^{-19} J and the value for a (3)→(1) transition, i.e. 4.84 × 10^{-19} J, falls in it, and so fits into the scheme; in the figure the line is marked F_2. Working backwards, the corresponding angle is 24.25°.

(c) Finally, as the F_2 [(3)→(1)] line is established, we would expect to see a first-order diffraction line resulting from the same atomic transition. So, there should be a line (G) at $\theta_G = \sin^{-1}[(hc)/(d \times 4.84 \times 10^{-19})] = 11.80°$. The completed table is therefore

Line	A	B	C_2	D	E	F_2	G
θ (degrees)	8.34	11.16	11.48	18.48	35.49	24.25	11.80
$n\lambda$ (nm)	290	387	398	634	1161	821	409
$E(n=1)$ (10^{-19} J)	6.83	5.11	4.97	3.12	1.71	–	4.84
$E(n=2)$ (10^{-19} J)	13.66	10.23	9.95	6.24	3.42	4.84	–

S 117   In this problem we simply follow instructions!
(a) From the thin lens formula, for a distant object, u^{-1} is very small and v tends to f. Since the magnification is given by v/u, for a fixed (but distant) u the size of the final image is proportional to f.

(b) The figure shows the required construction.

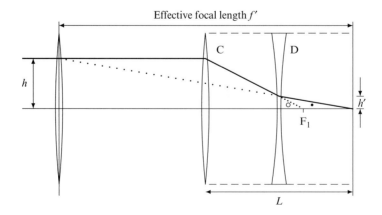

The left-hand lens is not a physical lens, but shows where a single convex lens would need to be placed in order to produce the same image as the actual telephoto combination of C and D; the camera casing, of length L is indicated by the dashed lines.

The path of a light ray from a distant object is shown by the heavy line: it is parallel to the axis until it reaches the lens C at the front of the camera; it is then refracted by C towards the point marked F_1, the focal point of the objective; before it reaches F_1, it is refracted by D towards a point further along the axis than F_1, where it forms an image on the recording medium (film or solid-state sensor).

The final direction of travel, when extrapolated backwards, intercepts the original paraxial ray at the point where a single converging lens would have to be placed in order to give the same final image of the distant object. This point is a distance f' from the recording medium at the back of the camera, and so determines the effective focal length of the telephoto combination.

(c) As indicated in the figure, let h be the ray's initial distance from the axis, and h' that when it is refracted by D. Measuring all lengths in mm, and taking C and D to mean the centres of the lenses, the relevant distances along the axis are $CF_1 = 76$, $CD = 60$, $DF_1 = 16$. For the diverging lens D, the focal length is -25 and the (virtual) object it focuses onto the sensor is at $u_2 = -16$. Therefore the distance v_2 from D to the sensor is given by

$$-\frac{1}{16} + \frac{1}{v_2} = -\frac{1}{25} \quad \Rightarrow \quad \frac{1}{v_2} = \frac{25 - 16}{400} \quad \Rightarrow \quad v_2 = 44.4,$$

and $L = 60 + 44.4 = 104.4$

Now consider the ratio h'/h in relation to the two right-angled triangles defined by the acute angles marked by $\circ$ and $\bullet$ in the figure.

$$\frac{16}{76} = \frac{h'}{h} = \frac{44.4}{f'} \quad \Rightarrow \quad f' = 211.$$

The magnification gain due to the telephoto lens is therefore $211/104.4 \approx 2$.

(d) The same result can be obtained directly from the formula

$$\frac{1}{f'} = \frac{1}{f_1} + \frac{1}{f_2} - \frac{d}{f_1 f_2},$$

i.e.

$$\frac{1}{f'} = \frac{1}{76} - \frac{1}{25} + \frac{60}{76 \times 25} \quad \Rightarrow \quad f' = 211,$$

but without any indication as to why it happens!

S 118   Before we superimpose the pressure amplitudes arising from the two sources, we need to determine their relative phases at the point of reception. The figure shows the paths of two such nearly-parallel pressure waves that will be detected in the direction θ.

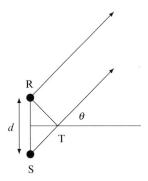

The phase of the wave originating from source S lags behind that from source R because of the extra distance ST that it has to travel, and the phase difference between the two must include this, as well as that of π due to the anti-phasing of the two sources. The length of the extra path ST is $d \sin \theta$ and, since the wavelength of the sound is $\lambda = 340/100 = 3.40$ m, this equates to a phase shift of

$$\phi = 2\pi \times \frac{d \sin \theta}{3.40} \text{ rad.} \qquad (*)$$

We must determine d from the additional information provided, as follows.

Let the signal from one source, R, be $A \cos \omega t$, then that from the other, S, is $A \cos(\omega t + \pi + \phi) = -A \cos(\omega t + \phi)$. When the two are superimposed, we have[20]

$$B(t) = A \cos \omega t - A \cos(\omega t + \phi) = 2A \sin\left(\omega t + \frac{\phi}{2}\right) \sin\left(\frac{\phi}{2}\right).$$

The amplitude of the received signal is therefore $2A \sin \frac{1}{2}\phi$, and since this is $0.1A$, we must have that $\sin \frac{1}{2}\phi = 0.05$ and that $\phi = 0.100$ rad. As this happens when $\theta = \pi/4$, it follows that

$$d = \frac{3.40 \, \phi}{2\pi \, \sin(\pi/4)} = 0.077 \text{ m} = 7.7 \text{ cm.}$$

For a general angle θ, $(*)$ becomes

$$\phi = 2\pi \times \frac{0.077 \sin \theta}{3.40} \text{ rad} = 0.1422 \sin \theta \text{ rad,}$$

---

[20] If the identity used for the second equality is unfamiliar to you, see the note at the end of this solution.

giving a maximum value for the amplitude $2A \sin \frac{1}{2}\phi$ of $0.1419A$ when $\theta = \pm\pi/2$; when θ is zero or π, $\phi = 0$ and there is no received sound wave.

Note To reduce the complexity of the equations involved, but without losing any generality, we can consider $\cos 2\alpha - \cos 2\beta$ as follows:

$$
\begin{aligned}
\cos 2\alpha - \cos 2\beta &= \cos[(\alpha + \beta) + (\alpha - \beta)] - \cos[(\alpha + \beta) - (\alpha - \beta)] \\
&= [\cos(\alpha + \beta)\cos(\alpha - \beta) - \sin(\alpha + \beta)\sin(\alpha - \beta)] \\
&\quad - [\cos(\alpha + \beta)\cos(\alpha - \beta) + \sin(\alpha + \beta)\sin(\alpha - \beta)] \\
&= -2\sin(\alpha + \beta)\sin(\alpha - \beta) \\
&= 2\sin(\alpha + \beta)\sin(\beta - \alpha).
\end{aligned}
$$

In the actual application above, $\alpha = \frac{1}{2}\omega t$ and $\beta = \frac{1}{2}(\omega t + \phi)$.

S 119 For transmitters in a single row running North–South, we would expect any given direction with a particular property to be one of four with the same property, i.e. directions (measured anticlockwise from North) of the form ϕ, $\pi - \phi$, $\pi + \phi$ and $2\pi - \phi$. To have *precisely* six directions clearly requires at least two such groups of four, but with some overlap between them.

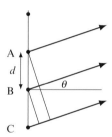

In the figure, transmitters A and C are in phase, and B has a phase lag of ϕ with respect to them. For the signals from A and C to interfere constructively in a direction θ, the usual Young's slit condition applies, but with a slit spacing of $2d$. Thus,

$$
\frac{2d\sin\theta}{\lambda} 2\pi = 2\pi m, \text{ where } m \text{ is an integer.}
$$

For the radiation from B to reinforce A and C in the same direction θ, we must also have

$$
\frac{d\sin\theta}{\lambda} 2\pi + \phi = 2\pi n, \text{ where } n \text{ is an integer.}
$$

For both conditions to hold requires $\phi = (2n - m)\pi$, and, since m and n are both integers, ϕ has to be a multiple of π.

For $\phi = \pi$: with $m = n = 1$, $\sin\theta = \lambda/2d$; with $m = 3$ and $n = 2$, $\sin\theta = 3\lambda/2d$; etc. Each of the strong-signal values of θ corresponds to *four* physical directions, one in each quadrant of the compass. Consequently, if there are two physically possible values satisfying $0 < \theta_1 < \theta_2 < \pi/2$, there would, in general, be eight such directions – already too many, without even considering larger values of m and n. To get exactly six, two pairs of the eight directions must coincide; this is only possible if θ_2 is exactly equal to $\pi/2$, i.e.

$$\frac{3\lambda}{2d} = \sin\theta_2 = \sin\frac{\pi}{2} = 1.$$

This requires that the spacing $d = \frac{3}{2}\lambda$. Further, it implies that $\theta_1 = \sin^{-1}\frac{1}{3} = 19.5°$. The six strong-signal directions are as shown in the figure below.

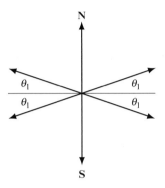

S 120   It is not immediately obvious how to fill in the missing values, but we write the wave in its most general form

$$y(x, t) = A\cos(kx + \omega t + \phi),$$

where k is its wavenumber, equal to $2\pi/\lambda$, ω its angular frequency, equal to $2\pi/T$, and ϕ an unknown phase that allows us to use a purely cosine function. So, for example,

$$9.059 = y(7, 4) = A\cos(7k + 4\omega + \phi). \tag{$*$}$$

By itself this does not take us very far, *but*, using the identity for the sum of two cosines (given at the beginning of this chapter), we can construct another expression involving the same RHS. Consider the sum of

$$y(6, 4) = A\cos(6k + 4\omega + \phi) = 11.340,$$
$$\text{and } y(8, 4) = A\cos(8k + 4\omega + \phi) = 6.687,$$
$$\text{i.e. } y(6, 4) + y(8, 4) = 2A\cos(7k + 4\omega + \phi) \times \cos k = 18.027.$$

Substituting from $(*)$ into the final equality gives

$$2 \times 9.059 \times \cos k = 18.027 \quad \Rightarrow \quad \cos k = 0.99498 \quad \Rightarrow \quad k = \pm 0.100\,\text{m}^{-1}.$$

A similar calculation using $y(7,3)$, $y(7,4)$ and $y(7,5)$ leads to

$$w = \cos^{-1}\frac{17.757}{2 \times 9.059} = \cos^{-1} 0.98008 = \pm 0.200\,\text{s}^{-1}.$$

With k and w both known, we can deduce that the speed of the wave, given by w/k, is 2.00 m s^{-1}.

We now turn to finding a, b, c and d. By applying the same techniques to the first and last column of the table, and to its first and last rows, we obtain four relationships between the unknowns.

$$a + c = 2 \times 11.340 \times \cos w = 2 \times 11.340 \times 0.98008 = 22.228,$$
$$b + d = 2 \times 6.687 \times \cos w = 2 \times 6.687 \times 0.98008 = 13.107,$$
$$a + b = 2 \times 13.508 \times \cos k = 2 \times 13.508 \times 0.99498 = 26.880,$$
$$c + d = 2 \times 4.249 \times \cos k = 2 \times 4.249 \times 0.99498 = 8.455.$$

Eliminating two of the unknowns produces a repeated equation for the difference between the other two. However, we note that, with our calculated values for k and w,

$$6k + 4w = 1.4 = 8k + 3w,$$

and therefore $y(6,4)$ *must* be equal to $y(8,3)$, i.e. to b. So, $b = 11.340$, and the others are then found to be $a = 15.540$, $d = 1.767$ and $c = 6.687$ – this final value could have been found in the same way as b was. The completed table is therefore

t/s	x/m	6	7	8
3		15.540	13.508	11.340
4		11.340	9.059	6.687
5		6.687	4.249	1.767

Where the same value of y appears as two different entries in the table, the one with the earlier time has a larger value of x, showing that the wave is travelling in the *negative* x-direction. That leaves just the amplitude of the wave!

We have several expressions of the form $A\cos z$, and to extract A we need one for $A\sin z$, with the same value of z. The second identity given at the start of this chapter holds the key. Rather than add two readings, we *subtract* them: consider

$$11.340 = y(8,3) = A\cos(1.4 + \phi),$$
$$9.059 = y(7,4) = A\cos(1.5 + \phi),$$
$$13.508 = y(7,3) = A\cos(1.3 + \phi).$$

Applying the identity to the final two results,

$$4.449 = 2A\sin(1.4 + \phi)\sin(0.1) \quad \Rightarrow \quad A\sin(1.4 + \phi) = 22.282.$$

Finally, combining this with the first of the three results gives

$$A^2\sin^2(1.4 + \phi) + A^2\cos^2(1.4 + \phi) = (22.282)^2 + (11.340)^2 \Rightarrow A = 25.0\,\text{cm}.$$

S 121   This question contains only one numerical value – a lot will need to be deduced from the qualitative description and the known properties of polaroid sheets.

In the figure below and in our calculations, we will denote by A_n and I_n the amplitude and intensity of the light *emerging* from the nth polaroid sheet, with the sheets numbered in the order they are introduced in the question, i.e. the final transmitted intensity is I_2 (and not I_3). It should be remembered that all three sheets are identical, except for their orientations; we take sheets 1 and 2 to be polarized vertically and horizontally, respectively.

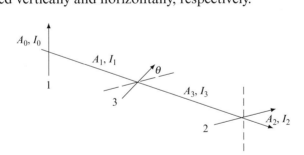

Since the initial light is unpolarized, it can be considered as being an uncorrelated superposition of two amplitudes, each of magnitude $A_0/\sqrt{2}$, transversely polarized in orthogonal directions. Consequently, we might expect I_1 to be one-half of I_0. However, it isn't; and so there must be some absorption in the polaroid sheet. As $I_n = |A_n|^2$ for all n, there must be an amplitude reduction introduced by each sheet, even for light of the 'correct' polarization. From what happens at sheet 1, this reduction factor is

$$\sqrt{\frac{32\%}{50\%}} = 0.8, \quad \text{implying that } A_1 = \frac{0.8}{\sqrt{2}}A_0.$$

As the third sheet initially transmits no light, and the light incident upon it is vertically polarized, its polarization planes must be parallel to those of sheet 2. The angle θ through which it is turned is as shown in the figure.

When light polarized in a particular direction is incident on a sheet whose polarization planes make an angle ϕ with that direction, the amplitude of the transmitted light is a factor $0.8 \cos \phi$ smaller than the incident amplitude. Thus

$$A_3 = 0.8 \, A_1 \cos\left(\tfrac{1}{2}\pi - \theta\right) = 0.8 \, A_1 \sin \theta, \quad \text{and} \quad A_2 = 0.8 \, A_3 \cos \theta.$$

Putting all of these results together, we have finally that

$$
\begin{aligned}
I_2 &= \left| 0.8 \cos \theta \, 0.8 \sin \theta \, \frac{0.8}{\sqrt{2}} A_0 \right|^2 \\
&= \left| 0.32 \sin 2\theta \, \frac{0.8}{\sqrt{2}} \right|^2 I_0 \\
&= (0.32)^3 \sin^2 2\theta \, I_0 = 3.3 \times 10^{-2} \sin^2 2\theta \, I_0
\end{aligned}
$$

gives the intensity of the transmitted light as a function of θ.

S 122   We need to determine the tension T in the ring when it is spinning, and to do so we consider the forces acting upon a part of it that subtends a small angle 2θ at its centre, as in the figure.

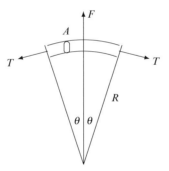

The components of T that act radially have to balance the centrifugal force F experienced by the ring segment, i.e.

$$2T \sin \theta = F = (\rho A \, 2\theta R) R \omega^2,$$

where A is the cross-sectional area of the ring. Since θ is (vanishingly) small, this becomes

$$T = \rho R^2 A \omega^2.$$

Since the material has a finite Young modulus, it will be stretched, and if the ring radius increases by ΔR, the corresponding strain will be

$s = (2\pi\Delta R)/(2\pi R)$. The stress that produces this strain is T/A, and so

$$\frac{\rho R^2 A\omega^2}{A} = \frac{T}{A} = Es = E\frac{\Delta R}{R},$$

giving ΔR as $\rho R^3 \omega^2 / E$.

S 123   The figure shows a small sector of the cross-section of the hose, one that subtends an angle $d\theta$ at the axis of the hose. The outward force on this piece of hose, due to an excess internal pressure p is $pr\,d\theta$ per unit length of the hose (which is measured perpendicular to the plane of the figure). This has to be balanced by the tension in the reinforcing nylon string.

Although the string is helically wound, for our *estimate* we will assume that each small piece of it is perpendicular to the axis of the hose; this approximation will be reviewed later.

If the tension in the string is T, then the net inward force on the strip shown in the figure is $2T\sin\frac{1}{2}\theta \approx T\,d\theta$ for each of the n turns per unit length of the hose. The maximum tension that can be withstood is given by

$$T = \pi(2.5 \times 10^{-4})^2 \times 1.0 \times 10^8 = 20\,\text{N},$$

and so we need

$$20\,n\,d\theta = pr\,d\theta = 4 \times 1.0 \times 10^5 \times 10^{-2}\,d\theta \quad \Rightarrow \quad n = 200\,\text{turns m}^{-1}.$$

Of these 200 turns, half will be clockwise and half anticlockwise, but the total length of nylon needed is approximately $2\pi r n = 12.6$ m.

With 100 turns per metre in each helix, successive turns are 1 cm apart. As this is for a string length of slightly more than 2π cm, the angle each piece of the string makes with the assumed direction perpendicular to the hose axis is about $\sin^{-1}(1/6)$; the cosine of this angle is close to unity, and so the simplifying approximation should be a reasonable one.

S 124   This is a question that hardly needs any physics or mathematical knowledge, except perhaps how to maximize a quadratic function.

The time for the first part of the cycle is $T_1 = V/F$, and for the second, $T_2 = V/(P - F)$. Thus the total time for the cycle is

$$T = T_1 + T_2 = \frac{V}{F} + \frac{V}{P - F} = \frac{PV}{F(P - F)}.$$

This is minimal when $F(P-F)$ is maximal, i.e. when $F=\frac{1}{2}P$ (either by inspection or using simple calculus). The minimal value of T is therefore $4V/P$, which answers the last part of the question.

When $F=\frac{1}{2}P$, $T_1 = 2V/P = T_2$, but

$$\text{if } F<\tfrac{1}{2}P \text{ then } T_1 > \tfrac{2V}{P}, \quad T_2 < \tfrac{2V}{P} \text{ i.e. } T_1 > T_2,$$
$$\text{if } F>\tfrac{1}{2}P \text{ then } T_1 < \tfrac{2V}{P}, \quad T_2 > \tfrac{2V}{P} \text{ i.e. } T_1 < T_2.$$

Thus by timing each of the two parts of the rinse cycle, it can be determined whether F is less than or greater than its optimal value, and an adjustment made in the correct sense (though it could, of course, overcompensate). Some guidance as to how much adjustment is needed can be obtained from the fact (which you may care to prove) that

$$F = \frac{P}{2}\left(\frac{2T_2}{T_1+T_2}\right),$$

i.e. that F should be changed by a factor $(T_1+T_2)/2T_2$ to get the optimal setting.

S 125 First of all, we note that the rainwater filling the normal channel does not come into the calculation – although both are increased, the pressure and gravitational load it generates are countered by the channel's walls and base, whether or not there is a flood. The pressure exerted on the barrier by any floodwater depends only upon how far below the free surface it is, and *not* upon what is below it.

Measuring all vertical heights y upwards from the level OA, and denoting the density of rainwater by ρ_0, the hydrostatic pressure p is given by $p(y)=\rho_0 g(b-y)$, and the total horizontal force tending to push the barrier sideways is

$$F = \int_0^b p(y)\,dy = \int_0^b \rho_0 g(b-y)\,dy = \rho_0 g(b^2 - \tfrac{1}{2}b^2) = \tfrac{1}{2}\rho_0 g b^2.$$

Since the mass of the barrier is $\frac{1}{2}ab\rho$, condition (a) in the Manual is met provided

$$\mu\,\frac{1}{2}ab\rho g > \frac{1}{2}b^2\rho_0 g, \quad \text{i.e.} \quad \frac{a}{b} > \frac{\rho_0}{\mu\rho}.$$

The (clockwise) couple produced by a depth dy of the water, which tends to tip the barrier over, is $p(y)y\,dy$ and the total couple is therefore

$$\int_0^b \rho_0 g(b-y)y\,dy = \rho_0 g\left(\tfrac{1}{2}b^3 - \tfrac{1}{3}b^3\right) = \tfrac{1}{6}\rho_0 g b^3.$$

The couple opposing such tilting is due to the weight of the barrier, and since the centre of mass of a triangle is one-third of its height (a in this case) above its base, this restoring couple has magnitude $\frac{1}{2}ab\rho g \times \frac{2}{3}a = \frac{1}{3}a^2 b\rho g$. Thus condition (b) set by Health & Safety is met if

$$\frac{1}{3}a^2 b\rho g > \frac{1}{6}\rho_0 g b^3 \quad \text{i.e.} \quad \frac{a^2}{b^2} > \frac{\rho_0}{2\rho}.$$

The density of concrete is given as 2,400 kg m^{-3}, and that of (rain)water is 1,000 kg m^{-3}; further, since $\mu = 0.7$, the two conditions become

$$\frac{a}{b} > \frac{\rho_0}{\mu\rho} = 0.595,$$

$$\frac{a^2}{b^2} > \frac{\rho_0}{2\rho} = 0.2083 \quad \Rightarrow \quad \frac{a}{b} > 0.456.$$

The first condition is the more stringent, and requires a to be at least 60% of the barrier height. As might have been anticipated, the values of c and d are irrelevant.

S 126

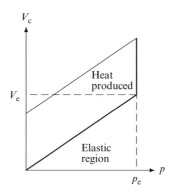

The buffer material behaves elastically until the yield point is reached, and so a plot of the amount of *compressed* material V_c as a function of the applied pressure is initially a straight line joining the origin to the point (p_e, V_e), where $p_e = 10^7$ N m^{-2} and V_e is 8% of the original volume of the two blocks, taken together. These points are shown on the above figure. The numerical value of V_e is

$$V_e = 0.08 \times 2 \times 20 \times 10^{-3} \times 0.2 = 6.4 \times 10^{-4} \, \text{m}^3.$$

Now the maximum elastic energy the two buffers can store (and return) is given by the area of the marked triangle (more formally, the $\int p\,dV$) which is $\frac{1}{2} \times 10^7 \times 6.4 \times 10^{-4} = 3.2 \times 10^3$ J. It follows that the maximum impact velocity v_i of the mass that will result in an elastic collision is v_e, where

$$\frac{1}{2}16v_e^2 = 3.2 \times 10^3 \quad \Rightarrow \quad v_e = 20 \, \text{m s}^{-1}.$$

Beyond this elastic limit, the work done as the plane of action of the critical pressure p_e moves through the material appears as heat. Both the work and the heat are represented in the figure by an area above the thick line.

So, in summary, $v_r = v_i$ for $v_i \leq v_e$, but $v_r = v_e = 20$ m s^{-1} for $v_i > v_e$.

S 127 We first note that the cross-sectional area of the rod is $A = \pi r^2$ and that of the cylinder is $\pi[(2r)^2 - (\sqrt{2}r)^2] = 2\pi r^2 = 2A$.

Before the heating takes place, the tension in the rod is $F = \frac{1}{2}\sigma_y \times A$. For the washer to be in equilibrium, this must be balanced by the reaction from the compressed cylinder which therefore has a stress of $F/2A = \frac{1}{4}\sigma_y$ in it; it's strain is $\sigma_y/4E$, though this is irrelevant, as it is compensated for by the positions of the nuts when they are tightened.

When the cylinder has been heated by T, there is an *additional* linear strain in it of αT, resulting in a net reaction force at the washer of $2AE(\alpha T + \sigma_y/4E)$. If σ is the resulting stress in the rod (transmitted by the washer), then

$$A\sigma = 2AE\left(\alpha T + \frac{\sigma_y}{4E}\right).$$

For σ to equal σ_y requires

$$\alpha T = \frac{\sigma_y}{4E} \quad \Rightarrow \quad \alpha = \frac{\sigma_y}{4ET}.$$

This result could have been obtained more directly by equating the change of force on the washer to the increased force, and hence stress, in the rod, but it is instructive to consider all the forces, stresses and strains involved.

Though not of any special significance, we note that the thermal expansion of the cylinder, were it free to expand, would exactly compensate for the compression involved when the device was assembled.

S 128 Since the density of the sphere is $3\rho_0$, we must expect that it will sink to at least the depth where the liquid density has that value, i.e. to at least $2D$. Further, since there is some viscosity, this must be the depth at which the sphere finally comes to rest.

Though hardly necessary, the figure shows some relevant facts.

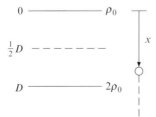

We measure the distance *downwards* from the surface by x, so that the liquid density is described everywhere by $\rho(x) = \rho_0(1 + x/D)$. At a depth x, the net upward force (upthrust minus weight) provides an acceleration of the sphere in the negative x-direction. If V is its volume,

$$3V\rho_0\ddot{x} = -V\rho_0\left(1 + \frac{x}{D}\right)g + 3V\rho_0 g,$$

$$\ddot{x} = -\frac{x}{3D}g + \frac{2}{3}g.$$

This is the equation for simple harmonic motion, but not about the origin. Guided both by this equation, and our previous observation about the sphere's final resting place, we change to a new origin and write $x = y + 2D$, when the equation becomes

$$\ddot{y} = -\frac{y}{3D}g - \frac{2}{3}g + \frac{2}{3}g = -\frac{g}{3D}y.$$

This *is* SHM, about $x = 2D$ with a period of $2\pi\sqrt{3D/g}$.

The amplitude of the oscillation is determined by the initial conditions; as the sphere starts from rest at $x = \frac{1}{2}D$, that represents a turning point of the motion, which therefore has amplitude $2D - \frac{1}{2}D = \frac{3}{2}D$. It follows that the deepest the sphere sinks, on completing half of its first oscillation, is $2D + \frac{3}{2}D = \frac{7}{2}D$.

Because of the small, but non-zero, viscosity, subsequent oscillations will have smaller amplitudes; eventually the damped oscillations will die away to nothing and, as noted earlier, the sphere will settle at a depth of $2D$.

S 129   It could be confusing to work out the part the boat plays in each scenario, so we establish a formula for the change in water level that applies to all three cases. We do this by considering all the objects that pass through the free surface of the water.

Let us denote by U the volume of all objects in the pond beneath the water surface; the water's contribution to this will remain unaltered throughout. So far as changes are concerned, we denote by ΔV_B the volume of that part of the boat that rises through the water surface when something is thrown overboard, and by ΔV_O the volume of water displaced by a jettisoned object when it is in the pond.

By Archimedes, ΔV_B must be equal to the volume of water that weighs the same as the object thrown overboard. For each of the scenarios, the rise in water level h is given by

$$h = \frac{\Delta U}{A} = \frac{1}{A}(-\Delta V_B + \Delta V_O).$$

(*a*) As noted above ΔV_B here is equal to M_S/ρ_W, whilst ΔV_O is simply M_S/ρ_S. Since $\rho_S > \rho_W$, h is negative, and the pond level *falls* by

$$\frac{M_S}{A}\left(\frac{1}{\rho_W} - \frac{1}{\rho_S}\right).$$

(*b*) This time, ΔV_B is equal to M_T/ρ_W, but, because the timber floats, and not all of it is underwater, the form of ΔV_O is different from that in (*a*); it is, in fact, also equal to M_T/ρ_W. The consequence is that $h=0$, and there is *no change* in the pond level.

(*c*) We first establish the implication of the fact that the joined stone and timber sink to the bottom of the pond. This implies that their average density is greater than that of water, i.e.

$$\frac{M_S + M_T}{M_S/\rho_S + M_T/\rho_T} > \rho_W,$$

or, more conveniently,

$$M_S/\rho_W + M_T/\rho_W > M_S/\rho_S + M_T/\rho_T. \qquad (*)$$

Clearly, $\Delta V_B = (M_S + M_T)/\rho_W$ and the extra underwater volume ΔV_O is just the sum of the individual volumes. Hence

$$\Delta U = -\frac{M_S + M_T}{\rho_W} + \frac{M_S}{\rho_S} + \frac{M_T}{\rho_T}.$$

However, we have from $(*)$ that the RHS of this is negative, and shows that the pond level *falls* by

$$\frac{M_S}{A}\left(\frac{1}{\rho_W} - \frac{1}{\rho_S}\right) - \frac{M_T}{A}\left(\frac{1}{\rho_T} - \frac{1}{\rho_W}\right).$$

S 130 At first sight there may not appear to be enough data provided for the question to be answered. However, let the rod have length L, cross-sectional area A, density ρ, mass $M = AL\rho$, and Young modulus E.

From the rod's extension under a load, we have

$$\text{stress} = \text{Young modulus} \times \text{strain},$$

$$\frac{10^3\, Mg}{A} = E\frac{0.4 \times 10^{-3}}{L},$$

$$\frac{10^3\, Mg\, \rho L}{M} = \frac{0.4 \times 10^{-3} E}{L},$$

$$\frac{E}{\rho} = 2.5 \times 10^6\, L^2 g\, \mathrm{m}^2\, \mathrm{s}^{-2}.$$

When the rod is clamped horizontally at its mid-point, that point cannot move and must be a node for standing longitudinal waves. The free ends, on the other hand, can be antinodes of longitudinal displacement. The lowest-frequency mode of vibration is the one with the smallest number of nodes, here just one, and a wavelength λ such that the rod's length is equal to the distance between successive antinodes, i.e. $L = \lambda/2$.

The lowest possible frequency for standing waves is therefore

$$f = \frac{v}{\lambda} = \frac{1}{2L}\sqrt{\frac{E}{\rho}} = \frac{1}{2L}\sqrt{2.5 \times 10^6 \, L^2 g} = 2.48 \,\text{kHz}.$$

It is perhaps a little surprising that we do not need explicit values for M, L, ρ, and E. Of course, combinations of some of them are hidden/wrapped up in the 0.4 mm static extension. Further, in some sense, longitudinal waves are simply dynamic versions of such extensions, and can therefore be expected to be governed by the same physical properties of the rod and the material from which it is made.

S 131   Before we can calculate either of the elasticity properties, we need to establish the equilibrium separation x_0 of the atoms.

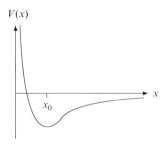

The figure gives a rough indication of the variation of $V(x)$ with x, and x_0 is the separation at which the potential is a minimum, or, more physically, the force F acting on an atom (given by $-dV/dx$) is zero:

$$0 = F = -\frac{dV}{dx} = \frac{6A}{x^7} - \frac{12B}{x^{13}} \quad \Rightarrow \quad x^6 = \frac{12B}{6A}, \quad \text{i.e. } x_0 = \left(\frac{2B}{A}\right)^{1/6}.$$

To calculate the modulus of elasticity we need to find the (restoring) force acting when an atom is displaced from its equilibrium position by, say, y. The strain will then be y/x_0 and the force will be

$$F(x_0 + y) = -\left.\frac{dV}{dx}\right|_{x_0 + y} = -\frac{dV}{dx} - y\frac{d^2V}{dx^2} + \cdots,$$

where the derivatives in the Taylor expansion are evaluated at $x = x_0$.

The first derivative of V is, of course, zero when $x = x_0$. The second derivative, evaluated at $x = x_0$, is

$$\frac{d^2 V}{dx^2} = -\frac{7 \times 6A}{x^8} + \frac{13 \times 12B}{x^{14}} \tag{*}$$

$$= \frac{1}{x^8}\left(-42A + \frac{156B}{(12B/6A)}\right)$$

$$= \left(\frac{A}{2B}\right)^{8/6}(-42A + 78A) = \frac{18A}{2^{1/3}}\left(\frac{A}{B}\right)^{8/6}.$$

The modulus of elasticity (the analogue of the Young modulus for a one-dimensional structure) is E, given by

$$E = \frac{F}{\text{strain}} = \frac{F}{(y/x_0)} = \left(\frac{2B}{A}\right)^{1/6}\frac{18A}{2^{1/3}}\left(\frac{A}{B}\right)^{8/6} = 18\left(\frac{A^{13}}{2B^7}\right)^{1/6}.$$

The atomic chain will break when the applied force becomes larger than the force needed to sustain the current extension, i.e. when the force is at its maximum. This will occur at the value of x at which $dF/dx = d^2V/dx^2 = 0$, which, from (*), means that

$$-\frac{42A}{x^8} + \frac{156B}{x^{14}} = 0.$$

This can be arranged to read

$$x = \left(\frac{13}{7}\right)^{1/6}x_0.$$

The corresponding breaking *strain* is

$$s = \frac{1}{x_0}\left[\left(\frac{13}{7}\right)^{1/6}x_0 - x_0\right] = \left(\frac{13}{7}\right)^{1/6} - 1 = 0.11.$$

Note This result depends only on the inverse powers of x involved, and not on A or B; this is not true of the breaking *force*, given by $6Ax^{-7} - 12Bx^{-13}$ when $x^6 = 26B/7A$, and has a value of $0.60(A^{13}/B^7)^{1/6}$.

S 132   For both situations the determining factor is the stress at the top of the cable, as indicated in the figure.

(a) The left-hand figure shows this situation. Let the cross-section of the cable be A; then the mass of the cable is $\ell A \rho$ and the total load to be supported is $Mg + \ell A \rho g$. This must not produce a stress greater than σ_0. We therefore have

$$Mg + \ell A \rho g = W,$$

$$\frac{Mg + \ell A \rho g}{A} = \frac{W}{A} = \sigma_0,$$

$$\text{i.e. } A = \frac{Mg}{\sigma_0 - \ell \rho g}.$$

Re-substituting this into the first equation gives

$$Mg + \ell \rho g \frac{Mg}{\sigma_0 - \ell \rho g} = W \quad \Rightarrow \quad M = \frac{\sigma_0 - \ell \rho g}{\sigma_0} \frac{W}{g}$$

as the maximum load that can be carried on the fully extended cable.

(b) The right-hand figure shows the material formed into a cable of length ℓ and cross-sectional area A, where $V = \ell A$. If we measure the distance up the cable from its free end by x, then, in order to support the material below it, the tension there must be $T(x) = \rho g A x$, and the stress at the same point is $T/A = \rho g x$. The local strain is therefore $\rho g x / E$.

The elastic energy stored in a small length dx of the cable is then given by

$$dU = \frac{1}{2} \times \text{stress} \times \text{strain} \times \text{volume element} = \frac{1}{2} \rho g x \frac{\rho g x}{E} A \, dx,$$

and the total stored energy in the cable is

$$U = \int_0^\ell \frac{1}{2} \rho g x \frac{\rho g x}{E} A \, dx = \frac{1}{2} \frac{\rho^2 g^2}{E} \frac{A \ell^3}{3}.$$

But, as the stress at the top of the cable must not exceed σ_0, A is determined by $V \rho g / A = \sigma_0$, which is equivalent to $\ell \rho g = \sigma_0$. And so the maximum stored elastic energy is

$$U = \frac{\sigma_0^2}{6E} A \ell = \frac{\sigma_0^2 V}{6E},$$

and this is attained if the cable length is made equal to $\sigma_0 / \rho g$.

S 133 The figure shows the frame with the soap film in place, but because of the different scalings necessarily used for the 'horizontal' and 'vertical' dimensions (5 : 1), the rubber threads do not appear to take up their actual shapes, which are *circular* arcs of (as yet) undetermined radius r. The corresponding endpoints of the two rods subtend an (also currently unknown) angle 2ϕ at the centres of the corresponding circles.

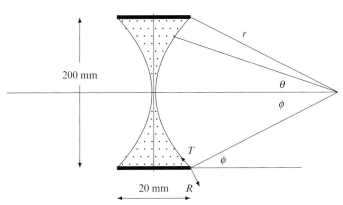

Although gravity does not come into our analysis, for convenience we will refer to 'horizontal' and 'vertical' directions as they appear in the figure. Because of the symmetry of the configuration about the 'horizontal' axis, all 'vertical' forces cancel in pairs. Further, because small differences are involved, we will retain more significant figures than usual in numerical work.

At each point of a thread, the surface tension acting upon it does so at right angles to the local direction of the thread; because the magnitude of this radial force is the same at all points, the thread is formed into an arc of a circle. So, for a length $r\,d\theta$ situated at a typical angle θ to the 'horizontal' axis, a force with horizontal component $\gamma r \cos\theta\, d\theta$ acts towards the 'vertical' axis. There are two sides to the soap film and so the total force in this direction is

$$F_{ST} = 2 \int_{-\phi}^{\phi} \gamma r \cos\theta\, d\theta = 4\gamma r \sin\phi.$$

This force is balanced by the reactions R of the frame to the tension T in the thread; these act in the directions of the tangents to the thread at its endpoints, i.e. they make an angle of $\frac{1}{2}\pi - \phi$ with the 'horizontal'.[21] Since T and R are equal and opposite forces, the net contribution of the two reactions is $2T\cos(\frac{1}{2}\pi - \phi) = 2T\sin\phi$.

As the forces due to surface tension and frame reaction must balance, we have

$$4\gamma r \sin\phi = 2T\sin\phi, \qquad \text{i.e.} \quad 2\gamma r = T = 15(2\phi r - 0.2)\,\text{N},$$

since each thread has been stretched from an initial length of 0.2 m to an equilibrium length of $2\phi r$.

---

[21] The figure shows this direction, and it also appears to be close to the tangent to the curve as drawn. However, because of the different scalings in the two directions, there is no reason to expect this, and it is down to chance.

We must now use the configuration's geometry to determine r and ϕ:

$$r \sin \phi = 0.1 \quad \text{and} \quad r(1 - \cos \phi) = 0.01.$$

From standard geometric identities, it follows that

$$0.1 = \frac{0.01}{0.1} = \frac{1 - \cos \phi}{\sin \phi} = \frac{2 \sin^2 \phi/2}{2 \sin \phi/2 \, \cos \phi/2} = \tan\left(\frac{\phi}{2}\right), \text{ i.e. } \phi = 0.1993 \text{ rad.}$$

Further, $r \sin \phi = 0.1$ now gives $r = 0.5050 \,\text{m}$, that the thread extension is 1.33×10^{-3} m, that $T = 0.01996$ N, and, finally, that $\gamma = 0.01976 \approx 20 \,\text{mN m}^{-1}$.

S 134   Let the overall thickness of the composite slab be L, with the individual slab thicknesses of $L/2$ or $L/3$, as appropriate, and take the two outer surface temperatures as T and 0. The simple figure below shows the two cases.

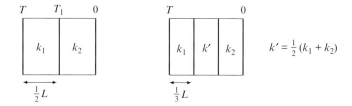

If the interface temperature in the two-component structure is T_1, as in the figure, then the temperature gradients in the two halves are $(T - T_1) \div (L/2)$ and $(T_1 - 0) \div (L/2)$. Since heat does not accumulate at the interface, the two halves must carry the same flux F:

$$\frac{k_1(T - T_1)}{L/2} = F = \frac{k_2(T_1 - 0)}{L/2} \quad \Rightarrow \quad T_1 = \frac{k_1 \, T}{k_1 + k_2}.$$

Putting this value for T_1 into the second equality, shows that

$$F = \frac{2k_2}{L} \frac{k_1 \, T}{k_1 + k_2} \quad \text{and so} \quad H_2 = \frac{2 \, k_1 \, k_2}{L(k_1 + k_2)}.$$

Writing this in terms of H_2^{-1} gives

$$\frac{1}{H_2} = \frac{L(k_1 + k_2)}{2 \, k_1 \, k_2} = \frac{L}{2}\left(\frac{1}{k_1} + \frac{1}{k_2}\right). \tag{$*$}$$

Equation $(*)$ is very similar to that for resistors in parallel, or for capacitors in series. For the former analogy, a 'resistor' has the value $R_i \sim k_i/\ell$, where $\ell = L/2$ is the thickness of each layer; for the latter, a 'capacitor' has the value $C_i \sim k_i/\ell$. In both cases, H_2 is the value of the equivalent single resistor or capacitor.

For the three-part composite, we extend the formula for H in the same way as we would if we were adding a further parallel resistor or series capacitor. Either way, the heat transfer coefficient H_3 is given by

$$\frac{1}{H_3} = \frac{L}{3}\left(\frac{1}{k_1} + \frac{1}{k_2} + \frac{2}{k_1 + k_2}\right). \qquad (**)$$

Given the internal symmetry of the right-hand sides of $(*)$ and $(**)$, it is likely that substitution of a single set of arbitrary numbers for k_1 and $k_2 \neq k_1$ would decide which of H_2 and H_3 is the smaller.[22]
For a more general proof, set

$$\frac{1}{LH_3} - \frac{1}{LH_2} = \frac{1}{3}\left(\frac{1}{k_1} + \frac{1}{k_2} + \frac{2}{k_1 + k_2}\right) - \frac{1}{2}\left(\frac{1}{k_1} + \frac{1}{k_2}\right)$$

$$= \frac{2}{3}\frac{1}{k_1 + k_2} - \frac{1}{6}\left(\frac{1}{k_1} + \frac{1}{k_2}\right)$$

$$= \frac{4k_1k_2 - k_2(k_1 + k_2) - k_1(k_1 + k_2)}{6k_1k_2(k_1 + k_2)}$$

$$= -\frac{(k_1 - k_2)^2}{6k_1k_2(k_1 + k_2)} < 0.$$

Thus $H_3^{-1} < H_2^{-1}$, implying $H_2 < H_3$ and that the two-part composite is a better heat insulator than the three-part structure.

S 135 At first sight it might seem surprising that the lock gates can open by themselves. After the passage of the boat into the canal, the lock will be filled with fresh water to a height f, and, if the lock has been refilled by opening the harbour gates, it contains a mixture of fresh and salt water to a height s. Neither of these lock conditions could provide the force necessary to open the harbour gates, which have a depth s of (denser) salt water behind them.

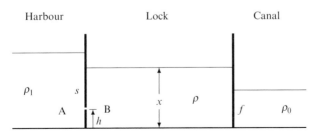

---

[22] For example, if, arbitrarily, we set k_1 and k_2 equal to 2 and 4 units (either way round) then $H_2^{-1} = 0.375L$ and $H_3^{-1} = 0.361L$, showing that $H_2 < H_3$.

These considerations make it clear that the height x of the water in the lock, must, at some point in the filling process, become greater than s. And there is no reason why it should not be so, provided salt water continues to flow through the by-pass pipe – and all that this requires is that the pressure p_A at A is still greater than that, p_B, at B (see the figure).

The gates will open when the force on them from the lock side exceeds that from the harbour side. To evaluate such forces, we will need the result that the force per unit length on a vertical wall of a tank of liquid of density ρ and depth b is $\frac{1}{2}\rho g b^2$.[23]

We first need the density of the water in the lock. Given that it starts with a depth f of fresh water, when the total depth has risen to x the lock water has a density

$$\rho = \frac{f\rho_0 + (x-f)\rho_1}{x}.$$

It follows that the hydrostatic pressures on the two sides of the valve are then given by

$$p_A = \rho_1 g(s-h) \quad \text{and} \quad p_B(x) = \rho g(x-h) = \left[\frac{f\rho_0 + (x-f)\rho_1}{x}\right]g(x-h).$$

As noted, the harbour gates will open when the force on the lock side exceeds that on the harbour side, that is when x is marginally greater than X, where

$$\frac{1}{2}\left[\frac{f\rho_0 + (X-f)\rho_1}{X}\right]gX^2 = \frac{1}{2}\rho_1 g s^2.$$

For x to reach X (exceeding s whilst it does so, but subsequently dropping back to that value) it is necessary that $p_B(X) < p_A$ (so that, whilst $s < x \le X$, the by-pass valve continues to add salt water to the lock, even though the lock water surface is above that in the harbour).

Denoting the dimensionless ratio $(\rho_1 - \rho_0)/\rho_1$ by 2λ to simplify the notation, the force balance and the pressure inequality can be written as

$$(X - 2\lambda f)X = s^2 \quad \text{and} \quad (X - 2\lambda f)(X-h) < X(s-h).$$

We can see directly from the first of these that X *must* be greater than s. The equation itself has the explicit solution

$$X = \lambda f + \sqrt{\lambda^2 f^2 + s^2}.$$

---

[23] If you are not familiar with this formula, see solution S 125 appearing on page 243.

Expanding the inequality, and then simplifying it using both the equality and its solution gives

$$X^2 - 2\lambda fX - hX + 2\lambda fh < sX - hX,$$

$$s^2 + 2\lambda fh < s\left(\lambda f + \sqrt{\lambda^2 f^2 + s^2}\right),$$

$$s^2 - \lambda f(s - 2h) < s\sqrt{\lambda^2 f^2 + s^2}. \tag{*}$$

The LHS of (*) is $< s^2$ provided $s - 2h > 0$, and the RHS is clearly $> s^2$. So, we conclude that if the valve is sufficiently low down that $h < \frac{1}{2}s$, the harbour gates will open by themselves.

In fact, equation (*) is less stringent than this and, depending on the values of s, f and λ, slightly greater values of h could result in the gates opening. Since $\lambda \ll 1$, and both f and h are less than s, a binomial expansion of the RHS of (*) shows that h can exceed $\frac{1}{2}s$ by as much as $\frac{1}{4}\lambda f$. Try it!

S 136   Firstly, we note that two-thirds of the area of each end face of rod 3 is not in contact with anything, and can play no part in transmitting forces from one rod to another; the system behaves as if the cross-sectional area were A throughout.

Secondly, the stresses (and hence the forces that are transmitted to neighbouring rods) must be equal in all of the rods, otherwise the junctions would have net forces acting on them and could not be in equilibrium. In other words, the stress must be the same throughout, and the answers to (a) and (b) will be the same.

Thirdly, because of the symmetry of the set-up, if junction J moves by x to the right, junction K moves x to the left, i.e. $y = -x$.

Considering junction J, rod 1 will have extended by x, thus reducing the compressional reaction it produces there, but there will be extra compression of $L\alpha_1 T$ as a result of the reaction from rod 3 to the additional force on it caused by the heating and expansion of rod 1. Thus, using 'stress equals E times strain',

$$\sigma_J = E_1 \left(-\frac{x}{L} + \alpha_1 T\right).$$

The stress produced at J by rod 3, which has been compressed by $2x$, as well as by having to compensate for the temperature rise, is given by

$$\sigma_J = E_3 \left(\frac{2x}{L} + \alpha_3 T\right).$$

Equating these two values:

$$E_3\left(\frac{2x}{L} + \alpha_3 T\right) = E_1\left(-\frac{x}{L} + \alpha_1 T\right),$$

$$x(2E_3 + E_1) = LT(E_1\alpha_1 - E_3\alpha_3),$$

$$x = \frac{LT(E_1\alpha_1 - E_3\alpha_3)}{2E_3 + E_1}.$$

The stress, common to all three rods, is

$$\sigma = \sigma_J = E_1\alpha_1 T - \frac{E_1 T(E_1\alpha_1 - E_3\alpha_3)}{2E_3 + E_1}$$

$$= \frac{E_1 E_3(2\alpha_1 + \alpha_3)}{2E_3 + E_1} T.$$

Although x can be positive or negative, depending on the sign of $(E_1\alpha_1 - E_3\alpha_3)$, the stress σ is always positive.

S 137   As shown in the figure, let x be the (perpendicular) distance from the base of the channel, so that the free surface of the oil is $x = D$.

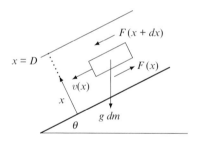

Before we can compute the overall flow rate, we must first establish the velocity profile $v(x)$. To do this, consider the forces acting on a 'slab' of oil of (infinitesimal) thickness dx, width W and unit length in the direction of flow.

If $F(x)$ is the 'drag force' acting on it due to the oil between $x = 0$ and $x = x$, then the difference in F between the lower and upper surfaces of the slab must be balanced by the relevant component of the weight of the slab; if it were not, the oil would either accelerate or decelerate. Thus

$$F(x) = F(x + dx) + (W \times 1 \times dx)\rho g \sin\theta.$$

Since dx is infinitesimal, we can replace $F(x + dx) - F(x)$ by $(dF/dx) \times dx$, obtaining

$$\frac{dF}{dx} + W\rho g \sin\theta = 0.$$

But, from the definition of η,

$$F = \eta(1 \times W)\frac{dv}{dx} \quad \Rightarrow \quad \frac{dF}{dx} = \eta W \frac{d^2v}{dx^2},$$

to give

$$\eta\frac{d^2v}{dx^2} + \rho g \sin\theta = 0.$$

To obtain $v(x)$ this has to be integrated twice. The first integration gives

$$\eta\frac{dv}{dx} + \rho g \sin\theta\, x = c_1.$$

However, since the air provides no viscous drag, dv/dx must be zero at $x = D$, thus requiring $c_1 = \rho g \sin\theta\, D$. We then have

$$\eta\frac{dv}{dx} = \rho g \sin\theta (D - x), \quad \text{and hence } \eta v(x) = \rho g \sin\theta (Dx - \tfrac{1}{2}x^2) + c_2.$$

This time, the base of the channel provides the boundary condition; $v(0) = 0$ implies that $c_2 = 0$, and we have finally that the velocity profile is

$$v(x) = \frac{\rho g \sin\theta}{\eta}\left(Dx - \frac{1}{2}x^2\right).$$

In order to obtain the total volume flow rate, we still need to integrate this over a cross-section of the channel:

$$V = \int_0^D W\,dx\, v(x)$$

$$= \frac{W\rho g \sin\theta}{\eta}\int_0^D \left(Dx - \frac{1}{2}x^2\right)dx$$

$$= \frac{W\rho g \sin\theta}{\eta}\left(\frac{1}{2}D^3 - \frac{1}{6}D^3\right)$$

$$= \frac{W\rho g \sin\theta\, D^3}{3\eta}.$$

S 138 As well as the notation indicated by the figure below, let the cross-sectional area of the blade be A, and its density be ρ. The limiting criterion for the engine speed is that the maximum stress, which clearly occurs at the axis of rotation, should not exceed the given value.

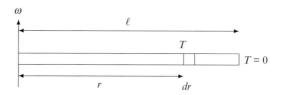

To determine the variation of the stress T with radial distance r, consider a small element of length dr, at the two ends of which the stresses are $T = T(r)$ and $T + dT = T(r + dr)$; dT will be negative, since T is maximal at $r = 0$, and zero at the blade's tip $r = \ell$. Balancing the tensions and the centripetal force for the small length:

$$T - (T + dT) = \rho A \, dr \times \omega^2 r \quad \Rightarrow \quad \frac{dT}{dr} = -\rho A \omega^2 r.$$

This can be integrated to give

$$T(0) - T(r) = \tfrac{1}{2} \rho A \omega^2 r^2,$$

and, in particular, at $r = \ell$ where $T(\ell) = 0$,

$$\frac{T(0)}{A} = \frac{\rho \omega^2 \ell^2}{2}.$$

The LHS of this equation must not exceed 400 MPa, and sets the maximum for ω at

$$\omega = \sqrt{\frac{400 \times 10^6 \times 2}{4510 \times (0.9)^2}} = 468.0 \,\text{rad s}^{-1} = 4469 \,\text{r.p.m.}$$

To find the total blade extension due to centrifugal forces, let the length dr be stretched to $dr + dx$. Then, we have

$$\frac{dx}{dr} = \text{strain} = \frac{\text{stress}}{E} = \frac{T(r)}{EA}.$$

This can now be integrated over the range $0 \leq r \leq \ell$:

$$\int_0^\ell dx = \frac{1}{EA} \int_0^\ell \left[T(0) - \frac{\rho A \omega^2 r^2}{2} \right] dr$$

$$= \frac{\rho \omega^2}{2E} \int_0^\ell (\ell^2 - r^2) \, dr,$$

$$x(\ell) = \frac{\rho \omega^2}{2E} \left(\ell^3 - \frac{\ell^3}{3} \right)$$

$$= \frac{\rho \omega^2 \ell^3}{3E} = \frac{2T(0)}{A} \frac{\ell}{3E}$$

$$= \frac{2 \times 0.9 \times 400 \,\text{MPa}}{3 \times 117 \,\text{GPa}} = 2.1 \,\text{mm}.$$

S 139 Let us denote the resistances of a 100 W bulb, A's 200 W bulb, and B's 50 W bulb, by R_0, R_A and R_B respectively. If the r.m.s. supply voltage is V, then,

$$R_0 = \frac{V^2}{100}, \qquad R_A = \frac{V^2}{200}, \qquad R_B = \frac{V^2}{50},$$

and the total resistance in the modified circuit is

$$R = V^2 \left(\frac{1}{200} + \frac{1}{50} \right) = \frac{V^2}{40}.$$

The current I that flows in it is $V \div (V^2/40) = 40/V$ amps, and the power taken from the mains is $P = VI = 40$ W.

In A's room the ceiling light receives

$$P_A = R_A I^2 = \frac{V^2}{200} \left(\frac{40}{V} \right)^2 = 8 \text{ W},$$

whilst B's bulb operates at

$$P_B = R_B I^2 = \frac{V^2}{50} \left(\frac{40}{V} \right)^2 = 32 \text{ W},$$

They each pay for $\frac{1}{2} \times 40 = 20$ W. B is clearly the brighter student – in more ways than one!

A bonus for B was that he or she was expecting A to install a 100 W bulb, and that the total resistance would be

$$R = V^2 \left(\frac{1}{100} + \frac{1}{50} \right) = \frac{3 V^2}{100},$$

leading to a total power from the mains of 33.3 W, of which two-thirds would be used in B's room. The cost would be that of 16.7 W. So, in practice, B got 32 W for the price of 20 W, whilst only expecting to get 22.2 W for the price of 16.7 W, a gain of 20%.

It's not clear what each expected to save by their initial agreement – A may even have thought of a 50% cost for 200 W, rather than the theoretical 25% cost for 25 W each – but both of them were rather in the dark about what would happen, with A in a near blackout.

S 140 The figure shows the circuit when only a single resistor is connected to the battery terminals.

The current drawn from the battery is straightforwardly $I = V/(r + R)$, and so the power delivered to R is

$$P_R = R \frac{V^2}{(r + R)^2},$$

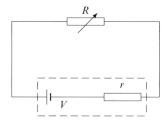

and this is maximized when $dP_R/dR = 0$, i.e. when

$$(r + R)^2 \times 1 - R \times 2(r + R) = 0, \text{ i.e. when } R = r.$$

This standard result – that the external delivered power is maximized when the external resistance is matched to the internal one – can now be applied to more complicated situations.

Resistance S, being connected across the battery is in parallel with R. To maximize the external power dissipation, we should aim to make the effective resistance of the combined S and R configuration equal to r, in accord with the above result, i.e. make

$$\frac{RS}{R + S} = r \quad \Rightarrow \quad R = \frac{Sr}{S - r}. \tag{*}$$

This is possible provided $S > r$; the current through the circuit will then be $V/(r + r)$. Externally, the current divides between the parallel resistances in inverse ratio to their values. Thus the current through S is $R/(R + S) \times (V/2r)$ and the power dissipated in it is

$$P_S = S \frac{R^2}{(R + S)^2} \frac{V^2}{4r^2} = \frac{S R^2 V^2}{4 R^2 S^2} = \frac{V^2}{4S}.$$

That dissipated in R is similarly found to be $V^2/4R$.

If $S < r$, a maximizing setting for R is not possible, as it would have to be negative [see (*)]. For a general setting of R, it can be shown that

$$P_S = \frac{RS(R + S) V^2}{(RS + rR + rS)^2},$$

which increases monotonically with R, from 0 at $R = 0$ to $S V^2/(S + r)^2$ as $R \to \infty$, i.e. is effectively not there.

The practical answer is to cut one of the leads to S and insert R, set to $r - S$, in series with it. Then the load again matches the battery resistance, and the delivered power is as good as it can be!

S 141 The figure shows the situation schematically, but before we can find the power dissipated in the 50 Ω resistor, we need to find the fault resistance r, its location, and the resistance ρ per km of each core. For the two exploratory measurements and the final power calculation, care will be needed to make sure that the resistances included in the three circuits are the correct ones.

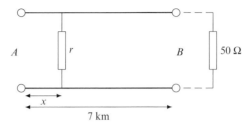

If the fault is located x km from A, then we have from the resistance measurements that

$$2x\rho + r = 64,$$
$$2(7 - x)\rho + r = 70,$$

and, from the potential divider reading, that

$$\frac{r}{r + 2x\rho} = \frac{15}{16}.$$

The three equations have the straightforward solution: $r = 60\,\Omega$, $\rho = 1\,\Omega$ per km, and $x = 2$ km.

We now need the equivalent resistance of the whole circuit when the 50 Ω resistor is connected at B. The resistor and cable cores beyond the fault are equivalent to a resistor of $50 + 2(5 \times 1) = 60\,\Omega$. This is in parallel with the 60 Ω resistance due to the fault, making the equivalent of a 30 Ω resistor, which, in series with the 2 ohms of each core connected to A, produces a total equivalent resistance of 34 Ω.

When a voltage V is connected across A, the current that is drawn is $V/34$ amps, and the power supplied is $P = V^2/34$ W. Since the two cores are effectively connected by two 60 Ω resistors in parallel, half of the current goes through each. In particular, that through the resistor at B is $I_B = V/68$, and the power dissipated in the load is $50\,I_B^2$. Thus

$$\frac{\text{power in load}}{\text{power supplied}} = \frac{50\,V^2}{(68)^2} \frac{34}{V^2} = 0.37 = 37\%.$$

S 142 If the potential of the balloon remains constant, then so must the current flowing through the resistor, i.e. the balloon's capacitance must be made to change in such a way that charge is lost at a constant rate.

Let the charge on the balloon be $Q(t)$ and the balloon's radius be $r(t)$. Then, at any time it has capacitance $C(t) = 4\pi\epsilon_0 r(t)$ and a potential V relative to earth of $Q(t) \div C(t)$. If V is to remain constant, we must have $Q(t) = VC(t)$ for all t.

Now, the current through the resistor is provided by the charge leaving the balloon and so is $I = -dQ/dt$. The p.d. across the resistor, RI, is also the p.d. between the balloon and earth. Therefore,

$$\frac{V}{R} = I = -\frac{dQ}{dt} = -4\pi\epsilon_0 \frac{dr}{dt} V,$$

showing that for the potential to remain constant the rate of radius change must be

$$\frac{dr}{dt} = -\frac{1}{4\pi\epsilon_0 R} \quad \Rightarrow \quad r = a - \frac{t}{4\pi\epsilon_0 R},$$

where a is the given initial balloon radius. The deflation will last until $r \approx 0$, i.e. for a period $T = 4\pi\epsilon_0 aR$.[24]

With r reducing at the required uniform rate of $1/(4\pi\epsilon_0 R)$, the (steady) current flowing through the resistor is $I = V/R$, and so energy is dissipated in it at the rate RI^2. Over the complete deflation period, the total heat produced is

$$H = RI^2 T = R\frac{V^2}{R^2} 4\pi\epsilon_0 a R = C(0) V^2,$$

independent of the value of R, as might be expected.

However, this result may seem to be at odds with the fact that the energy stored in a capacitor is (normally) given as $\frac{1}{2}CV^2$, putting an upper limit on what is available for subsequent dissipation. However the two charging situations are not the same. Normally, the capacitance C_0 of the capacitor being charged is fixed, and the potential difference through which any incremental charge has to be moved increases from zero to the final potential during the charging process. Here, the potential is fixed at the (higher) final value, and the capacitance increases, bringing with it the additional charges $V dC$ necessary to maintain the constant potential. Formally

$$E = \int V \, dQ = \int_0^{C_0} V V \, dC = V^2 C_0.$$

---

[24] This has the form of capacitance $\times$ resistance, and so is a type of time constant, but different from that for an exponential decay.

In other words, all the charge has to be put onto the capacitor against the full final potential, whereas for the normal fixed-capacitance capacitor the corresponding potential is, on average, only one half of this value.

S 143 This relatively simple configuration can be transformed into a 'potential divider', but it must be remembered that, since the charges on connected plates must be equal but opposite, and $V = Q/C$, potential differences are *inversely* proportional to capacitances.

Firstly, the parallel capacitances are equivalent to one of $80\,\mu\text{F}$; then, $80\,\mu\text{F}$ and $120\,\mu\text{F}$ in series are equivalent to a single capacitance C_e of $(80 \times 120)/(80 + 120) = 48\,\mu\text{F}$. The total stored energy is therefore

$$E = \tfrac{1}{2}C_e V^2 = \tfrac{1}{2} \times 48 \times 10^{-6} \times (250)^2 = 1.5\,\text{J}.$$

The division of 250 V between the $80\,\mu\text{F}$ and $120\,\mu\text{F}$ capacitors in series is 150 V to 100 V, respectively, i.e. the *larger* p.d. is across the parallel pair of capacitors. The energy stored in the $50\,\mu\text{F}$ capacitor is therefore $\tfrac{1}{2} \times 50 \times 10^{-6} \times (150)^2 = \tfrac{9}{16}\,\text{J}$.

S 144 Although the devices involved here have unusual characteristics, the basic circuit rules

- Elements connecting the same pair of points have the same potential difference across them.
- The total current leaving a junction is equal to the total current entering it.

still apply.

Since A and B connect the same pair of points, the potential differences across them must be equal, i.e.

$$aI_A^2 = V_A = V_B = bI_B^2. \qquad (*)$$

Further, since all of the current I $(= I_A + I_B)$ passes through C, we must have

$$V_A + c(I_A + I_B)^2 = V.$$

This equation can be rewritten, using $(*)$, in terms of V and either I_A or I_B:

$$aI_A^2 + cI_A^2 \left(1 + \sqrt{a/b}\right)^2 = V, \qquad \text{or} \qquad aI_B^2 + cI_B^2 \left(1 + \sqrt{b/a}\right)^2 = V.$$

Thus

$$I_A = \frac{V^{1/2}}{\left(a + c + 2c\sqrt{a/b} + ca/b\right)^{1/2}} = \frac{b^{1/2}\,V^{1/2}}{(ab + bc + 2c\sqrt{ab} + ca)^{1/2}}.$$

Similarly,

$$I_B = \frac{a^{1/2} V^{1/2}}{(ab + bc + 2c\sqrt{ab} + ca)^{1/2}}.$$

The power taken from the source is $P = VI = V(I_A + I_B)$, given explicitly by

$$P = \frac{(a^{1/2} + b^{1/2}) V^{3/2}}{(ab + bc + ca + 2c\sqrt{ab})^{1/2}}. \qquad (**)$$

The most obvious deviation from the normal 'fixed resistance' result is that the power taken is proportional to $V^{3/2}$, rather than to V^2, implying that the resistance is proportional to $V^{1/2}$. This is what we might expect for a single device X connected to a voltage source V:

$$\frac{V}{I_X} = R_X = xI_X \quad \Rightarrow \quad R_X = x\sqrt{\frac{V}{x}} = \sqrt{xV}.$$

We can use $(**)$ to give the power taken by two devices in series by letting $a \to \infty$:

$$P \to \frac{a^{1/2} V^{3/2}}{a^{1/2}(b + c)^{1/2}} = \frac{V^2}{(b + c)^{1/2} V^{1/2}},$$

giving an effective resistance of $\sqrt{(b + c)V}$.

For two devices in parallel, we let $c \to 0$, effectively removing C from the circuit, and obtain

$$P \to \frac{(a^{1/2} + b^{1/2}) V^{3/2}}{(ab)^{1/2}} = \frac{(a^{1/2} + b^{1/2}) V^2}{(abV)^{1/2}},$$

giving as the combination formula

$$\frac{1}{R} = \frac{1}{\sqrt{bV}} + \frac{1}{\sqrt{aV}}.$$

In summary, the devices act as resistances $(xV)^{1/2}$ in parallel, but in series they add in quadrature.

S 145 Initially, it may not be clear what determines the maximum current, as any current, if continued for long enough, will provide the energy needed to raise the wire's temperature to its melting point. However, even though its conduction losses are to be ignored, the fuse will suffer radiation losses.

For any particular current through it, the wire's temperature will rise to that needed for the radiation from it ($\propto T^4$) to match the heat generated ($\propto RI^2$). Clearly, the loss cannot be further increased once the wire's melting point T_0 has been reached; at this point the fuse 'blows'.

So, the only properties of copper that matter for this particular question are T_0 and the resistivity ρ; the other three pieces of data provided are irrelevant. Quantitatively, in the standard notation, for a fuse wire of length L we have

$$\frac{\rho L}{\pi r^2} I^2 = RI^2 = \text{rate of heat generation} = \text{radiation rate} = 2\pi r L \sigma T_0^4,$$

where σ is the Stefan constant. This simplifies to

$$I = \sqrt{\frac{2\pi^2 r^3 \sigma T_0^4}{\rho}} = \sqrt{\frac{2\pi^2 \times 10^{-9} \times 5.7 \times 10^{-8} \times (1084 + 273)^4}{1.8 \times 10^{-8}}} = 460 \,\text{A}.$$

Don't try this at home!

S 146   If the capacitor were ideal, i.e. able to hold its charge when isolated, there would be no reason to expect that the reading in (b) would differ from that in (a), and we would not expect to have to constantly top up the charge on it, in the way described in (c). Clearly charge leaks from one plate to the other, even when it is isolated, meaning that the conduction path between the plates does not have infinite resistance. Therefore, the capacitor is more properly represented by a capacitor C and a resistor R in parallel, as illustrated in the figure below.

Result (c) shows that the leakage current, i.e. the current that circulates around the CR loop when the capacitor is charged, is 120 μA when the voltage drop across the capacitor is 15 V and hence that the resistance is $R = 15/(120 \times 10^{-6}) = 1.25 \times 10^5 \,\Omega$.

Observation (a) shows that the initial voltage across the capacitor, when it was isolated,[25] was 16.0 V. The subsequent discharge through R follows the form

$$V(t) = V(0) e^{-t/CR}, \text{ and so } 14.1 = 16.0 e^{-30/CR},$$

---

giving

$$CR = 30 \div \ln \frac{16.0}{14.1} = 237\,\text{s} \quad \Rightarrow \quad C = \frac{237}{1.25 \times 10^5} = 1.90\,\text{mF}.$$

Thus a truer representation of the nominal 2.5 mF capacitor is a capacitor of 1.90 mF in parallel with a resistor of $1.25 \times 10^5\,\Omega$.

S 147 Since the current I drawn from the battery is required, we take that as one of the defining currents; for the other we use I_1, the current through the 10 Ω resistor.

We need to establish the value of R for which Z first starts to conduct current; when it does, the latter will be vanishingly small, and consequently there will be no significant p.d. across the 5 Ω resistor. Under these conditions (with voltages measured relative to the negative terminal of the battery) $V_B = 0$ and therefore $V_A = 5$.

It follows that, at this setting, $I = I_1 = 5/10 = 0.5$ A. Thus $5 = V_A = 10 - RI = 10 - 0.5R$, giving $R = 10\,\Omega$; this is the critical value that divides the two regimes, $R > 10\,\Omega$ with Z non-conducting, and $R < 10\,\Omega$ with a current flowing through Z.

For $R > 10$, we have straightforwardly that $I = 10\,\text{V} \div (10 + R)\,\Omega = 10/(10 + R)$ A. For $R < 10$, three equations describe the situation:

$$V_A = 10I_1,$$
$$V_A - 5 = 5(I - I_1),$$
$$10 - V_A = RI.$$

Elimination of V_A and I_1 gives

$$RI = 10 - \frac{10}{3}(1 + I) \quad \Rightarrow \quad I = \frac{20}{3R + 10}\,A.$$

A graphical representation of these results is shown in the figure below; note the change of slope at the point $(10, 0.5)$.

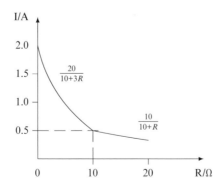

S 148 Once we have established the elements of M, we will be able to connect $\mathbf{x}$ for the final pair of terminals with that for the first pair by $\mathbf{x}_{\text{first}} = MMM\mathbf{x}_{\text{final}}$. In both cases the ratio of V_n to I_n gives the value of the resistance that is connected, or would be measured, across that pair of terminals.
(a) Equating the net (clockwise) voltage drops around the two obvious loops to zero gives us

$$0 = -V_1 + RI_1 + RI_1 - RI_2,$$
$$0 = V_2 + RI_2 + RI_2 - RI_1.$$

From these equations it follows that

$$I_1 = \frac{1}{R}V_2 + 2I_2, \quad \text{and hence that} \quad V_1 = 2V_2 + 4RI_2 - RI_2.$$

In matrix form:

$$\begin{bmatrix} V_1 \\ I_1 \end{bmatrix} = \begin{bmatrix} 2 & 3R \\ R^{-1} & 2 \end{bmatrix} \begin{bmatrix} V_2 \\ I_2 \end{bmatrix}.$$

(b) With the three units in cascade, and referring to the intermediate junctions as A and B, we have $\mathbf{x}_{\text{first}} = M\mathbf{x}_A$, $\mathbf{x}_A = M\mathbf{x}_B$, and $\mathbf{x}_B = M\mathbf{x}_{\text{final}}$, i.e. $\mathbf{x}_{\text{first}} = MMM\mathbf{x}_{\text{final}}$. With $\mathbf{x}_{\text{final}}$ having elements RI and I, for some (irrelevant) I, and $R_{\text{m}} = V_1/I_1$ the measured resistance, the explicit matrix equation reads

$$\begin{bmatrix} V_1 \\ I_1 \end{bmatrix} = \begin{bmatrix} 2 & 3R \\ R^{-1} & 2 \end{bmatrix} \begin{bmatrix} 2 & 3R \\ R^{-1} & 2 \end{bmatrix} \begin{bmatrix} 2 & 3R \\ R^{-1} & 2 \end{bmatrix} \begin{bmatrix} RI \\ I \end{bmatrix}$$

$$= \begin{bmatrix} 2 & 3R \\ R^{-1} & 2 \end{bmatrix} \begin{bmatrix} 7 & 12R \\ 4R^{-1} & 7 \end{bmatrix} \begin{bmatrix} RI \\ I \end{bmatrix}$$

$$= \begin{bmatrix} 26 & 45R \\ 15R^{-1} & 26 \end{bmatrix} \begin{bmatrix} RI \\ I \end{bmatrix}.$$

Finally

$$R_{\text{m}} = \frac{V_1}{I_1} = \frac{71RI}{41I} = \frac{71}{41}R.$$

Mathematical note. The matrix M has unit determinant $(ad - bc)$, whatever the value of R, and consequently so do all multiples M^n of it. This can be checked explicitly for $n = 1$, 2 and 3 using the values calculated above.

S 149 To estimate ρ, we need to calculate the kinetic energy loss at each collision and the total number of collisions that occur in the wire in unit time; this can then be equated to $RI^2 = (\rho L/A)I^2$.
The total current flowing through the wire is given by $I = Ane v_{\text{d}}$, where n is the electron density and v_{d} is the drift velocity of the electrons. Further, since

the electron's acceleration is assumed uniform, its speed just before a collision is $2v_d$. Thus the rate of heat generation for one electron is

$$\frac{1}{\tau}\frac{1}{2}m_e(2v_d)^2 = \frac{1}{2}m_e\left(\frac{2I}{neA}\right)^2\frac{1}{\tau}.$$

For all the electrons in a volume of size $L \times A$, the total heat production rate is

$$\frac{2m_e}{n^2e^2\tau}\frac{I^2}{A^2}nAL, \text{ and this must be equal to } \frac{\rho L}{A}I^2,$$

giving

$$\rho = \frac{2m_e}{ne^2\tau}.$$

From the given data, we find that $n = 8.9 \times 10^3 \div (63.5 \times 1.66 \times 10^{-27}) = 8.44 \times 10^{28}$ m^{-3}. and so

$$\rho = \frac{2 \times 9.11 \times 10^{-31}}{8.44 \times 10^{28}(1.60 \times 10^{-19})^2 \times 4 \times 10^{-14}} = 2.1 \times 10^{-8} \ \Omega\,\text{m}.$$

Note This estimate is within 25% of the actual value, but little significance can be attached to it, as no independent basis for the given value of τ has been provided.

S 150 The figure below shows, on the left, the series connection of two inductors, and, on the right, their parallel connection.

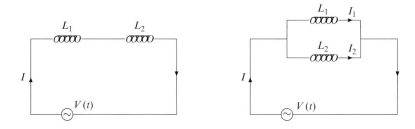

(a) With the two inductors in series the same varying current $I(t)$ flows through each. Across each of the inductors there is a voltage drop (or gain) $V_n(t)$ equal to $L_n\, dI/dt$, and the total drop around the circuit must be equal to the voltage $V(t)$ of the source:

$$V(t) = V_1(t) + V_2(t) = L_1\frac{dI}{dt} + L_2\frac{dI}{dt} = (L_1 + L_2)\frac{dI}{dt},$$

showing that the equivalent single inductor is $L_e = L_1 + L_2$, i.e. they combine in the same way as resistors.

The right-hand figure shows a parallel arrangement, and, by inspection, the voltage drop across *each* of the inductors is $V(t)$. The currents, however, are not equal, but I_1 and I_2, where $I_1 + I_2 = I$. Hence,

$$L_1 \frac{dI_1}{dt} = V(t) = L_2 \frac{dI_2}{dt},$$

and

$$\frac{dI}{dt} = \frac{dI_1}{dt} + \frac{dI_2}{dt} = V(t) \left(\frac{1}{L_1} + \frac{1}{L_2} \right) = \frac{V(t)}{L_e},$$

where

$$\frac{1}{L_e} = \frac{1}{L_1} + \frac{1}{L_2}.$$

This is the direct analogue of the combination rule for parallel resistors.

(b)

We take the charge on one of the plates of the capacitor as $Q = Q_0 \cos \omega t$. It then follows that

$$V_C = \frac{Q_0 \cos \omega t}{C}, \quad I = \frac{dQ}{dt} = -\omega Q_0 \sin \omega t, \quad V_L = L\frac{dI}{dt} = -L\omega^2 Q_0 \cos \omega t.$$

Now, equating the total voltage drop around the circuit to the applied voltage, we have $V_L + V_C = V$, or more explicitly,

$$-L\omega^2 Q_0 \cos \omega t + \frac{Q_0 \cos \omega t}{C} = V_0 \cos \omega t,$$

which can be rearranged as

$$Q_0 = \frac{C V_0}{1 - (\omega^2/\omega_0^2)},$$

where $\omega_0^2 = 1/LC$. As Q_0 shows resonance at $\omega = \omega_0$, so does its time derivative, namely the current $I(t)$.

(c) With up to two inductances L and one or two capacitances C it is possible to engineer a resonance frequency $1/\sqrt{xLyC} = \sqrt{xy}\,\omega_0$, where x and y can each take one of three values, $\frac{1}{2}$, 1 and 2. The question wording excludes cases in which $xy = 1$. The solutions, as circuits, are:

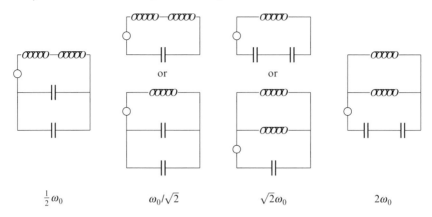

$\frac{1}{2}\omega_0$ $\omega_0/\sqrt{2}$ $\sqrt{2}\,\omega_0$ $2\omega_0$

S 151   A bit of detective work, in drawing up a 'list of suspects', is needed here! For each of the measurements, two of the three resistors d, e and f are connected in series, and the third is in parallel with the combination. Thus we have the three equations

$$AB \qquad \frac{f(d+e)}{d+e+f} + a + b = 7\tfrac{3}{13},$$

$$AC \qquad \frac{e(d+f)}{d+e+f} + a + c = 6\tfrac{9}{13},$$

$$BC \qquad \frac{d(e+f)}{d+e+f} + b + c = 10\tfrac{1}{13}.$$

Since all of a, b, ..., f are integers, and the maximum that $d + e + f$ can be is 15 (and, in particular, not 26, 39, ...), we must have that $d + e + f = 13$. Consequently, the resistance between any pair of terminals is of the form

$$R = n_1 + n_2 + \frac{n_3(13 - n_3)}{13}. \qquad (*)$$

We now draw up a table listing the various possibilities for the fractional part of this expression. We test, in turn, each of the resistance values as candidates for being n_3.

n_3	1	2	3	4	5	6
$n_3(13 - n_3)$	12	22	30	36	40	42
Compatible with	–	AC	–	–	BC	AB
n_3 is resistor	–	e	–	–	d	f
$n_1 + n_2$ from $(*)$	–	$a + c = 5$	–	–	$b + c = 7$	$a + b = 4$

With the $2\,\Omega$, $5\,\Omega$ and $6\,\Omega$ resistors accounted for, the last line of the table shows that a, b and c must be $1\,\Omega$, $3\,\Omega$ and $4\,\Omega$, respectively. In summary,

$$a=1\,\Omega,\ b=3\,\Omega,\ c=4\,\Omega,\ d=5\,\Omega,\ e=2\,\Omega,\ f=6\,\Omega.$$

S 152 When the supply voltage V_s is greater than that on plate P of the capacitor, the diode conducts and charges the latter up to 10 V, the maximum value of V_s. As V_s drops, the diode becomes reversed biased and ceases to conduct. The voltage V_P also drops as the capacitor discharges through the resistance R with a time constant given by $CR = 50 \times 10^{-6} \times 10 \times 10^3 = 0.5$ s. This is much longer than the period $T_0 = f^{-1} = 0.02$ s of the a.c. supply. Consequently, V_P does not fall very far before it is recharged to 10 V by the next cycle of the 50 Hz supply. This is illustrated in the figure.

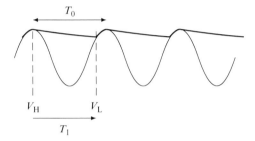

The voltage measured across AE ranges between those at the points marked as V_H and V_L in the figure. Clearly $V_H = 10$ V, and V_L can be calculated as approximately

$$V_L = 10\,e^{-0.02/0.5} = 10\,e^{-0.04} = 9.608 \text{ V}.$$

Thus the peak-to-peak voltage variation in the (nearly) d.c. output at A is 0.39 V (with an average value close to 9.80 V). Since the value of 0.02 used above should really be replaced by T_1 from the next paragraph, and the latter is a little less than 0.02, the value 0.39 V is a slight overestimate of the variation.

If δ is the interval in each cycle during which the diode is conducting, then, as can be seen from the figure $\delta = T_0 - T_1$, where T_1 is the time interval between V_H and V_L. Taking V_H as the time $t=0$, we see that T_1 is the interval after which the source voltage and that on plate P are again equal, with T_1 satisfying

$$e^{-T_1/RC} = \cos\left(2\pi \frac{T_1}{T_0}\right).$$

This is a 'transcendental' equation for T_1 and there is no routine way to solve it, but, since we know that $RC \gg T_0 \gg \delta$, we can find an approximate value for

δ by writing $T_1 = T_0 - \delta$ and making series expansions of functions with small arguments.

$$e^{-T_1/RC} = \cos\left(2\pi \frac{T_1}{T_0}\right),$$

$$e^{-T_0/RC} e^{\delta/RC} = \cos\left(2\pi - \frac{2\pi\delta}{T_0}\right) = \cos\frac{2\pi\delta}{T_0},$$

$$\left(1 - \frac{T_0}{RC} + \cdots\right)\left(1 + \frac{\delta}{RC} + \cdots\right) = 1 - \frac{1}{2}\left(\frac{2\pi\delta}{T_0}\right)^2 + \cdots.$$

$$\text{Since } \frac{\delta}{RC} \ll \frac{T_0}{RC}, \qquad \frac{T_0}{RC} \approx 2\pi^2 \left(\frac{\delta}{T_0}\right)^2,$$

$$\left(\frac{\delta}{T_0}\right)^2 \approx \frac{T_0}{2\pi^2 RC} = \frac{1}{50\pi^2}.$$

Thus $\delta/T_0 \approx 0.045$, i.e. $\delta \approx 0.045 f^{-1}$.

S 153 (a) Using $+$ to represent a series connection and $\|$ to represent a parallel one:

(i) *A* to *B*:

$$5 + (2\|18) = 5 + \frac{2 \times 18}{2 + 18} = 6.8\,\Omega.$$

(ii) *B* to *C*: $10\|10 = 5\,\Omega.$

(iii) *B* to *C* with *AC* shorted:

$$10\|(2 + (5\|8)) = 10\|\left(2 + \frac{5 \times 8}{5 + 8}\right) = 10\|\frac{66}{13} = \frac{10 \times (66/13)}{10 + (66/13)} = 3.37\,\Omega.$$

(b) The $8\,\Omega$ resistor is irrelevant, as it is connected directly across the battery system, and therefore does not affect the current in the $4\,\Omega$ resistor.

We can analyse the circuit by considering three clockwise current loops, with the obvious choice of suitable loops being as shown in the figure below.

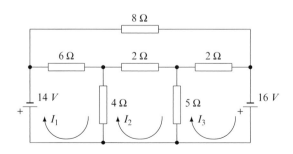

For each of the loops, we apply Kirchoff's (second) law – which says that 'the sum of the voltage drops across each of the components in a closed loop is equal to the algebraic sum of the voltage sources within that loop'.

Since we are dealing here only with resistors, each voltage drop is of the form 'resistance × the algebraic sum of all currents flowing through that resistor'. Thus, for the three loops,

$$-14 = 6I_1 + 4(I_1 - I_2) = 10I_1 - 4I_2,$$
$$0 = 4(I_2 - I_1) + 2I_2 + 5(I_2 - I_3) = -4I_1 + 11I_2 - 5I_3,$$
$$16 = 5(I_3 - I_2) + 2I_3 = -5I_2 + 7I_3.$$

Elimination of I_1 from the first two equations, followed by the elimination of I_3 from the result of this and the third equation, leads to $I_2 = 1$ A. Back substitution then gives $I_1 = -1$ A, making the current in the $4\,\Omega$ resistor $I_2 - I_1 = 2$ A clockwise. [$I_3 = 3$ A.]

(c) The voltage of the battery is irrelevant, so far as the question posed is concerned; it could only affect the amount of heat generated, not the optimum value of R.

The given circuit can be considered as a 'Wheatstone's bridge' circuit, a device for measuring an unknown resistance X, and illustrated in the figure.

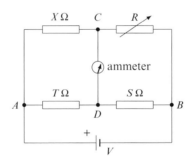

In the Wheatstone circuit, R is adjusted until no current flows through the ammeter connected between C and D. When this balance is achieved, the voltages V_C and V_D must be equal. But each is the 'tap-off' point of a potential divider for the voltage V connected across AB. Thus

$$\frac{R}{R+X} V = V_C = V_D = \frac{S}{S+T} V \quad \Rightarrow \quad \frac{X}{R} = \frac{T}{S}.$$

Applying the same principle to the given circuit, there will be no current (or heating) in the $5\,\Omega$ resistor when

$$\frac{1+2}{R\|10} = \frac{4}{12} \quad \Rightarrow \quad R\|10 = 9, \text{ i.e. } \frac{1}{R} + \frac{1}{10} = \frac{1}{9} \quad \Rightarrow \quad R = 90\,\Omega.$$

S 154   (a) The charging of the capacitor is through the $5\,\Omega$ and $2\,\Omega$ resistors in series, and so the time constant for this process is $\tau = 10 \times 10^{-6} \times (5 \times 10^3 + 2 \times 10^3) = 0.07$ s.

When K_2 is momentarily closed and V_1 thereby made positive, V_3 becomes $+10$ V, whatever it was before. The amplifier then starts to charge C, with the potential across the latter rising from zero to 10 V with a time constant τ of 0.07 s. The current needed to do this is $10\,e^{-t/\tau} \div 7 \times 10^3$ A.

But, in addition, the amplifier has to provide the current running to earth through the potential divider chain consisting of $5\,\mathrm{k}\,\Omega$ and $1\,\mathrm{k}\,\Omega$ in series. However, with K_1 closed, R_3 is bypassed and the relevant resistance of the chain is $5\,\mathrm{k}\,\Omega$; so the divider draws a current of $10 \div 5 \times 10^3 = 2$ mA. The total current drawn from the amplifier is therefore

$$I = 2 + \frac{10}{7}e^{-t/0.07} \ \mathrm{mA}.$$

Having K_1 closed also ensures that $V_2 = 0$, $V_1 > V_2$, and V_3 stays at $+10$ V.
(b) As 'several seconds' (i.e. many times τ) elapse before K_1 is opened, the capacitor is already fully charged to 10 V, and, as there is no longer any current flowing, $V_1 = 0$. This, together with $V_3 = +10$ V, is the initial state before K_1 is opened.

The opening of K_1 brings the $1\,\mathrm{k}\,\Omega$ resistor 'into play' and V_2 becomes $1/(1+5) \times 10 = +1.67$ V, i.e. greater than V_1. As a consequence V_3 becomes -10 V, and V_2 follows by becoming -1.67 V, virtually instantaneously as there is no capacitor to charge up (ignoring any inherent in the amplifier).

The resulting current flowing via the amplifier towards the (left-hand plate of the) capacitor (i.e. clockwise in the figure) is $(10/7)\,e^{-t/\tau}$ mA, where t is measured from the time of the switch. This results in a voltage across R_1 of $(5/7)10\,e^{-t/\tau}$ with earth at the higher potential; in other words, $V_1 = -(5/7)10\,e^{-t/\tau}$. This negative voltage decreases in magnitude as t increases, and after a certain time t_1 it is equal to $-(1/6)10$ V, i.e.

$$-\frac{5}{7}10\,e^{-t_1/\tau} = -\frac{1}{6}10,$$

$$e^{-t_1/0.07} = \frac{7 \times 1}{6 \times 5} \quad \Rightarrow \quad t_1 = 0.102\,\mathrm{s}.$$

When this happens, V_3 switches to $+10$ V and the current, and hence also the p.d. across R_1, reverse direction. The current again has magnitude $(10/7)\,e^{-t/\tau}$ mA, where t is now measured from the time of the most recent switch.

The sequence of events that previously happened with negative values of V_1 is now repeated, but with positive values. After a further period t_1, V_1 drops below $V_2 = +1.67$ V and V_3 returns to -10 V. The cycling then repeats indefinitely, with a period of about 0.20 s. The time variation of V_1 is sketched below (not to scale); V_3 produces a square-wave of ± 10 V with the same frequency, about 5 Hz.

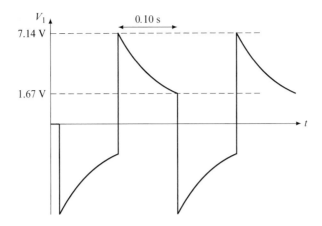

S 155   At the start of each cycle, the output voltage V, initially zero, increases as the capacitor C is charged through the resistor R according to the usual equation

$$V = V_0\left(1 - e^{-t/RC}\right).$$

However, when V reaches V_1 the device operates, and V drops rapidly to zero. This happens after a time T given by

$$V_1 = V_0\left(1 - e^{-T/RC}\right), \qquad\qquad (*)$$

which can be rearranged as

$$T = RC\ln\left(\frac{V_0}{V_0 - V_1}\right).$$

The process then repeats itself with period T.

   The figure shows the time variation of V, and, as the time-base must be linear starting from 0 V at $t = 0$, it is clearly the first part of the curve that must be used.

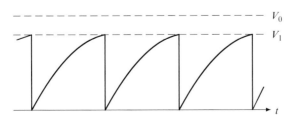

To examine the linearity of the curve in that region, we expand $(*)$ for the value of t at which non-linearity will be most significant, namely $t = T = 0.01$ s.

$$V_1 = V_0 \left\{ 1 - \left[1 - \frac{T}{RC} + \frac{1}{2} \left(\frac{T}{RC} \right)^2 - \cdots \right] \right\}$$

$$= V_0 \left[\frac{T}{RC} - \frac{1}{2} \left(\frac{T}{RC} \right)^2 + \cdots \right].$$

For 1% linearity, the quadratic term must be less than 1% of the linear one, i.e.

$$\frac{1}{2} \left(\frac{T}{RC} \right)^2 < 0.01 \frac{T}{RC} \quad \Rightarrow \quad RC > \frac{T}{0.02}, \text{ i.e. } RC > 0.5.$$

This limit on T/RC also implies that V_1 must be less than $0.02 V_0$. In summary, only a very small part of the capacitor recharging curve is suitable for the time-base, and a source voltage at least fifty times the trigger voltage of the discharge device is required.

S 156   In both cases, the final charge on the capacitor is $Q_0 = CN\mathcal{E}$ and the stored energy is $E_0 = \frac{1}{2} CN^2 \mathcal{E}^2$. There are a number of ways to calculate the wasted energy, but we choose one that makes explicit the rate at which heat is generated in the resistor.

Using method (a), the charge Q on the capacitor varies with time t according to

$$Q = Q_0 \left(1 - e^{-t/RC} \right) \text{ with a corresponding current } I = \frac{dQ}{dt} = \frac{Q_0}{RC} e^{-t/RC}.$$

The power wasted in the resistor is RI^2 and so the total energy wasted is

$$E_w = R \int_0^\infty \frac{Q_0^2}{R^2 C^2} e^{-2t/RC} \, dt = \frac{Q_0^2}{2C} = \frac{1}{2} CN^2 \mathcal{E}^2,$$

the same as the amount that is not wasted.

Using method (b), the amount of (further) charge transferred to the capacitor at each connection is $\Delta Q = C\mathcal{E}$, and the work done by the battery at the rth stage is $\Delta Q r \mathcal{E}$. Thus the total work done is[26]

$$E_t = C\mathcal{E}(\mathcal{E} + 2\mathcal{E} + \cdots + N\mathcal{E}) = C\mathcal{E}^2 \frac{1}{2} N(N+1).$$

The wasted energy using this method is

$$E_w = E_t - E_0 = \frac{1}{2} C\mathcal{E}^2 [N(N+1) - N^2] = \frac{1}{2} CN\mathcal{E}^2,$$

i.e. a factor of N less!

---

[26] Using the standard mathematical result that $\sum_1^N r = \frac{1}{2} N(N+1)$.

S 157 (a) We follow the instructions in the question. The sphere has capacitance $4\pi\epsilon_0 a$ and a total charge Q of $4\pi a^2 \sigma$. With Q fixed, the stored energy is $W = Q^2/2C = Q^2/(2 \times 4\pi\epsilon_0 a)$ and so for a small change da in the radius a,

$$dW = -\frac{Q^2}{8\pi\epsilon_0 a^2}\, da.$$

This must be equal to the work done by the pressure, $p\,\Delta V = p \times 4\pi a^2\, da$ in magnitude, and so

$$p = \frac{Q^2}{8\pi\epsilon_0 a^2}\frac{1}{4\pi a^2} = \frac{1}{2\epsilon_0}\frac{Q}{4\pi a^2}\frac{Q}{4\pi a^2} = \frac{\sigma^2}{2\epsilon_0}$$

is the outward pressure on the sphere as a result of it carrying a charge.

(b) The figure shows a section of the metal sphere that has radius $a\sin\theta$, where $\theta < \pi/2$, but is not necessarily small. It also shows the tensile force that acts around its perimeter, which has length $2\pi a\sin\theta$, and a cross sectional area of t times this.

$$F = Tt \text{ per unit length of the perimeter}$$

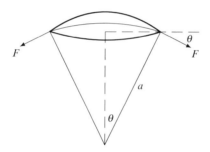

If the tensile stress in the metal is T, the net 'downwards' component of the total force acting on the perimeter is

$$F_\downarrow = 2\pi a\sin\theta \times t \times T \times \sin\theta.$$

The net 'upwards' force due to the gas pressure is

$$F_\uparrow = p \times \text{the horizontally projected area of the section} = p \times \pi(a\sin\theta)^2.$$

The two forces must be equal, and so $p = 2tT/a$.

(c) As in part (a), $Q = 4\pi a^2\sigma$, and so the electric field intensity at the sphere's surface, $r = a$, is

$$E = \frac{1}{4\pi\epsilon_0}\frac{4\pi a^2\sigma}{a^2} = \frac{\sigma}{\epsilon_0},$$

showing that ionizational discharge will occur when $\sigma = \epsilon_0 E_0$. To prevent prior rupture of the metal,

$$\frac{2tT}{a} > p = \frac{\sigma^2}{2\epsilon_0} = \frac{\epsilon_0^2 E_0^2}{2\epsilon_0}.$$

Hence, the required tensile breaking stress is

$$T > \frac{\epsilon_0 \, a \, E_0^2}{4t} = \frac{8.8 \times 10^{-12} \times 0.5 \times 9.0 \times 10^{12}}{4 \times 1.0 \times 10^{-6}} = 9.9 \times 10^6 \, \text{N m}^{-2}.$$

S 158   (a) If the middle plate had originally been placed centrally, then symmetry considerations would require that there be equal final charges on the two plates and therefore there could be no final p.d. between them. The actual voltage results from the *difference* ΔQ between the two final charges, with one charge being $Q_{\text{average}} + \frac{1}{2}\Delta Q$ and the other $Q_{\text{average}} - \frac{1}{2}\Delta Q$. The p.d. is that corresponding to a charge of $\frac{1}{2}\Delta Q$ on the capacitor formed by the two outer plates.

Quantitatively, the two (negative) charges are

$$Q_1 = \frac{\epsilon_0 A}{d_1} V = \frac{8.8 \times 10^{-12} \, 1.0 \times 10^{-2}}{1.0 \times 10^{-3}} \, 2000 = 1.76 \times 10^{-7} \, \text{C}$$

and

$$Q_2 = \frac{\epsilon_0 A}{d_2} V = \frac{8.8 \times 10^{-12} \, 1.0 \times 10^{-2}}{2.0 \times 10^{-3}} \, 2000 = 0.88 \times 10^{-7} \, \text{C}.$$

Thus $\frac{1}{2}\Delta Q = \frac{1}{2}(1.76 - 0.88) \times 10^{-7} = 4.4 \times 10^{-8}$ C, and the final potential difference is

$$V_{\text{final}} = \frac{\Delta Q}{2} \frac{d_3}{\epsilon_0 A} = \frac{4.4 \times 10^{-8} \, 3.0 \times 10^{-3}}{8.8 \times 10^{-12} \, 1.0 \times 10^{-2}} = 1500 \, \text{V}.$$

(b) At first sight there may seem to be too little information given, but let the area of each plate be A; the length of the box is then v/A. If d is the spacing between the plates, then $2Nd = v/A$, and the maximum voltage difference V that can be applied across the two sets of plates is $V = E_0 d$.

The energy stored in one pair of adjacent plates is

$$\frac{1}{2}CV^2 = \frac{1}{2}\frac{\epsilon_0 A}{d}(E_0 d)^2 = \frac{1}{2}\epsilon_0 A E_0^2 \frac{v}{2NA}.$$

As there are $2N$ capacitors, the maximum energy that can be stored without electrical breakdown is

$$E_{max} = \tfrac{1}{2}\epsilon_0 v E_0^2.$$

In fact, there was too much information, as the result is independent of N!

S 159 Although the charge carriers are electrons, we work with the conventional definitions of current and positive charges, so that Fleming's left hand rule can be applied directly.[27] Initially the current will be deflected in the y direction, resulting in excess charge collecting on one edge of the ribbon, and charge depletion on the other. This will generate an electric field across the ribbon whose effect is to oppose any further accumulation of deflected charge; equilibrium is reached when the two opposing forces acting on the carriers are equal in magnitude.

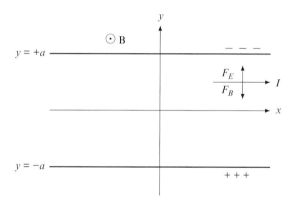

The electric field strength is $3.0 \times 10^{-6} \div 2.0 \times 10^{-2} = 1.5 \times 10^{-4}$ V m^{-1}, and the (lateral) force F_E on a carrier is e times this. If the magnitude of the carrier drift velocity is v, then the (lateral) force F_B experienced as a result of moving in the magnetic field is $Bev = e \times 0.75 \times v$ N. Equating the two balanced forces:

$$e \times 1.5 \times 10^{-4} = e \times 0.75 \times v \quad \Rightarrow \quad v = 2.0 \times 10^{-4}\,\text{m s}^{-1}.$$

---

[27] Fleming's left hand rule is applicable when a current flows at right angles to a magnetic field, and gives the direction of the thrust on (or motion of) the current carrier. If the thumb, and the first and second fingers of a left hand are held mutually orthogonal, then the **F**irst finger represents the **F**ield, the se**C**ond the **C**urrent, and the **T**humb the **T**hrust.

Since the actual carriers are electrons, the drift velocity is in the negative x-direction.

As shown in the figure, the magnetic force (on the notionally positive charge making up the current), given by Fleming's left hand rule, is towards the negative y-direction, resulting in excess positive charge on the edge $y = -a$ and hence an electric field and force in the positive y-direction, i.e. the edge $y = -a$ is at the higher potential.

S 160   In order to keep track of the dimensions of the quantities involved, we consider a cylinder of length L, rather than of unit length. Let the cross sectional area between the cylinders be A [equal to $\pi(b^2 - a^2)$, in an obvious notation], and the wire's radius be r. The total length used in the winding is then the volume to be filled divided by the wire's cross-sectional area, i.e. $\ell = AL/\pi r^2$. Its electrical resistance is therefore

$$R = \rho \times \frac{AL}{\pi r^2} \times \frac{1}{\pi r^2},$$

where ρ is the resistivity of the wire.

We now consider (one half of) a radial cross-section of the cylinder; it is a rectangle of length L and width $b - a$, and therefore contains $N = L(b - a)/\pi r^2$ cross-sections of the wire, i.e N turns.

The current required in the winding is $I = B/\mu_0 n = BL/\mu_0 N$, and the rate at which heat is generated there is

$$P = RI^2 = \frac{\rho\pi(b^2 - a^2)L}{\pi^2 r^4}\frac{B^2 L^2}{\mu_0^2}\frac{\pi^2 r^4}{L^2(b - a)^2} = \frac{\pi\rho(b + a)B^2 L}{\mu_0^2(b - a)}.$$

We conclude that the heat dissipated in the solenoid winding does *not* depend upon the radius (r) of the wire used.

S 161   For any particular value of a, the voltage should be such as to make the electric field just less than E_0 at the place where the field is strongest; this is clearly at the surface of the *inner* sphere where the charge density is greatest (or, the field lines are closest together). This precaution against breakdown imposes a link between a and the maximum value $V(a)$ of V that can be applied. This link can be found as follows:

$$Q = CV = 4\pi\epsilon_0\frac{ab}{b - a}V \implies \sigma = \frac{4\pi\epsilon_0}{4\pi a^2}\frac{ab}{(b - a)}V,$$

and hence, using the general connection $E = \sigma/\epsilon_0$ between a charge density and the local electric field intensity,

$$E = \frac{b}{a(b - a)}V. \tag{*}$$

As E must not exceed E_0, the link is that

$$V(a) = \frac{a(b-a)}{b} E_0.$$

We now turn to the stored energy, which, to avoid confusion, we denote by W.

$$W = \frac{1}{2}CV^2 = \frac{1}{2}\frac{4\pi\epsilon_0 ab}{(b-a)}\frac{a^2(b-a)^2}{b^2} E_0^2 = \frac{2\pi\epsilon_0 a^3(b-a)}{b} E_0^2.$$

This has to be maximized with respect to a; this happens when

$$\frac{d}{da}(a^3 b - a^4) = 3a^2 b - 4a^3 = 0 \quad \Rightarrow \quad a = \frac{3b}{4}.$$

From $(*)$, the corresponding appropriate potential difference is

$$V = b^{-1} \tfrac{3}{4}b \tfrac{1}{4}b E_0 = \tfrac{3}{16}b E_0.$$

Though it is not asked for, the corresponding value of W is $(27/128)\pi\epsilon_0 b^3 E_0^2$.

S 162 With $d = d_0 - \alpha Q$ and $Q = Q_0 + Q_1 \cos\omega t$, the voltage across the capacitor (of plate area A) is

$$V = \frac{Q}{C} = (Q_0 + Q_1 \cos\omega t)\frac{d_0 - \alpha(Q_0 + Q_1 \cos\omega t)}{\epsilon_0 A}. \qquad (*)$$

Since the RHS contains the square of $\cos\omega t$ there will be a component of frequency 2ω, the connection being the identity $\cos 2\omega t = 2\cos^2\omega t - 1$. To separate the constant and the various frequency components of V we multiply out the RHS of $(*)$ and use the identity.

$$\begin{aligned}
\epsilon_0 A V(t) &= d_0(Q_0 + Q_1 \cos\omega t) - \alpha(Q_0 + Q_1 \cos\omega t)^2 \\
&= (d_0 Q_0 - \alpha Q_0^2) + (d_0 Q_1 - 2\alpha Q_0 Q_1)\cos\omega t - \alpha Q_1^2 \cos^2\omega t \\
&= (d_0 Q_0 - \alpha Q_0^2 - \tfrac{1}{2}\alpha Q_1^2) + Q_1(d_0 - 2\alpha Q_0)\cos\omega t - \tfrac{1}{2}\alpha Q_1^2 \cos 2\omega t.
\end{aligned}$$

The component of frequency ω can be eliminated if Q_0 is chosen as $d_0/2\alpha$. The voltage across the capacitor is then

$$\begin{aligned}
V(t) &= \frac{d_0^2}{2\alpha} - \frac{d_0^2}{4\alpha} + \frac{1}{2}\alpha Q_1^2 - \frac{1}{2}\alpha Q_1^2 \cos 2\omega t \\
&= \frac{d_0^2}{4\alpha} + \frac{1}{2}\alpha Q_1^2 - \frac{1}{2}\alpha Q_1^2 \cos 2\omega t,
\end{aligned}$$

i.e. a constant component and one of frequency 2ω.

S 163   When the plates are in their final positions, we effectively have three capacitors connected in series. But such capacitors must have equal and opposite charges on their two plates, and this requirement has to be made compatible with the the the fact that each plate is isolated, and so can neither gain nor lose charge. To shorten references, we denote 10^{-9} C by Q.

Starting with plate A: it carries Q and, as all of its 'lines of force' must terminate on the upper face of C, that surface must carry $-Q$. However, as plate C carries a net charge of $+Q$, there must be $+2Q$ on its lower face. By a similar argument, there must be a charge of $-2Q$ on the upper face of D, and hence $+Q$ on its lower surface. Plate B carries the corresponding negative charge $-Q$ on its upper surface, as it did before the two capacitors were combined. The overall situation is as in the figure below.

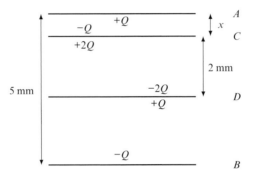

Starting from the given value for the original left-hand capacitor, we can say that, for a general value of x, measured in mm, the capacitance between a pair of plates separated by x is $10^{-10} \div x$ F. Thus, measured in volts, the relevant potentials are

$$V_{AC} = \frac{Q}{10^{-10}/x} = 10x, \quad V_{CD} = \frac{2Q}{10^{-10}/2} = 40,$$

$$V_{DB} = \frac{Q}{10^{-10}/(3-x)} = 30 - 10x,$$

i.e. $V_{CD} = 40$ V, whilst $V_{AB} = 10x + 40 + 30 - 10x = 70$ V. Perhaps a little surprisingly, both values are independent of x.

Whatever the value of x, when the CD pair of plates is slid between plates A and B, the positive charge on A is closer to the positive charge on C than to the negative charge on D, so there will be a net repulsion resulting from these two interactions. Similarly, the negative charge on B will experience a net repulsion as a result of its interaction with CD, because, for it, the negative charge on D is the closer of the two charge distributions. Although we cannot calculate its size, there will be a net repulsion, and the CD pair of plates will have to be pushed into place, and work will have to be done.

To find the total work needed, we calculate the difference in stored electrical energy between the final and initial states. Since the charges involved are all simple multiples of Q, we use the form $E = Q^2/2C$. For the initial state,

$$E_{\text{init}} = \frac{Q^2}{2 \times 10^{-10}/2} + \frac{Q^2}{2 \times 10^{-10}/5} = 3.5 \times 10^{-8} \, \text{J}.$$

For the final state, we have three capacitors (in series) to account for:

$$E_{\text{fin}} = \frac{Q^2}{2 \times 10^{-10}/x} + \frac{(2Q)^2}{2 \times 10^{-10}/2} + \frac{Q^2}{2 \times 10^{-10}/(3-x)}$$

$$= \frac{Q^2 x}{2 \times 10^{-10}} + \frac{4Q^2 \, 2}{2 \times 10^{-10}} + \frac{Q^2 \, (3-x)}{2 \times 10^{-10}}$$

$$= 4.0 \times 10^{-8} + 1.5 \times 10^{-8} = 5.5 \times 10^{-8} \, \text{J}.$$

The difference between E_{fin} and E_{init} of 2.0×10^{-8} J is a measure of the work needed to effect the amalgamation. Again, we note that the result does not depend upon the the value of x; this was to be expected as the stored final energy could have been calculated using the voltages found earlier – and they were independent of x.

S 164 (a) As, at the intermediate point, we simply have a capacitor with plate separation $\frac{1}{3}d$ connected to a source of voltage V, the first stage of the procedure is irrelevant, apart from giving the initial stored energy as

$$E_{\text{init}} = \frac{1}{2}CV^2 = \frac{1}{2}\frac{\epsilon_0 A}{d}V^2.$$

After the first stage the capacitor carries a charge $Q = CV = \epsilon_0 AV/(d/3)$, and, since the source is disconnected, this remains constant throughout the second stage. In particular, it is also the final charge on the capacitor, which therefore has final stored energy

$$E_{\text{fin}} = \frac{Q^2}{2C} = \frac{(3\epsilon_0 AV)^2}{d^2} \frac{d}{2\epsilon_0 A} = \frac{9}{2}\frac{\epsilon_0 A}{d}V^2.$$

Thus, the difference is a gain of $4\epsilon_0 AV^2/d$.

(b) Since the source is disconnected for the second stage, we do not have to consider work done on or by it; any mechanical work involved can be related directly to the electrostatic stored energy. Further, since Q is fixed, and known, for the whole of the second stage, it is easier to consider the capacitor's energy in the form $E = Q^2/2C$, with only C varying. For separation x, consider the effect of a (infinitesimally) small change dx:

$$C = \frac{\epsilon_0 A}{x} \quad \Rightarrow \quad dC = -\frac{\epsilon_0 A}{x^2} \, dx,$$

and the corresponding stored energy change is found from

$$E = \frac{Q^2}{2C} \quad \Rightarrow \quad dE = -\frac{Q^2}{2C^2} \, dC = \frac{Q^2}{2C^2} \frac{\epsilon_0 A}{x^2} \, dx.$$

But this must be equal to the mechanical work $F \, dx$ done by the separation force F. Therefore,

$$F = \frac{Q^2}{2} \left(\frac{x}{\epsilon_0 A} \right)^2 \frac{\epsilon_0 A}{x^2} = \frac{Q^2}{2 \epsilon_0 A},$$

showing that the force is independent of x. It's constant magnitude is given by

$$F = \left(\frac{3 \epsilon_0 A V}{d} \right)^2 \frac{1}{2 \epsilon_0 A} = \frac{9}{2} \frac{\epsilon_0 A}{d^2} V^2.$$

The above answers the question, as set, but we note that the total work done by the constant force is $W = \frac{2}{3} d \times F = 3 \epsilon_0 A V^2/d$. As expected, this is equal to the difference between the stored energy of $3 \epsilon_0 A V^2/2d$ at the start of the second stage and the final energy of $9 \epsilon_0 A V^2/2d$.

S 165 (a) After squaring both sides and cross-multiplying, we expand the squared term in parentheses and the simplify, as follows:

$$\frac{b}{[(b - a \cos \theta)^2 + (a \sin \theta)^2]^{1/2}} = \frac{a}{[(a \cos \theta - a^2/b)^2 + (a \sin \theta)^2]^{1/2}},$$

$$b^2[(a \cos \theta - a^2/b)^2 + (a \sin \theta)^2] = a^2[(b - a \cos \theta)^2 + (a \sin \theta)^2],$$

$$b^2 \left(a^2 - \frac{2a^3}{b} \cos \theta + \frac{a^4}{b^2} \right) = a^2(b^2 - 2ab \cos \theta + a^2),$$

$$a^2 b^2 - 2a^3 b \cos \theta + a^4 = a^2 b^2 - 2a^3 b \cos \theta + a^4.$$

This confirms the validity of the given equality.

(b) $\quad\quad\quad\quad \angle AOX = \theta$

For the two charges positioned as described in the question, and illustrated in the above figure, the potential ϕ_X at a point X a distance a from O is given by

$$\phi_X = \frac{1}{4\pi\epsilon_0}\left(\frac{q}{|BX|} - \frac{qa}{b|AX|}\right). \qquad (*)$$

From the geometry of the situation,

$$|BX|^2 = [(b - a\cos\theta)^2 + (a\sin\theta)^2] \text{ and } |AX|^2 = [(a\cos\theta - a^2/b)^2 + (a\sin\theta)^2],$$

and so it follows from result (a) that $b|AX| = a|BX|$. This, in turn, shows that the expression in the parentheses in $(*)$ is identically zero, and so therefore is ϕ_X. As this result follows for all values of θ, all points distant a from O have the same potential, namely zero.

(c) In order to investigate the applicability of the 'theorem', we need to compare the two situations, one with just the two charges, the other with a single charge outside an earthed sphere. The region under consideration is that *outside* and *on* a spherical surface centred on O and of radius a.

In both cases there is only one charge (that at B), and all the potentials on the boundaries – at infinity, on the conducting *earthed* physical sphere, and on the computed spherical surface that coincides with the sphere's location – are zero. So, the two systems satisfy the same charge placement and boundary conditions, and it follows from the 'theorem' that the forces in the specified region can be computed using either scenario.

Naturally, we use the much simpler two charge system to compute the result for the complex earthed conducting-sphere plus a single charge arrangement:

$$F = \frac{q \times (qa/b)}{4\pi\epsilon_0(b - a^2/b)^2} = \frac{q^2ab}{4\pi\epsilon_0(b^2 - a^2)^2}.$$

S 166 In order to determine the time constant for the charge decay, we need to know both the capacitance C of the pair of cylinders and the resistance R between them. The decay process is effectively that of discharging a capacitor through a resistor, for which the time constant $\tau = CR$.

Clearly, both the capacitance and the resistance depend upon the geometry and length of the cylinders, and we are given no information about these values. We therefore must assume values a and b for the cylinder radii $(b > a)$, and L for its length, even though we do not expect them to appear in the final answer.

Just as for a spherical conductor, the field outside a conducting cylinder depends only upon the total charge inside it, and not on its distribution within the cylinder. So, the field in the region $a < r < b$ is the same, whether the charge is on the axis or distributed (uniformly) over the inner cylinder. Consequently, if the charge on the smaller cylinder is Q the electric field in $a < r < b$ is given by

$$E(r) = \frac{Q}{2\pi\epsilon Lr}, \text{ with } \epsilon = \epsilon_r\epsilon_0.$$

Now the field is related to the potential by $E(r) = -dV(r)/dr$, and so

$$V(a) - V(b) = -\frac{Q}{2\pi\epsilon L}\int_b^a \frac{1}{r}\,dr = \frac{Q}{2\pi\epsilon L}\ln\frac{b}{a}.$$

Since $Q = CV$ defines capacitance,

$$C = \frac{2\pi\epsilon L}{\ln(b/a)}.$$

To calculate the resistance between the cylinders, we consider the resistance offered by a continuum of thin cylindrical shells of thickness dr, all connected in series (from $r = a$ to $r = b$); one is shown in the figure as a dark ring.

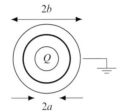

$$Q(t) = Q(0)e^{-t/\tau} \text{ with } \tau = CR$$

The contribution of one shell at radius r is $dR = \rho\,dr/(2\pi rL)$.[28] The full resistance is therefore

$$R = \frac{\rho}{2\pi L}\int_a^b \frac{1}{r}\,dr = \frac{\rho}{2\pi L}\ln\frac{b}{a}.$$

Finally, the time constant:

$$\tau = CR = \frac{2\pi\epsilon L}{\ln(b/a)}\frac{\rho\ln(b/a)}{2\pi L} = \epsilon\rho = \epsilon_r\epsilon_0\rho.$$

Despite what we noted earlier, it is still perhaps a little surprising that the time constant does not depend at all on the radii of the cylinders, nor on their length!

S 167   Because of its awkward shape and the lack of some dimensions, it is not possible to calculate the absolute value of the system's capacitance, nor, consequently, of the charge it carries.

However, we can find the reduction in the capacitance of the two cylinders when their overlap is reduced by Δx, provided any resulting drop in oil level can be ignored, or compensated for by feeding oil from a parallel reservoir. Although we do not know what it is, the total charge on the capacitor will remain constant (the voltage will not), and for this reason we consider the stored energy in the form $E = Q^2/2C$.

---

[28] In the usual formula, $R = \rho\ell/A$, the thickness 'ℓ' is here dr, and the face area 'A' is $2\pi r \times L$.

As is shown in the solution to Q 166 appearing on page 285, the capacitance per unit length of a pair of concentric cylinders that are long compared to the cylinder separation, and of radii a and b ($b > a$), is given by

$$C = \frac{2\pi\, \epsilon_r \epsilon_0}{\ln(b/a)}, \text{ here equal to } \frac{2\pi \times 4.5 \times 8.8 \times 10^{-12}}{\ln(52.5/50)} = 5.1 \times 10^{-9} \text{ F m}^{-1}.$$

So, when cylinder A is withdrawn by Δx, the change in capacitance is $\Delta C = -5.1 \times 10^{-9}\,\Delta x$ F.

The stored energy, and the change in it, are given by

$$E = \frac{Q^2}{2C} \quad \text{and} \quad \Delta E = -\frac{Q^2}{2C^2}\,\Delta C.$$

We may not know Q or C, but we do know that their ratio is $V = 30$ kV, and consequently that

$$\Delta E = -\frac{1}{2} \times 9.0 \times 10^8\,(-5.1 \times 10^{-9}\,\Delta x) = +2.3\,\Delta x.$$

The energy has increased as a result of the 'plunger' being withdrawn. Therefore, work must have been done in the process. We can conclude that the axial force between the cylinders must be such as to try to increase their overlap (downwards on cylinder A) and be of magnitude $2.3\,\Delta x/\Delta x = 2.3$ N.

S 168

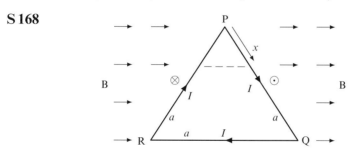

Being parallel to the magnetic field, side QR of the triangle has no electromagnetic force acting upon it. Sides PQ and RP are affected by the field, but they make angles of $\pi/3$ with it. As the components of the current that are perpendicular to the field flow in opposite directions in these two sides, one (PQ) experiences a force directed out of the plane of the figure, whilst the force on RP acts into the paper. They therefore produce a couple about a vertical axis. However, the contributions to that couple from equal-length sections of wire are not equal, because they are at varying distances from the axis.

Intuitively, we could assume that the total couple is one net force, $BIa\sin(\pi/3)$, multiplied by the average width of the triangle $a/2$; this gives the correct answer, but a more rigorous calculation is as follows.

Denote the distance from P down side PQ by x, as in the figure. The force acting on a length dx is

$$dF = IB\,dx\,\sin\frac{\pi}{3},$$

and the couple due to this force and the corresponding infinitesimal length on side PR is

$$dC = IB\sin\frac{\pi}{3}\,dx \times 2x\cos\frac{\pi}{3} = \frac{\sqrt{3}IB}{2}x\,dx.$$

The total couple is therefore

$$C = \frac{\sqrt{3}IB}{2}\int_0^a x\,dx = \frac{\sqrt{3}IBa^2}{4},$$

in line with the 'expected' result.

S 169 (a) The field strength is constant along AD, is directed out of the paper, and has strength $B_{\mathrm{AD}} = 2\mu_0/(10\pi \times 10^{-2})$ T. The force on side DA is therefore (using $F = BIL$)

$$B_{\mathrm{DA}} \times 3 \times 0.2 = \frac{8\pi \times 10^{-7}}{10\pi \times 10^{-2}}\,0.6 = 4.8 \times 10^{-6}\,\mathrm{N},$$

and from Fleming's left hand rule (see the footnote on page 279) is directed towards the straight wire. That on BC is $50/(100 + 50) = 1/3$ of this, and directed away from the wire. The net result is $(2/3) \times 4.8 \times 10^{-6} = 3.2 \times 10^{-6}$ N towards the wire. Any forces acting on the sides AB and CD cancel each other, as, although they sit in identical fields, the 3 A current flows in opposite directions in the two cases.

(b) For the individual side AB, the force is everywhere directed away from the centre of the rectangle, but, as the field strength varies along its length, an integration is necessary. For a small length dr that is distance r from the wire, the force is $2\mu_0/(2\pi r) \times 3\,dr$, and so the total force is

$$F = \frac{3\mu_0}{\pi}\int_{r_A}^{r_B}\frac{dr}{r} = \frac{3\mu_0}{\pi}\ln\frac{150}{50} = \frac{3 \times 4\pi \times 10^{-7}}{\pi}\ln 3 = 1.32 \times 10^{-6}\,\mathrm{N}.$$

An equal but opposing force acts on the side CB.

S 170 As the magnetic field only affects charges moving perpendicular to it, we resolve the electron's initial velocity into $v_\| = v\cos\theta$ parallel to the field and $v_\perp = v\sin\theta$ perpendicular to it. The latter causes the electron to follow a circular path in a (moving) plane perpendicular to the field, with the radius r of the circle given by

$$\frac{mv_\perp^2}{r} = Bev_\perp \quad \Rightarrow \quad r = \frac{mv_\perp}{Be}.$$

As $v_\parallel$ is unaffected by the field, the electron's path is a spiral, as shown in the figure.

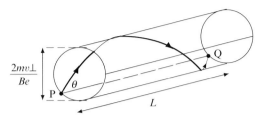

Although there is a range of values for $v_\perp$, the time taken for an electron to complete the circle, $\tau = 2\pi r/v_\perp = 2\pi m/Be$, is independent of it, i.e. all electrons are directly downstream from P after a time τ. During this time any particular electron will have travelled a distance

$$L = v_\parallel \tau = \frac{2\pi m v \cos \theta}{Be}.$$

The electron will pass within $(1 - \cos \theta)L \sim L\theta^2/2$ of the point Q, which lies a distance L directly downstream from P. Even for an electron with θ as large as $30°$, this is only $0.13L$. As all values of θ are 'small', all the electrons will pass Q at a distance that is very small compared to L.

S 171 Since the ring P is being supported against gravity, the force F it experiences *must* act upwards. To achieve this, either from knowing that oppositely directed currents repel each other (unlike oppositely charged particles), or from applying Fleming's left hand rule to determine the appropriate direction for B, we require the currents in P and Q to flow in opposite senses.

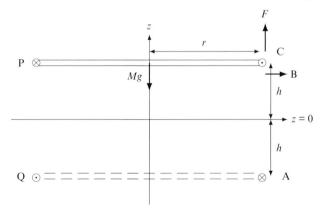

This is shown in the figure, where the current flowing into the paper at A produces the field B at the corresponding point C of ring P in the direction needed to make the force F act upwards, which it does if the current there is directed out of the paper, as shown.

As $r \gg h$, at any particular point, we can ignore the magnetic field due to the distant parts of the other ring, and also treat the field acting on any small part of one ring as if it were due to a parallel wire a distance $2h$ away. Thus the field flux is $B = \mu_0 I / 2\pi(2h)$, where I is the common magnitude of the two currents. The resulting force, taken over the complete ring, is therefore

$$F_t(h) = \frac{\mu_0 I}{4\pi h} \times I \times 2\pi r.$$

The ring will be in equilibrium when this is equal to the gravitational force, i.e.

$$\frac{\mu_0 I^2 r}{2h} = Mg \quad \Rightarrow \quad h_0 = \frac{\mu_0 I^2 r}{2Mg}.$$

For a more general h, the upward displacement force is

$$F_t(h) = \frac{\mu_0 I^2 r}{2h} - Mg = Mg \left(\frac{h_0}{h} - 1 \right),$$

and the resulting acceleration is

$$\ddot{h} = \frac{F_t(h)}{M} = g \left(\frac{h_0}{h} - 1 \right).$$

Now, if we set $h = h_0 + x$,

$$\ddot{x} = g \left(\frac{h_0}{h_0 + x} - 1 \right) = g \left[\left(1 + \frac{x}{h_0} \right)^{-1} - 1 \right]$$

$$= g \left(1 - \frac{x}{h_0} + \cdots - 1 \right) \approx -\frac{x}{h_0} g.$$

It follows that the ring's vertical position will execute approximately SHM with a frequency of $(2\pi)^{-1} \sqrt{g/h_0}$.

S 172   The figure shows a plan view of the tube, with the positive z-direction being out of the plane of the paper.

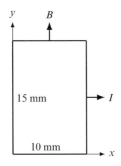

It does not matter what the mercury depth h is, in so far as the electromagnetic force on it is concerned; the total current is I and the magnetic field B acts upon it over a length of 10 mm. The total force acts upwards (Fleming's left hand rule) and has magnitude $2.0 \times 1.0 \times 10^3 \times 10^{-2} = 20$ N.

This force acts in the opposite direction to the weight of the mercury, which is

$$15 \times 10^{-3} \times 10 \times 10^{-3} \times 13.6 \times 10^3 \times 9.81 \times h = 20.0\,h\,\text{N}.$$

However, the maximum value of h is 0.90 m, giving a maximal gravitational force of 18.0 N. As this is less than the electromagnetic one, mercury will be ejected from the top of the tube, and will continue to be ejected as the total current flowing will not be reduced. Until, of course, the tube is empty, when mercury is all over the floor and the emergency services are called.

S 173 The figure shows the directions of the currents and the angle θ. It does not matter which current is I_1, and which is I_2, as we will only be concerned with their product.

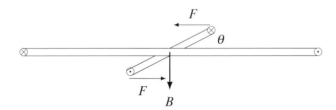

We are given the strength of the central field B, but the direction of it, for the current directions shown, has to be determined from the 'clockwise, when viewed along the current direction' rule (see Q 169). Applying it shows that *all* parts of the larger loop produce contributions that are in the direction shown in the figure.

We now turn to the forces and couple that act on the square loop. Each of its vertical sides carries the same current at right angles to the horizontal field, and so experiences a horizontal force F parallel to the plane of the fixed loop. However, as the currents in the two sides are in opposing directions, so are the forces, resulting in a couple of magnitude $F \times a \sin \theta$ acting on the square loop.

Taking into account the couple due to the torsion thread, the net couple C acting upon the square loop, and tending to increase θ is

$$C = \frac{\mu_0 I_1 I_2 a}{2R} \times a \sin \theta - k\theta.$$

For $\theta = 0$ to be a point of stable equilibrium, C has to be negative when $\theta \approx 0$. At that point $\sin \theta \approx \theta$ and so the required condition is that

$$k > k_0 = \frac{\mu_0 I_1 I_2 a^2}{2R}.$$

(i) If $k = (2/\pi)k_0$, the couple becomes

$$C = k_0 \sin \theta - \frac{2}{\pi}k_0\theta,$$

and, by inspection, this is equal to zero when $\theta = \frac{1}{2}\pi$. Moreover, for $\theta > \frac{1}{2}\pi$ the $\sin \theta$ term decreases whilst θ increases, and so θ does not increase further, and the square loop ultimately sets itself perpendicular to the circular loop.

(ii) For the yet weaker torsion thread $k = \frac{1}{2}k_0$, the square will move through more than $\frac{1}{2}\pi$, and the corresponding 'no net couple' equation is

$$\theta = 2 \sin \theta.$$

This is a 'transcendental' equation with no routine method of solution. It can be solved, to any given degree of accuracy by reasoned trial and error or certain numerical methods, and leads to $\theta = 108.6°$. This is the angle the square loop will turn through, though in practice it will reach it via a damped oscillation.

For intrigued mathematicians On the assumption that the solution does not differ much from $\theta = \frac{1}{2}\pi$ (and this can be checked retrospectively), the following solution using only school mathematics, is possible.

Suppose that the required solution is $\theta = \frac{1}{2}\pi + \delta$, where δ is small compared to unity. Then we have

$$\frac{\pi}{2} + \delta - 2\left(\sin\frac{\pi}{2}\cos\delta + \cos\frac{\pi}{2}\sin\delta\right) = 0,$$

$$\frac{\pi}{2} + \delta - 2\cos\delta = 0,$$

$$\frac{\pi}{2} + \delta - 2\left(1 - \frac{\delta^2}{2!} + \cdots\right) = 0,$$

$$\delta^2 + \delta + \left(\frac{\pi}{2} - 2\right) \approx 0.$$

The (positive) solution to this approximate quadratic equation is

$$\delta = \frac{-1 + \sqrt{1 - 4\left(\frac{\pi}{2} - 2\right)}}{2} = 0.3241 \text{ rad} = 18.6°.$$

So, $\theta = 90 + 18.6 = 108.6°$. Amazing!

S 174   Knowing r, to find τ all we need is the speed v of the meson. This will have to be determined from the given value of r.

We first note that since the force on the meson resulting from its motion in the magnetic field is perpendicular to that motion, no work is done on the meson, and its speed remains constant. The sideways force causes it to move into a circular orbit, with a radius given by

$$\frac{mv^2}{r} = Bev \quad \Rightarrow \quad r = \frac{mv}{Be}.$$

The expression to be substituted for m is that given in the question, with $m_0 = 207 \times 9.11 \times 10^{-31} = 1.886 \times 10^{-28}$ kg. The resulting equation is

$$0.198 = \frac{1.886 \times 10^{-28}\, v}{2.00 \times 1.60 \times 10^{-19} \times \sqrt{1 - (v/c)^2}}.$$

To solve this for v, we cross-multiply and square both sides.

$$(0.634 \times 10^{-19})^2 \left(1 - \frac{v^2}{c^2}\right) = (1.886 \times 10^{-28})^2 v^2 = (5.66 \times 10^{-20})^2 \frac{v^2}{c^2},$$

$$\left(1 - \frac{v^2}{c^2}\right) = \frac{(5.66 \times 10^{-20})^2}{(6.34 \times 10^{-20})^2} \frac{v^2}{c^2} = 0.797\frac{v^2}{c^2},$$

$$\frac{v^2}{c^2} = \frac{1}{1.797} \quad \Rightarrow \quad v = 0.746c.$$

Finally,

$$\tau = \frac{2\pi r}{v} = \frac{2\pi \times 0.198}{0.746 \times 3.00 \times 10^8} = 5.6 \times 10^{-9}\ \text{s}.$$

Note Since $\tau = 2\pi r/v = 2\pi m/Be$, it might seem that there is no need to find v. However, v is involved in the relativistic expression for m. It could be replaced in that expression by $2\pi r/\tau$, but then the equation for τ is no less complicated than that for v in the above solution.

S 175 Clearly, the sag in the wire vanishes when an upwards electromagnetic force just balances the wire's weight. Denoting its cross-section by A, and its density and length by ρ and L respectively, the wire's mass is $m = \rho LA$ and its resistance is $R = \rho_e L/A$, where ρ_e is the resistivity. With an applied voltage V_0 the current in the wire is V_0/R, and the force balance takes the form

$$B\frac{V_0 A}{\rho_e L} L = \rho LAg.$$

From this

$$\rho_e = \frac{BV_0}{\rho Lg} = \frac{1.0 \times 10^{-2} \times 88 \times 10^{-3}}{9.0 \times 10^3 \times 0.5 \times 9.81} = 2.0 \times 10^{-8}\ \Omega\,\text{m}.$$

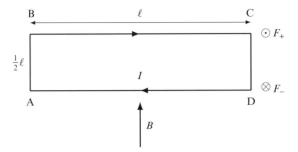

The figure shows a plan view of the newly formed rectangle lying on the rough table; its long sides are of length ℓ, where $3\ell = L$. As the long sides carry current in opposing directions, the electromagnetic forces acting on them are equal in magnitude but opposite in sense. They therefore produce a couple on the loop, acting about a horizontal axis perpendicular to the field, i.e. the axis is parallel to the long sides. With the circulating current and magnetic field as shown in the figure, the force F_+ on BC is upwards ($\odot$), whilst that on DA is F_- which acts vertically downwards ($\otimes$).

Since the table is rough, DA cannot slide – nor can it move downwards. Consequently, the only motion to consider is rotational with DA as the axis of rotation. When V is small, nothing happens; there is an upward force $VB\ell/R$ acting on BC, but gravity provides a couple of magnitude $\frac{1}{4}mg\ell$ which initially is greater than $VB\ell/R \times \frac{1}{2}\ell$, the couple resulting from the force on BC.

When V is large enough, the situation is reversed, and side BC rises from the table. This will happen when the force on BC is equal to $\frac{1}{2}mg$. This same force is needed even when the plane of the loop has been raised through an angle θ; this is because the arms of both couples, electromagnetic and gravitational, are reduced by the same factor, $\cos\theta$. At this critical value of V, the loop, once in motion, will continue to rise, may overshoot, but will finish up vertical.

Quantitatively, a force of mg is produced on a wire of length $L = 3\ell$ when $V = V_0$. Therefore to produce a force of $\frac{1}{2}mg$ using a wire of length ℓ requires a voltage of $3V_0/2$. This value is in the given range of $0 \leq V \leq 2V_0$. For $\frac{3}{2}V_0 < V \leq 2V_0$ nothing further will be apparent.

S 176

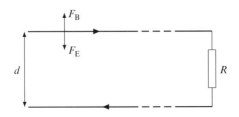

As the two cables act as a capacitor charged to $V = 33$ kV, the charges on them are of opposite signs, and the force between them is attractive. The size of the force per unit length has been given as $F_E = \sigma^2/(2\pi\epsilon_0 d)$, with $\sigma = C_1 V$ for a capacitance per unit length of C_1.

The magnetic-field induced force is, as shown in solution S 169, attractive for parallel currents and repulsive for anti-parallel ones; here, they are anti-parallel. Further, the repulsive force per unit length is $F_B = \mu_0 I^2/(2\pi d)$. Thus, there will be no net force when

$$\frac{C_1^2 V^2}{2\pi\epsilon_0 d} = \frac{\mu_0 I^2}{2\pi d},$$

$$R^2 = \frac{V^2}{I^2} = \frac{\epsilon_0\mu_0}{C_1^2}$$

$$= \frac{8.8 \times 10^{-12} \times 4\pi \times 10^{-7}}{1.0 \times 10^{-22}} = 1.11 \times 10^5 \ \Omega^2,$$

$$R = 331 \ \Omega.$$

With this (equivalent) load resistance, the required transmitted power is $P = V^2/R = 3.3$ MW, and does not depend on how far apart the cables are!

S 177 As we can assume that the deflections are all small, we can also assume that the transverse velocity acquired by an ion as a result of the electric field, does not significantly affect the time it spends in the region of length ℓ. Let an ion's initial (and approximate final) speed be V.

Considering first its motion in the z-direction: with both initially equal to zero, the velocity and position of the ion on exiting the fields are

$$v_z = \frac{eE}{M}\frac{\ell}{V} \quad \text{and} \quad z = \frac{eE}{2M}\frac{\ell^2}{V^2}.$$

When it travels through the field-free region it does so at an angle of $\approx v_z/V$ to the y-axis, and so its total z-displacement on the screen is

$$z = \frac{eE}{2M}\frac{\ell^2}{V^2} + \frac{eE\ell}{MV}\frac{L}{V} = \frac{eE\ell}{MV^2}\left(\frac{1}{2}\ell + L\right)$$

in the positive z-direction.

To obtain the motion in the x-direction, we need only replace eE by eBV, and so, at the screen,

$$x = +\frac{eB\ell}{MV}\left(\frac{1}{2}\ell + L\right).$$

Writing A for $e\ell(\frac{1}{2}\ell + L)/M$, the two results can be summarised by

$$x = \frac{AB}{V} \text{ and } z = \frac{AE}{V^2} \quad \Rightarrow \quad x^2 = \frac{AB^2}{E}z.$$

In words, for any given A (i.e. given ion mass) the fluorescence lies along a portion of a parabola (symmetric about the z-axis and passing through the origin), the range of V-values determining how large the portion is. Since A is inversely proportional to M, for any particular (non-zero) value of x, the corresponding value of z is larger if the mass is larger. Thus the ^{22}Ne parabola lies 'above' that for ^{20}Ne, as shown by the thickest line segments in the figure below.

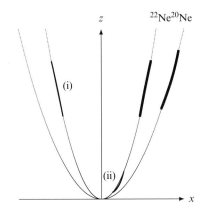

(i) If the direction of B is reversed, so is the x-deflection, motion in the z-direction being unaffected. This is shown for ^{22}Ne by the medium thickness line marked (i) in the figure.
(ii) If the mean energy of the ions is increased significantly, the fluorescence on each of the *same* pair of parabolas will move to a region that is nearer the origin, as indicated by symbol (ii) in the figure, again for ^{22}Ne.

S 178   In order to find the current through the axle we will need to determine (i) the emf generated, and (ii) the radial resistance of the disc.

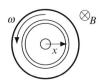

(i) In an obvious notation, the rate at which flux lines are being cut, expressed as the emf generated, is

$$\tfrac{1}{2}\omega R^2 \times B = \tfrac{1}{2}\, 10\pi\, (1.2)^2\, 10 = 226\,\text{V}.$$

(ii) Consider an annulus of radius x (see figure), thickness t, radial depth dx, and resistivity ρ. Its resistance is

$$dR = \frac{\rho\, dx}{2\pi x \times t},$$

and the total resistance of all the annuli (in series) is

$$R = \frac{\rho}{2\pi t} \int_{0.02}^{1.2} \frac{dx}{x} = \frac{\rho}{2\pi t}\, \ln 60 = 6.52 \times 10^{-6}\,\Omega.$$

When the leads are shorted, the total current that flows is

$$I = \frac{226\,\text{V}}{6.52 \times 10^{-6}\,\Omega} = 3.5 \times 10^7\,\text{A}.$$

Such an enormous current in such a large magnetic field will have serious mechanical repercussions – but that's another story!

S 179 From the data, the cross-sectional diameter of the ring is of the order of mm, and so is much smaller than the ring's radius. It follows that the only information needed about the magnet and its field is the radial magnetic flux density it produces at radius r; this determines the induced emf $\mathcal{E}$ as $2\pi r v B$.

Since the resistance of the ring is $2\pi r \rho_\text{e}/A$ the current that flows in the ring is

$$I = 2\pi r v B\, \frac{A}{2\pi r \rho_\text{e}} = \frac{A v B}{\rho_\text{e}},$$

in the sense shown in the figure in the question (Fleming's RHR). This current experiences an upwards (apply the LHR) force of $2\pi r I B$.

As the ring is not accelerating, there can be no net force acting on it, and so we must have that the upward electromagnetic force balances the downward gravitational one:

$$2\pi r\, \frac{A v B}{\rho_\text{e}}\, B = 2\pi r A \rho g,$$

$$v = \frac{\rho g \rho_\text{e}}{B^2}$$

$$= \frac{2.7 \times 10^3 \times 9.81 \times 2.8 \times 10^{-8}}{(0.8)^2} = 1.2\,\text{mm s}^{-1}.$$

Apart from g, only three actual numerical values are needed!

S 180   The flux through the loop is changing only during the interval between the times when its two sides of length b enter the field region. This is when current will flow in the loop, which will experience a retarding force and hence require work to be done on it to keep it moving. Once the loop is fully in the field region, if there were any current flowing in it the forces on opposite sides of the rectangle would cancel in pairs, no work would be necessary, yet heat would be continuously generated; it follows that there can be no such current.

The left hand figure below shows the situation when the loop is just half way into the field region.

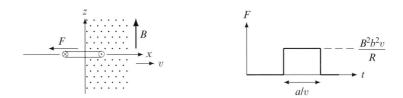

Whilst the loop is entering the field, the rate at which the magnetic flux through it is increasing is $bv \times B$. This gives the magnitude of the induced emf and Fleming's RHR shows that the direction of the consequent current flow is out of the paper, as shown in the figure. The magnitude of this current is $I = bvB/R$, and the 'leading edge' of the loop will experience a force of $F = B \times I \times b = B^2b^2v/R$ in the direction opposed to v, as given by the LHR.

This situation will last until the 'trailing edge' of the rectangle has entered the field, i.e. for a period of a/v. After this, there will be no current and hence no force. The right hand figure summarizes the time dependence of the force.

The total work done is 'force × distance moved in the direction of the force', and here this is $(B^2b^2v/R) \times a$. The total heat dissipated is 'power × time', and this is

$$H = RI^2 \times \frac{a}{v} = R\left(\frac{bvB}{R}\right)^2 \frac{a}{v} = \frac{B^2b^2va}{R};$$

as expected, the same as the total work done.

S 181   The situation is shown symbolically in the figure, with $D = 20$ mm, the mass $m = 10$ g and the magnetic flux density $B = 0.5$ T.

The two gear wheels revolve in *opposite* directions, with the larger one having an angular velocity ω, say; the smaller one then rotates at 5ω. However, the two radii to the contact point move in the same direction; consequently, the electric fields generated there reinforce each other. For the motion and field directions in the figure, Fleming's RHR shows that the axle of the smaller wheel is at the higher potential.

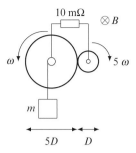

The rate of flux change through each wheel is the flux density multiplied by $\frac{1}{2} \times$ angular speed $\times$ (radius)^2, and so the net voltage generated is

$$V = \frac{1}{2}B\left[5\omega\,(10 \times 10^{-3})^2 + (50 \times 10^{-3})^2\,\omega\right] = 7.5 \times 10^{-4} \times \omega\ \text{V}.$$

When the mass is moving downwards with its terminal speed $v = a\omega$, where a is the radius of the larger wheel's axle, the rate of loss of gravitational energy will be equal to the electrical dissipation rate, V^2/R, i.e.

$$\frac{(7.5 \times 10^{-4})^2}{10 \times 10^{-3}}\,\omega^2 = mgv = 10 \times 10^{-3} \times 9.81 \times 1.0 \times 10^{-3} \times \omega.$$

This gives ω as $1.74\ \text{rad s}^{-1}$ and a terminal speed for the mass of a times this, i.e. $1.74\ \text{mm s}^{-1}$.

S 182 This is a straightforward application of the basic law of electromagnetism, $\mathcal{E} = -d\Phi/dt$.

Let B_0 be the minimum amplitude that can be detected, then $B = B_0 \cos \omega t$ and, in an obvious notation, the flux through the aerial is $\Phi = n A B_0 \cos \omega t$. The generated emf is therefore

$$\mathcal{E} = \frac{d\Phi}{dt} = \omega n A B_0 \sin \omega t.$$

As the aerial has negligible resistance, the full emf is delivered to the radio set at a power level of $\bar{\mathcal{E}}^2/R$. Since the induced emf is sinusoidal, $\bar{\mathcal{E}}^2 = \mathcal{E}^2/2$, and so we need

$$\text{noise power} = 10^{-12} < \frac{(2\pi f n A B_0)^2}{2R}.$$

Substituting values, and rearranging the inequality, gives

$$B_0 > \frac{\sqrt{2 \times 600 \times 10^{-12}}}{2\pi \times 10^5 \times 100 \times 0.25} = 2.2 \times 10^{-12}\ \text{T}.$$

S 183 The figure shows the arrangement, with the rails in the 'vertical' direction and the two wires 'horizontal'.

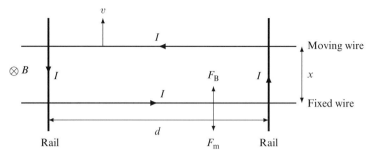

The moving wire in the magnetic field generates an emf, and so causes a current to circulate around the (growing) rectangular loop. As a result, the fixed wire, which is part of the loop, experiences a force F_B due to the applied field B; this force is in a direction parallel to the rails. It will also be acted upon by a force F_m caused by the azimuthal magnetic field associated with the current-carrying moving wire; this force will also be in a direction parallel to the rails.

The emf and current generated by the moving wire are straightforwardly given by

$$\mathcal{E} = -\frac{d\Phi}{dt} = dvB \quad \text{and} \quad I = \frac{\mathcal{E}}{(2d + 2x)\rho}.$$

For the field direction (into the paper) and wire motion shown in the figure, Fleming's RHR shows the current direction to be anticlockwise around the loop.

The force on the fixed wire due to the applied field is $F_B = BId$, and the LHR shows that it is directed in the same sense as v.

The second force affecting the fixed wire is that of repulsion between parallel wires carrying opposing currents. The force involved is that acting on a current I in a field of flux density $(\mu_0 I)/(2\pi x)$, directed out of the paper for the situation illustrated in the figure (see Q 169, if this is not familiar). Its strength is

$$F_m = \frac{\mu_0 I}{2\pi x} I d.$$

The net force on the fixed wire is zero when $x = x_0$, defined by $F_m = F_B$, i.e.

$$\frac{\mu_0 I^2 d}{2\pi x} = BId,$$

or, substituting more explicitly for I,

$$\frac{dvB}{2(d + x_0)\rho} = I = \frac{2\pi x_0 B}{\mu_0}.$$

This (quadratic) equation for x_0 can be rearranged as

$$x_0^2 + dx_0 - \lambda d = 0, \text{ with } \lambda = \frac{\mu_0 v}{4\pi\rho}.$$

With d a positive quantity and the constant term negative, this quadratic equation will have exactly one positive root, $\frac{1}{2}(\sqrt{d^2 + 4\lambda} - d)$.

S 184   Let the particle's mass be m, and denote the *average* flux density *within* the orbit by $\widehat{B}$, which must be carefully distinguished from B, the flux *at* the orbit. The total flux enclosed by the orbit is $\Phi = \pi r^2 \widehat{B}$. Consequently, the electric field generated by a changing field strength, with no change in r, is

$$E = \frac{1}{2\pi r} \pi r^2 \frac{d\widehat{B}}{dt},$$

and the increase in particle velocity in time interval dt is

$$dv = \frac{Ee}{m} dt = \frac{er}{2m} \frac{d\widehat{B}}{dt} dt = \frac{er}{2m} d\widehat{B}. \tag{$*$}$$

In the steady state, the centripetal force F satisfies

$$\frac{mv^2}{r} = F = evB. \tag{$**$}$$

To maintain the same radius for the orbit, we need the infinitesimal changes on both sides of this equation to be equal. The electromagnetic force involves two variables, B and v, both of which change. So, we need

$$\frac{2mv}{r} dv = dF = ev\, dB + eB\, dv.$$

From ($*$) we can substitute for dv, and use ($**$) to eliminate B from the final term. We then obtain

$$ev\, d\widehat{B} = ev\, dB + eB \frac{er}{2m} d\widehat{B} = ev\, dB + \frac{1}{2} ev\, d\widehat{B}.$$

It then follows that the required condition is $B = \frac{1}{2}\widehat{B}$.

S 185   Although the voltage sources in this situation are not standard cells, the usual laws of current conservation at junctions, and a sum of the form '$\Sigma R_k I_k$' round any closed circuit being equal to the sum of the voltage sources included, is still applicable.

Let the current in the main circular loop be I and that part which is diverted through the voltmeter be i, as indicated in the figure.

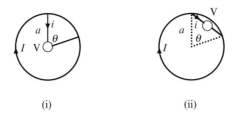

(i)   (ii)

Then, for both situations, (i) and (ii), we consider two closed *clockwise* circuits, (a) around the complete circle, and (b) around the loop formed by the voltmeter, its leads, and the part of the circle of length $a\theta$ between the junction points. Denoting the area enclosed by circuit (b) by A_b, we have

$$a\rho(2\pi - \theta)I + a\rho\theta(I - i) = \pi a^2 \dot{B},$$
$$-Ri + a\rho\theta(I - i) = A_b \dot{B}.$$

A minor rearrangement of these equations gives

$$2\pi a\rho I - a\rho\theta i = \pi a^2 \dot{B}, \qquad\qquad (*)$$
$$a\rho\theta I - (R + a\rho\theta)i = A_b \dot{B}. \qquad\qquad (**)$$

Since we need only the value of Ri, we eliminate I by multiplying $(*)$ by θ, and $(**)$ by 2π and subtracting:

$$-a\rho\theta^2 i + 2\pi(R + a\rho\theta)i = (\pi a^2\theta - 2\pi A_b)\dot{B}. \qquad\qquad (\dagger)$$

Considerations so far cover both situations, (i) and (ii), but now we consider them individually.

(i) Here $A_b = \frac{1}{2}a^2\theta$ and the RHS of ($\dagger$) is equal to zero, and so, therefore, is i; the voltmeter will read 0 V.

(ii) In this case A_b is the difference between the area $\frac{1}{2}a^2\theta$ of the sector, and that of the isosceles triangle of side a and included angle θ, i.e. $\frac{1}{2}a^2 \sin\theta$. Substituting this into ($\dagger$) gives

$$\left[2\pi(R + a\rho\theta) - a\rho\theta^2\right] i = \left(\pi\theta - \tfrac{1}{2}2\pi\theta + \tfrac{1}{2}2\pi \sin\theta\right) a^2 \dot{B},$$

showing that the voltmeter reading will be

$$V = Ri = \frac{\pi \sin\theta \, a^2 R\dot{B}}{2\pi R + a\rho\theta(2\pi - \theta)}.$$

As might be expected, this gives $V = 0$ when $\theta = \pi$.

S 186 In these situations, in which a motion generates an emf, Fleming's right hand rule (RHR) is the appropriate one to apply.[29]

(a) The emf is generated when the conducting tidal (salt)water flows across the (downwards) vertical component, $B_V = B \sin 66°$, of the Earth's field. When the tide is coming in, the North bank is positive and the emf is

$$\mathcal{E} = B_V \times 2 \times 250 = 5 \times 10^{-5} \times 0.9135 \times 2 \times 250 = 23\,\text{mV}.$$

On an outgoing tide the South bank is at the higher potential.

(b) $720\ \text{km h}^{-1} = 200\ \text{m s}^{-1}$. We give only the numerical expression for the emf in the form $Bv\ell$; if necessary, the corresponding physical quantities can be established from the values used. We note that the jet's length is irrelevant, as it is parallel to the direction of flight and therefore cannot have a voltage developed across it.

(i) $\mathcal{E} = 6 \times 10^{-5} \times 200 \times 80 = 0.96$ V across the wings, with the starboard (right) wing tip positive.

(ii) The jet is flying parallel to the Earth's field and there are no generated emfs.

(iii) $\mathcal{E} = 3 \times 10^{-5} \times 200 \times 8 = 48$ mV across the height, with the bottom positive.

(iv) The emf generated from the vertical component of the Earth's field is independent of the direction of flight, but the vertical emf generated is not.

- $\mathcal{E}_H = 5 \times 10^{-5} \sin 66° \times 200 \times 80 = 0.73$ V across the wings, with the starboard (right) wing tip positive.
- $\mathcal{E}_V = 5 \times 10^{-5} \cos 66° \times 200 \cos 45° \times 8 = 0.23$ mV across the height, with the top positive.

S 187

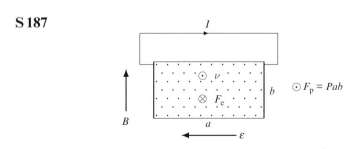

---

[29] Fleming's right hand rule is applicable when a conductor moves at right angles to a magnetic field, and gives the direction of the emf generated in the conductor. If the thumb, and the first and second fingers of a right hand are held mutually orthogonal, then the **F**irst finger represents the **F**ield, the s**E**cond the **E**mf direction, and the thu**M**b the **M**otion. The second 'instruction' is sometimes replaced by 'the se**C**ond the **C**urrent direction'.

The figure shows a cross-section of the pipe, with the conducting walls and field direction 'vertical', and the mercury flow direction out of the paper. The force F_p due to the pressure difference also acts out of the paper.

This is a system with 'feedback', since the mercury flow in the magnetic field generates an emf, and hence a current, which, when interacting with the field, produces a force, which in turn affects the mercury flow velocity. This clearly leads to a self-consistency condition which will establish the value of v.

We start by considering the situation when there is no field present. The only force is that due to the pressure and that is Pab in magnitude. As this produces a flow velocity v_0, we have $v_0 \propto Pab$, and so for a more general net force F the resulting flow speed is

$$v = \frac{F}{Pab} v_0.$$

Next we find the emf generated when the field is present and the flow velocity is v. This is given by the field strength multiplied by the rate at which area perpendicular to the field changes, $\mathcal{E} = -d\Phi/dt = Bva$. The direction of the electric field it generates is given by Fleming's RHR and is as shown in the figure.

The current arising as a result of $\mathcal{E}$ is

$$I = \frac{\mathcal{E}}{R} \quad \text{where} \quad R = \frac{\rho a}{b\ell}, \text{ i.e. } I = \frac{Bvb\ell}{\rho}.$$

It flows in the same direction as $\mathcal{E}$, that is from right to left in the figure.

As this current flows in the presence of the field, there is a force $F_e = BIa$ acting on it. Fleming's LHR shows that its direction is into the plane of the paper, as shown in the figure, and, as to be expected, opposing the mercury flow that is ultimately the original cause of its existence.[30] As a result of this force, the net force acting on the mercury is reduced, and the new value of v satisfies

$$v = \frac{v_0}{Pab} \left(Pab - B \frac{Bvb\ell}{\rho} a \right)$$

$$= v_0 - \frac{v_0 B^2 \ell}{\rho P} v.$$

This, when rearranged, gives v as

$$v = \frac{v_0}{1 + \dfrac{v_0 B^2 \ell}{\rho P}}.$$

It should be noted that v is independent of the sign (direction) of B.

---

[30] This is an example of the widely applicable Le Chatelier's Principle, perhaps best known in connection with chemical equilibrium.

S 188 A qualitative answer may be something like the following:

Because we do not know the resistances of the coil windings, we cannot find the shapes of the pulses that arise when the contact either opens or closes, causing the capacitor to charge or discharge. But we do know that both events will lead to a change in the magnetic flux produced by the primary coil (connected to the battery) and consequently give rise to an emf in the secondary. The former will be of the order of 12 V, but the latter will be very much larger, as the secondary will have many more turns than the primary, and so, although they will have essentially the same flux density through them, the secondary will have many times more flux linkage.

In the secondary, there will be high voltage pulses (off opposite signs) delivered to the two spark plugs, and a large electric field will be produced in the narrow gap between the central conductor and the earthed casing of each plug. If this electric field is strong enough, there will be electrical breakdown of the air-petrol mixture, a spark will be generated, and the mixture will explode.

Both cylinders will receive a spark at each make or break, but only the one containing the (nearly) fully compressed vapour will make effective use of it – the other will hold only spent gases and be near the end of the exhaust part of its cycle.

Since the spark plug gap is 0.75 mm and an electric field intensity of 3×10^6 V m^{-1} is required, the emf in the secondary coil must be *at least* $7.5 \times 10^{-4} \times 3 \times 10^6 = 2.25$ kV. For an ideal voltage transformer, the turns ratio is equal to the ratio of secondary to primary voltages, and so here this ratio needs to be at least $2.25 \times 10^3 \div 12 \approx 200$.

Again, as we do not know the values of relevant resistances, we cannot calculate the current taken by the discharge, but we can say that its average value over one spark must be about 50 mJ $\div$ 2.25 kV ≈ 2 μA. Whilst both the current and voltage are changed by a transformer, their product, i.e. the power, is not – at least in the idealized case. The energy taken from the battery for each spark is 50 mJ, and so with four sparks per engine revolution, the average power taken at 2000 r.p.m. is

$$4 \times 5 \times 10^{-2} \times \frac{2000}{60} \approx 6\,\text{W},$$

leading to an average current of $6 \div 12 = 0.5$ A. In practice both the turns ratio and the battery current will be greater than the values calculated here.

S 189 We do not know the distribution of energies of the water molecules in the wet washing, but we can assume that it contains a Boltzmann factor of the form $\exp(-E/kT)$, where E is the energy needed for the process under consideration, here the release of one water molecule.

Now, $1 + 1 + 16 = 18$ kg of water contain 6.0×10^{26} molecules and requires 18×2.3 MJ to evaporate. The energy per molecule is therefore

$$E = \frac{18 \times 2.3 \times 10^6}{6.0 \times 10^{26}} = 6.9 \times 10^{-20} \text{ J per molecule.}$$

The rate at which the washing dries will be determined by the fraction of the water molecules with energy E. If T is the temperature required to increase that fraction by a factor of ten, then it must satisfy

$$e^{-E/kT} = 10\, e^{-E/kT_0},$$

$$\frac{E}{k}\left(\frac{1}{T_0} - \frac{1}{T}\right) = \ln 10,$$

$$\frac{1}{300} - \frac{1}{T} = \frac{1.4 \times 10^{-23}}{6.9 \times 10^{-20}} \ln 10,$$

$$T = 349 \text{ K.}$$

The required airing cupboard temperature is therefore $76°$C.

S 190   This is a problem in which we need to introduce additional symbolic quantities in order to carry out the required calculations. Denote the Sun's radius by $R_\odot$, and its distance from Earth by D.

The figure below shows the lens and two light rays from one edge of the Sun which are brought to a focus on one edge of the (inverted) image.

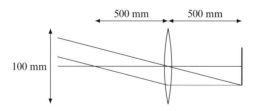

Since the Sun is far away, the two rays are virtually parallel and the image is formed in the focal plane of the lens.[31] It then follows that the angles subtended at the centre of the lens by the object and image are equal, i.e.

$$\frac{R_\odot}{D} = \frac{r}{f}, \qquad\qquad (*)$$

where r is the radius of the disc.

---

[31] The same conclusion is reached by setting $u = \infty$ in the lens formula.

The total power of the Sun's (black body) radiation is $\sigma T_\odot^4 \times 4\pi R_\odot^2$, and so that captured by the lens is

$$P = \frac{\pi(d/2)^2}{4\pi D^2} \sigma T_\odot^4 \, 4\pi R_\odot^2.$$

When the the disc has reached its achievable maximum temperature, this absorbed power must be reradiated by the (two-sided) disc, which will have a temperature T given by

$$P = \sigma T^4 \times 2\pi r^2.$$

Equating these two expressions for P gives (after some cancelling)

$$\frac{R_\odot^2 d^2}{8 D^2 r^2} T_\odot^4 = T^4.$$

Using (∗) to simplify the LHS of this equation yields

$$\frac{d^2}{8f^2} T_\odot^4 = T^4 \quad \Rightarrow \quad T = \sqrt{\frac{d}{\sqrt{8}f}} \, T_\odot = 1595 \text{ K}$$

as the maximum achievable temperature for the disc.

S 191 The figure defines the notation as well as giving a diagrammatic representation of the process.

$P_1 = 200$ MW
$W = 100$ MW
$P_2 = 100$ MW

It is clear that the efficiency of the power station is $\eta = 100/(100 + 100) = 0.5$ If the boilers' temperature is T_1, the theoretical maximum efficiency is

$$\eta' = \frac{W}{P_1} = \frac{T_1 - T_2}{T_1},$$

where $T_2 = 283$ K.

Now, η *must* be less than η', and so

$$0.5 < \frac{T_1 - 283}{T_1} \quad \Rightarrow \quad T_1 > 566 \text{ K} = 293°\text{C}.$$

Thus the boilers' temperature must be 293°C or higher.

S 192

(a) In the equilibrium state, whatever the size of the new heat flux ϕ', it has to be the same for all four gaps; this is so, even if the sheets are not equally spaced (as in the figure). Now the net flow in any one gap is equal to the difference between the powers radiated by the two sheets that define it, and so

$$\sigma T_{i+1}^4 - \sigma T_i^4 = \phi' \quad \text{for } i = 0, 1, 2, 3.$$

If these four equations are added, we obtain

$$\sigma(T_4^4 - T_0^4) = 4\phi'. \tag{*}$$

But the LHS of this equality is the expression for the original heat flux ϕ, before the extra plates were inserted, i.e. $\phi' = \frac{1}{4}\phi$. Equation (*) also provides a value for ϕ'/σ, namely $\frac{1}{4}[(373)^4 - (283)^4] = 3.236 \times 10^9 \text{ K}^4$.

The intermediate temperatures are now given by

$$T_i = \left[(283)^4 + i \times 3.236 \times 10^9\right]^{1/4}.$$

They are $T_1 = 40.4°C$, $T_2 = 63.9°C$ and $T_3 = 83.3°C$.

(b) With n additional sheets, $n + 1$ gaps are created and the analogue of (*) is

$$\sigma(T_4^4 - T_0^4) = (n + 1)\phi',$$

indicating that the heat flow to the water is reduced by a factor of $n + 1$.

S 193 We can use the maximum temperature information to determine the constant of proportionality in Newton's law, and then integrate the equation to find $T = T(t)$ during the cooling phase.

As an equation, Newton's law takes the form

$$\frac{d(CT)}{dt} = -k(T - T_0), \tag{*}$$

for some constant k. However, since T_1 is the *maximum* obtainable temperature, at this temperature the cooling loss rate must just balance the power provided by the heater:

$$k(T_1 - T_0) = P.$$

This provides an expression for k that can be substituted into $(*)$, which then reads

$$\frac{dT}{dt} = -\frac{1}{C}\frac{P}{T_1 - T_0}(T - T_0).$$

The variable temperature T only appears in one factor on the RHS of this, and the equation is easily rearranged in integral form:

$$\int_{T_1}^{(T_1+T_0)/2} \frac{dT}{T - T_0} = -\frac{P}{C(T_1 - T_0)}\int_0^\tau dt,$$

$$\left[\ln(T - T_0)\right]_{T_1}^{(T_1+T_0)/2} = -\frac{P}{C(T_1 - T_0)}(\tau - 0),$$

$$\ln\left[\frac{\frac{1}{2}(T_1 - T_0)}{(T_1 - T_0)}\right] = -\frac{P}{C(T_1 - T_0)}\tau,$$

$$\tau = \frac{C(T_1 - T_0)}{P}\ln 2.$$

S 194 When a molecule travels from the higher to the lower temperature wall, comes into thermal equilibrium with the latter, and then returns, the net energy transported is $\frac{3}{2}k\Delta T$. The frequency with which it does that is $\bar{c}/2d$, where d is the wall separation. If N is the total number of molecules in a volume Ad, i.e. those impinging upon area A of the walls, then the flux is given by

$$\phi = \frac{1}{A}\left(\frac{3}{2}k\Delta T\frac{\bar{c}}{2d}N\right).$$

But N can be related to the gas pressure through

$$\tfrac{1}{3}Nm\bar{c}^2 = pV = pdA,$$

meaning that the heat flux can be written as

$$\phi = \frac{1}{A}\left(\frac{3}{2}k\Delta T\frac{\bar{c}}{2d}\frac{3pdA}{m\bar{c}^2}\right)$$

$$= \frac{9k}{4m}p\frac{\bar{c}}{c_{rms}^2}\Delta T = \frac{9k}{4m}p\left(\frac{8}{3\pi}\right)^{1/2}\frac{1}{c_{rms}}\Delta T$$

$$= \frac{9k}{4m}p\left(\frac{8}{3\pi}\right)^{1/2}\sqrt{\frac{m}{3kT}}\Delta T = p\left(\frac{9k}{2\pi mT}\right)^{1/2}\Delta T.$$

S 195 The physics of this problem is relatively straightforward, but obtaining an equation for T that *can* be solved requires some insight.

Clearly, in a steady state, all the heat conducted down the bar is equal to the *net* radiation emitted by the end at temperature T, taking account of the radiation received from the surroundings. If A is the cross-sectional area of the bar,

$$\lambda \frac{T_1 - T}{\ell} A = \sigma(T^4 - T_0^4) A.$$

When this is rearranged as an equation for T it becomes the quartic equation

$$\ell \sigma T^4 + \lambda T - \lambda T_1 - \ell \sigma T_0^4 = 0, \tag{*}$$

for which there is no routine solution method.

If T is 'not too different from T_0', we can write $T = T_0 + \Delta T$ with $\Delta T \ll T_0$. The factor T^4 can then be replaced by

$$T^4 = T_0^4 + 4T_0^3 \Delta T + 6T_0^2 (\Delta T)^2 + 4T_0 (\Delta T)^3 + (\Delta T)^4,$$

with each of the last four terms a factor of the order of $\Delta T/T_0$ smaller than the one before it. The first term cancels with a term already present in (*), and so we keep the second term (linear in ΔT), but ignore those containing higher powers of ΔT. The result is

$$4\ell \sigma T_0^3 \Delta T + \lambda T - \lambda T_1 \approx 0,$$

$$4\ell \sigma T_0^3 (T - T_0) + \lambda T - \lambda T_1 \approx 0,$$

$$T(\lambda + 4\sigma \ell T_0^3) \approx \lambda T_1 + 4\sigma \ell T_0^4,$$

from which

$$T \approx \frac{\lambda T_1 + 4\sigma \ell T_0^4}{\lambda + 4\sigma \ell T_0^3}.$$

S 196 The basic mechanism is that heat flows out of the water, resulting in the freezing of some of it to form ice. All of the flow is through ice, and it is ice that is formed, and so the density and conductivity of water are irrelevant. Further, the temperature gradient in the ice will decrease with time as the ice gets thicker, whilst the air and water temperatures, and hence their difference $\Delta \theta$, remain unchanged; the heat flow rate will, therefore, also decrease.

Let the ice thickness be x and consider for unit area A the change dx that occurs during an infinitesimal time interval dt. The heat that flows out of the water is $dH = \lambda_{ice} \times A \times (\Delta \theta / x) \times dt$. But this must equal the latent heat of fusion of the additional ice formed, i.e. $dH = \rho_{ice} A \, dx \times L$, where L is the specific latent heat of fusion. Equating these two expressions for dH and rearranging as a differential equation:

$$\frac{dx}{dt} = \frac{\lambda_{\text{ice}} \, \Delta\theta}{L\rho_{\text{ice}}} \frac{1}{x}.$$

This integrates simply and, with $x = 0$ at $t = 0$, gives

$$\frac{1}{2}x^2 = \frac{\lambda_{\text{ice}} \, \Delta\theta}{L\rho_{\text{ice}}} \, t.$$

Thus the time for a 10 cm layer of ice to form is

$$t = \frac{L\rho_{\text{ice}}}{2\lambda_{\text{ice}} \, \Delta\theta} x^2$$

$$= \frac{3.3 \times 10^5 \times 917}{2 \times 2.3 \times 2} (0.1)^2 = 3.3 \times 10^5 \text{ s} \approx 91 \text{ hr}.$$

S 197 This problem appears to be short of relevant data, and so let us denote the radii of the Sun and Mars by R and r, respectively, and the separation of the two bodies by d. We will have to suppose that Mars behaves as if it were a black body, both when absorbing energy from the Sun, and when radiating it isotropically; we also need to assume that it has no internal heat sources. This situation is shown in the figure.

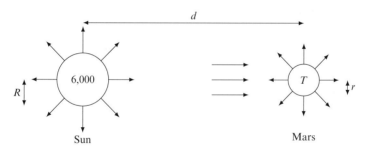

The total energy radiated by the Sun is $4\pi R^2 \sigma T_{\odot}^4$, where $T_{\odot} = 6000$ K. The fraction of this that falls on Mars (and is assumed to be totally absorbed) is $\pi r^2 / 4\pi d^2$. In the steady state, all this energy must be radiated away from the Mars surface, which is (assumed to be) behaving like a black body at temperature T. Therefore,

$$\frac{\pi r^2}{4\pi d^2} 4\pi R^2 \sigma T_{\odot}^4 = 4\pi r^2 \sigma T^4.$$

After multiple cancellations, we are left with

$$\frac{R^2}{4d^2} T_{\odot}^4 = T^4 \quad \Rightarrow \quad T = T_{\odot} \sqrt{\frac{R}{2d}}.$$

But R/d is given as one half of $0.35°$, and so

$$T = 6000 \sqrt{\frac{0.35}{4} \frac{\pi}{180}} = 234 \, \text{K}.$$

So, this estimate of the temperature on Mars is totally independent of the size of Mars, and only involves R and d as their ratio!

S 198   Work can be extracted from the system if the two bodies are brought to a common temperature; let this be T_0, as indicated in the figure.

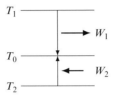

The maximum possible amount of work can be extracted only if all processes are carried out reversibly, so that the entropy of the system as a whole does not increase.

As the temperature at which each small amount of heat is gained or lost varies throughout the process, an integration is necessary. For the two bodies the changes in entropy S are

$$\Delta S_1 = \int_{T_1}^{T_0} \frac{C\,dT}{T} \quad \text{and} \quad \Delta S_2 = \int_{T_2}^{T_0} \frac{C\,dT}{T}.$$

The sum of these two changes must be zero, i.e.

$$C \ln \frac{T_0}{T_1} + C \ln \frac{T_0}{T_2} = 0 \quad \Rightarrow \quad \ln \frac{T_0^2}{T_1 T_2} = 0 \quad \Rightarrow \quad T_0^2 = T_1 T_2.$$

Thus, for ideal efficiency of extraction, the final common temperature must be the geometric mean of the two initial values.

The work extracted from the first body is $W_1 = C(T_1 - T_0)$, and that supplied to the second one is $W_2 = C(T_0 - T_2)$. Thus the total extractable work is

$$W = W_1 - W_2 = CT_1 + CT_2 - 2CT_0 = C[T_1 + T_2 - 2(T_1 T_2)^{1/2}].$$

This same conclusion could have been found directly from the initial and final heat contents of the two bodies, once the required final temperature was established.

S 199 This problem can be analysed using either (i) a (pragmatic) physics approach, or (ii) a more complex mathematical method.

(i) Consider the temperature $T(r)$ of that part S_r of the uranium sphere that is a centred sphere of radius r ($r < a$). From the heat conduction equation, the total rate of heat flowing radially outwards from its surface is

$$\frac{dH}{dt} = \text{area} \times \text{thermal conductivity} \times \text{temperature gradient} = 4\pi r^2 \lambda \frac{dT}{dr}.$$

However, on physical grounds, in the steady state this must be equal to the total rate of heat generation within S_r, i.e. to $\frac{4}{3}\pi r^3 H$. Thus

$$4\pi r^2 \lambda \frac{dT}{dr} = \frac{4}{3}\pi r^3 H \quad \Rightarrow \quad \frac{dT}{dr} = \frac{H}{3\lambda} r.$$

This is the same equation as that marked (†) in method (ii) below, and the rest of the solution follows the argument given there.

(ii) The more mathematically demanding approach is as follows.

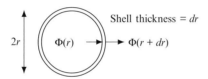

The figure shows a spherical shell of inner radius r and thickness dr. The equation governing the temperature T as a function of r will be obtained by equating the difference in (radial) heat flows across the two surfaces, to the heat that is generated between them.

The heat flow into the inner surface is

$$\Phi(r) = 4\pi r^2 \times \lambda \times \frac{dT}{dr}.$$

That flowing out of the outer surface is $\Phi(r + dr)$, which can be written, using a Taylor expansion[32] as

$$\Phi(r + dr) = \Phi(r) + \frac{d\Phi}{dr} dr + \cdots.$$

---

[32] Taylor expansions have the general form

$$f(a + z) = f(a) + f'(a)z + \frac{1}{2!}f''(a)z^2 + \cdots + \frac{1}{n!}f^{(n)}(a)z^n + \cdots.$$

The difference between these two must, for arbitrarily small dr, be equal to the heat generated in the shell, i.e. to $4\pi r^2\, dr \times H$:

$$\frac{d}{dr}\left(4\pi r^2 \lambda \frac{dT}{dr}\right) dr = 4\pi r^2 H\, dr,$$

$$\frac{d}{dr}\left(r^2 \frac{dT}{dr}\right) = \frac{H}{\lambda} r^2,$$

$$r^2 \frac{dT}{dr} = \frac{H}{3\lambda} r^3 + c_1,$$

$$\frac{dT}{dr} = \frac{H}{3\lambda} r + \frac{c_1}{r^2}.$$

The constant of integration c_1 must be zero, because if it were not, it would imply that $dT/dr \to \infty$ as $r \to 0$ and a consequent infinite heat flow away from the sphere's centre. Thus,

$$(\dagger) \qquad\qquad \frac{dT}{dr} = \frac{H}{3\lambda} r \quad \Rightarrow \quad T(r) = \frac{H}{6\lambda} r^2 + c_2.$$

The constant c_2 is the temperature at the centre of the sphere ($r=0$), and so the required temperature difference is

$$T(a) - T(0) = \frac{Ha^2}{6\lambda} = \frac{5.5 \times 10^3 \times (0.1)^2}{6 \times 46} = 0.20\,\text{K}.$$

S 200   The figure shows the essential p-V diagram for one (Otto) cycle of the petrol engine; in practice, several refinements are necessary, including an emptying of the cylinder (on the diagram a two-way line at constant pressure joining (V_1, p_1) to (V_2, p_1) and back) to provide fresh oxygen for the next cycle.

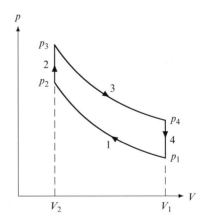

The work done by the air (step 3), and on the air (step 1), are each given by an integral of the form $\int p\,dV$, and the net useful work W available from the engine is represented on the diagram by the difference between these two integrals, i.e. by the area inside the closed curve. Along an adiabatic curve pV^γ is constant, and so

$$pV^\gamma = p_2 V_2^\gamma \text{ on curve 1, and } pV^\gamma = p_3 V_2^\gamma \text{ on curve 3.}$$

We now calculate the area representing W.

$$W = \left(\int_{V_2}^{V_1} p\,dV \right)_{\text{curve 3}} - \left(\int_{V_2}^{V_1} p\,dV \right)_{\text{curve 1}}$$

$$= \int_{V_2}^{V_1} \frac{p_3 V_2^\gamma}{V^\gamma}\,dV - \int_{V_2}^{V_1} \frac{p_2 V_2^\gamma}{V^\gamma}\,dV$$

$$= (p_3 - p_2) V_2^\gamma \int_{V_2}^{V_1} \frac{1}{V^\gamma}\,dV$$

$$= \frac{(p_3 - p_2) V_2^\gamma}{1 - \gamma} (V_1^{1-\gamma} - V_2^{1-\gamma})$$

$$= \frac{(p_3 - p_2)}{\gamma - 1} (V_2 - V_1^{1-\gamma} V_2^\gamma)$$

$$= \frac{(p_3 - p_2)}{\gamma - 1} V_2 \left(1 - \frac{V_2^{\gamma-1}}{V_1^{\gamma-1}} \right)$$

$$= (p_3 - p_2) V_2 \frac{1 - s^{\gamma-1}}{\gamma - 1}.$$

All of the externally provided energy H is supplied during step 2, and can be measured by the increase in internal energy of the air during that stage. It is most obviously measured by the change in temperature T of the air, but, since $pV = nRT$ for an ideal gas, we can use the change in pV instead. Thus $H = (p_3 - p_2) V_2$, showing that the efficiency of the cycle is

$$\eta = \frac{W}{H} = \frac{1 - s^{\gamma-1}}{\gamma - 1}.$$

S 201 In both cases, we need to establish the difference between the upthrust experienced by the balloon and its own weight. If the balloon initially contains n_0 moles of air and its volume is V_0, then $p_0 V_0 = n_0 R T_0$ at the start of either process.

With the valve *closed*: When the temperature is increased to T, the pressure increases only marginally above p_0, but the volume of the balloon increases to $V = (T/T_0)V_0$ and the density decreases to $(T_0/T)\rho_0$, the total mass remaining unchanged. However, the upthrust is now $\rho_0 V$ and so the lift force available is

$$L_{\text{closed}} = (\rho_0 V - \rho_0 V_0)g = \rho_0 V_0 g \left(\frac{T}{T_0} - 1 \right) = \rho_0 V_0 g \left(\frac{T - T_0}{T_0} \right).$$

With the valve *open*: When the temperature is increased to T, the pressure remains at p_0 and the volume remains unchanged, but air is driven out and the number of moles remaining drops to $n = (T_0/T)n_0$; the balloon's weight drops accordingly. The lift force available is

$$L_{\text{open}} = \rho_0 V_0 g - \frac{T_0}{T} \rho_0 V_0 g = \rho_0 V_0 g \left(\frac{T - T_0}{T} \right).$$

Since $T > T_0$, we conclude that $L_{\text{open}} < L_{\text{closed}}$ and that it is better to have the valve closed during the heating process.

S 202   Consider a vertical column of stable air of unit cross-section. A layer of air in the column of depth dx has a weight $\rho(x)g\, dx$, and this is supported by the pressure difference between its upper and lower surfaces. Thus,

$$\rho(x)g\, dx = p(x) - p(x + dx) = -\frac{dp}{dx}\, dx. \qquad (*)$$

To use this equation we need the connection between p and ρ. Since $pV = nRT$ and $n = M/N_A m$, where M is the mass of the gas and m that of one molecule, we have

$$pV = nRT = \frac{M}{N_A m} RT \quad \Rightarrow \quad \rho = \frac{M}{V} = \frac{N_A m}{RT} p = \frac{m}{kT} p.$$

Equation $(*)$ can therefore be written as

$$\frac{dp}{dx} = -\frac{gm}{kT} p.$$

We now need T as a function of x. With the linear decrease and the given figures, the rate of decrease is $(15.0 - 7.5) \div 1.15 = 6.52$ K km^{-1}. Thus, in general $T(x) = (288 - 0.00652x)$ K, where x is measured in metres. The differential equation connecting p and x can now be written explicitly as

$$\frac{1}{p}\, dp = -\frac{gm}{k(288 - 0.00652x)}\, dx,$$

which integrates to

$$\ln \frac{p(x)}{p(0)} = \frac{gm}{0.00652k} \ln \left(\frac{288 - 0.00652x}{288} \right),$$

with $m = 28.8 \times 1.66 \times 10^{-27}$ kg.

When $p(x) = \frac{1}{2}p(0)$,

$$-\ln 2 = 5.21 \ln\left(\frac{288 - 0.00652x}{288}\right) \quad \Rightarrow \quad x = 5500\,\text{m}.$$

This is the height at which the pressure is one half of that at sea-level.

S 203 The figure shows the initial (solid) and final (dashed) positions of the piston, and defines the quantity y.

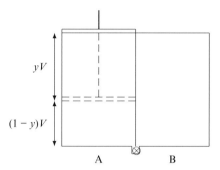

There are only two unknowns in this problem, y and T_f, once n the number of moles of gas involved has been set. Before the valve is opened, we have, as usual, $pV = nRT$, and, after the operation has been completed, the corresponding equation of state is

$$p(2 - y)V = nRT_f. \tag{$*$}$$

The final equation must come from the energy balance between the work done on the gas and its internal energy:

$$pyV = \frac{3}{2}nR(T_f - T).$$

Substituting into $(*)$ from the other two equations gives

$$2nRT - \frac{3}{2}nR(T_f - T) = nRT_f,$$

which, after some cancellation, can be rearranged as

$$T_f = \frac{2T + \frac{3}{2}T}{1 + \frac{3}{2}} = \frac{7}{5}T.$$

The corresponding value for y is $y = \frac{3}{5}$, but that is not called for.

S 204   Since the system pressure increases monotonically with its temperature, we need only consider the situations at 5°C and 100°C. Let the volume of
the gas at these two temperatures be V_1 and V_2 respectively. The initial pressure of the gas is 1 atmosphere plus that due to the water column of height h,
i.e.

$$p_1 = 1.00 \times 10^5 + (10^3 \times 9.81 \times 15) = 2.47 \times 10^5 = 2.47\,\text{atm}.$$

Since for a fixed mass of gas, pV/T is constant,

$$\frac{2.47\,V_1}{278} = \frac{(3.0 + 1.0)\,V_2}{373} \quad \Rightarrow \quad V_2 = 0.8285\,V_1.$$

But, we also have that the volume reduction of air in the bottle must be equal
to the volume expansion of the water in the system:

$$V_1 - V_2 = V \times \alpha\,\Delta T = 0.20 \times 2.00 \times 10^{-5} \times (100 - 5) = 3.8 \times 10^{-4}\,\text{m}^3.$$

Hence

$$V_1 - 0.8285\,V_1 = 3.8 \times 10^{-4} \quad \Rightarrow \quad V_1 = 2.2 \times 10^{-3}\,\text{m}^3.$$

This is the absolute minimum size of the bottle, as we have made no allowance
for the water that enters it when the valve is initially opened – nor for the
requirements of Health & Safety regulations!

S 205   Let the mass of gas be denoted by M. The equation governing this
situation is the force balance that keeps the balloon from falling down, or going
through the ceiling.

$$\text{excess pressure} \times \text{area} = \text{upthrust} - \text{weight},$$

$$(p_h - p_a)A = (\rho_a - \rho_h)Vg = Mg\left(\frac{\rho_a}{\rho_h} - 1\right),$$

$$A = \frac{Mg}{(p_h - p_a)}\left(\frac{\rho_a}{\rho_h} - 1\right).$$

Now, for any ideal gas

$$pV = nRT = \frac{M}{N_A m}RT \quad \Rightarrow \quad \rho = \frac{M}{V} = \frac{N_A m p}{RT},$$

from which it follows that

$$\frac{\rho_a}{\rho_h}\frac{p_h}{p_h} = \frac{m_a}{m_h} = \frac{28.8}{2}.$$

For the given pressures this becomes

$$\frac{\rho_a}{\rho_h} = \frac{1.00}{1.05}\frac{28.8}{2} = 13.71.$$

Putting this into the expression for A yields

$$A = \frac{1.00 \times 10^{-4} \times 9.81}{0.05 \times 10^5} \times 13.71 = 2.7 \times 10^{-6}\,\mathrm{m^2}$$

as the area in contact with the ceiling.

S 206   Three principal assumptions that are made when deriving the ideal gas equation for a monatomic gas are (i) that interactions with the containing boundary are simply those of an elastic recoil, (ii) that the size of the molecules themselves can be neglected, and (iii) that there are no (weak) mutual forces between the molecules.

The measurements given do not change the boundary conditions, and so cannot give information about (i). However, as the gas density is reduced, the finite size of the molecules becomes less important, since they occupy a smaller and smaller fraction of the volume within the bounding surface, i.e. the conditions approach those of assumption (ii). At the same time, the average distance between molecules is increased and any mutual forces that are present will decrease in strength, making assumption (iii) more likely to be a valid one.

If the gas were ideal, the ratio r of the two measured pressures should be independent of the gas density. To test this, we rewrite the table, but with an additional column giving that ratio.

p at T_0 (atm)	p at temperature T (atm)	$r = p(T)/p(T_0)$
1.013	1.3795	1.3610
0.6243	0.8494	1.3606
0.2021	0.2749	1.3602
$\rightarrow 0$	$\rightarrow 0$	1.3600

There are only three readings, but there does appear to be a trend; for each reduction of about 0.4 atm in $p(T_0)$, r reduces by 0.0004. For the 'ideal' situation we need $p(T_0) \rightarrow 0$, i.e. a further reduction of 0.2 atm, giving a further reduction in r of 0.0002. Thus our best estimate of $p(T)$ for use in the ideal gas equation is 1.3600; this is shown in the row that has been added to the table. The best estimate of T is therefore

$$T = 1.3600 \times 273.15 = 371.48\,\mathrm{K} = 98.33°C.$$

S 207 (a) For a ball that moves with speed c at right angles to the particular side, the impulse it gives to that side at each impact is $2mc$, and this occurs with a frequency $c/2L$. The total force on that side is therefore

$$F = \frac{1}{2}\sum_1^{1000} \frac{c_i}{2L} 2mc_i = \frac{1}{L} \times \sum_1^{1000} \frac{1}{2}mc_i^2 = \frac{E}{L} = 50 \text{ N}.$$

(b) Only one half of the balls will be moving in a direction that makes escape a possibility, and they will only meet the relevant side with a frequency of $\bar{c}/2L$. Of those reaching the relevant side, a fraction d/L will escape. The rate of reduction of numbers is therefore given by

$$-\frac{dN}{dt} = \frac{1}{2}\frac{d}{L}\frac{\bar{c}}{2L}N \quad \Rightarrow \quad N(t) = N(0)e^{-\lambda t},$$

where $\lambda = d\bar{c}/4L^2$. We do not know $\bar{c}$ and will have to use $\sqrt{\bar{c^2}}$ instead; from the total energy

$$\bar{c^2} = \frac{2E_m}{m} = \frac{2 \times 100}{1000 \times 0.001} = 200 \text{ m}^2 \text{ s}^{-2} \quad \Rightarrow \quad \bar{c} \approx 14.1 \text{ m s}^{-1},$$

giving $\lambda = 8.81 \times 10^{-3} \text{ s}^{-1}$.

(i) For $N(t_1) = 900$, $t_1 = -\ln(0.9)/(8.81 \times 10^{-3}) = 12$ s.

(ii) For $N(t_2) = 50$, $t_1 = -\ln(0.05)/(8.81 \times 10^{-3}) = 340$ s.

(c) Calculation (a) was a direction evaluation based on the mechanics of the system, and so, within statistical fluctuations, is 'exact'.

Both of the results in (b) can only be approximations because of the way the value for $\bar{c}$ was determined. However, estimate (i) for 900 ball-bearings should be reasonably good; the number of balls is still high and the situation is close to the calculable initial state.

This is not the case for estimate (ii), and, in addition, the average velocity $\bar{c}$ of the remaining balls will be significantly lower than that assumed, because the faster balls are more likely to have already escaped as they impact the relevant side more frequently. Moreover, with only a small number of balls left, collisions will be much less frequent, and the assumption of instantly equalized numbers parallel to each of the sides becomes unjustified. Both of these effects make a more realistic estimate likely to be much greater than the 340 s given above.

S 208   The figure shows the 'one-molecule gas' and the general notation.

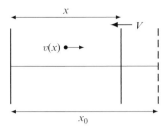

(a) When the nth collision with the moving wall occurs, denote the wall separation by x_n, and the speed of the molecule after that collision by v_n.

The relative speed of the molecule and moving wall before the nth collision is $v_{n-1} + V$, and it must be the same after the elastic impact. So,

$$v_n - V = v_{n-1} + V.$$

During the interval between the $(n-1)$th and the nth impact, which lasts for $t_n = (x_n + x_{n-1})/v_{n-1}$, the moving wall has reduced the separation by Vt_n. Therefore,

$$x_n = x_{n-1} - \frac{x_n + x_{n-1}}{v_{n-1}} V,$$

$$x_n(v_{n-1} + V) = x_{n-1}(v_{n-1} - V),$$

$$x_n(v_n - V) = x_{n-1}(v_{n-1} - V),$$

$$x_n v_n \approx x_{n-1} v_{n-1} \approx x_0 v_0.$$

The last line, which shows the constancy of the product vx, is justified because $v_0 \gg V$ and all other $v_n > v_0$.

(b) The rise in temperature of the one molecule has to be evaluated by finding the work done to move the mobile wall, and then equating it to the kinetic (thermal) energy of the 'gas'.

As all molecular speeds are very much greater than V, with wall collisions occurring at a high frequency, we may treat x and v as continuous variables for the present purpose. At each collision with the moving wall the impulse on the wall is $\Delta p = 2mv$, and these collisions occur with a frequency $v/2x$. The average force on the piston (and on the stationary wall) is therefore

$$F = 2mv \times \frac{v}{2x} = \frac{mv^2}{x} = \frac{m\, x_0^2 v_0^2}{x\, x^2} = \frac{m x_0^2 v_0^2}{x^3}.$$

Now the total work done $(-\int F\,dx)$ is converted to molecular kinetic energy of magnitude $\frac{3}{2}k\Delta T$. It follows that the temperature rise of the 'one-molecule gas' is

$$\Delta T = -\frac{2}{3k} \int_{x_0}^{x} \frac{mx_0^2 v_0^2}{x^3} \, dx$$

$$= \frac{mx_0^2 v_0^2}{3k} \left[\frac{1}{x^2} - \frac{1}{x_0^2} \right]$$

$$= \frac{mv_0^2}{3k} \left[\left(\frac{x_0}{x} \right)^2 - 1 \right].$$

S 209   Since the escaping molecules have a greater average energy $(2kT)$ than that of the population they left $(\frac{3}{2}kT)$, the temperature of those remaining will fall. If $N(t)$ is the number of molecules remaining after time t, then at each escape dN is negative – but so is dE That is, although molecules are escaping,

$$\frac{dE}{dN} = +2kT,$$

where E is the total internal energy of the remaining gas.

That internal energy can be written as $E = \frac{3}{2} NkT$, and so

$$+2kT = \frac{dE}{dN} = \frac{3}{2} k \left(N \frac{dT}{dN} + T \right),$$

leading to

$$\frac{3}{2} kN \frac{dT}{dN} = \left(2 - \frac{3}{2} \right) kT,$$

$$3 \frac{dT}{T} = \frac{dN}{N},$$

$$\ln T^3 = \ln N + k \quad \Rightarrow \quad \frac{T^3}{N} \text{ is a constant.}$$

As $T \to \frac{1}{2} T_0$, the number of remaining molecules $\to N_0/8$.

Note. If the differential equation is mistakenly written with a minus sign, the conclusion is reached that as N reduces, the temperature soars as $N^{-3/7}$.

S 210   (a) The connection between C_p and C_V is

$$C_p \Delta T = C_V \Delta T + \int p \, dV.$$

But, since p is constant and $pV = RT$, we have $p \, dV = R \, dT$ and $\int p \, dV = R \Delta T$. Substituting this into the above, and cancelling through by ΔT gives the stated result:

$$C_p = C_V + R. \qquad\qquad (*)$$

(b) Using the expressions for dQ, dU and dT given in the question, we have, successively,

$$0 = dQ = dU + p\,dV$$

$$= C_V \left(\frac{V\,dp + p\,dV}{R} \right) + p\,dV$$

$$= (C_V + R)p\,dV + C_V V\,dp.$$

$$0 = \frac{C_V + R}{C_V} \frac{dV}{V} + \frac{dp}{p} \quad \text{[on dividing through by } C_V p V]$$

$$= \gamma \frac{dV}{V} + \frac{dp}{p}, \quad \text{[using the result } (*) \text{ and the definition of } \gamma]$$

$$\text{constant} = \gamma \ln V + \ln p \quad \Rightarrow \quad pV^\gamma \text{ is constant for an adiabatic change.}$$

(c) No comment!

Note. Equation $(*)$, as it stands, is only valid for one mole of gas, i.e. the specific heats are 'per mole'; it can be extended to n moles and calculation (b) is slightly altered. The final adiabatic equation is valid for an arbitrary quantity of gas. For a monatomic ideal gas, $\gamma = \frac{5}{3}$.

S 211 The original measurements can be used to determine the work function ϕ for sodium, as follows:

energy of a single photon = work function + the maximum K.E. of a photoelectron.

The final term is equal to the electron charge $\times$ the stopping potential, and so, algebraically,

$$hf = \frac{hc}{\lambda} = \phi + eV_{\text{s}},$$

and numerically (with all energies measured in electron-volts),

$$\frac{6.6 \times 10^{-34} \times 3.0 \times 10^8}{450 \times 10^{-9} \times 1.60 \times 10^{-19}} = \phi + 0.50,$$

giving $\phi = 2.26$ eV.

To consider the other scenarios we draw up a table comparing the actual stopping potential V_{a} with the critical value V_{c} needed to suppress the photocurrent, calculated from

$$V_{\text{c}} = \frac{hc}{e\lambda} - \phi = \frac{1.24 \times 10^{-6}}{\lambda} - 2.26.$$

Case	λ	V_a/V	V_c/V	Photocurrent?
(i)	300	1.50	1.87	yes
(ii)	500	0.25	0.22	no
(iii)	750	−1.00	−0.61	no*

*Although arithmetically it looks as if there should be a photocurrent, physically there will not be, because the (quantized) energy in a single photon is insufficient to allow a photoelectron to leave the sodium.

S 212   The equations describing the decay rate for the X nuclei, and the supply and decay rates for type Y are

$$\frac{dx}{dt} = -\lambda_1 x,$$

$$\frac{dy}{dt} = \lambda_1 x - \lambda_2 y. \tag{$*$}$$

In addition, we require $x(0) = N$ and $y(0) = 0$.

The proposed expression for $y(t)$ is

$$y(t) = \frac{\lambda_1 N}{\lambda_2 - \lambda_1} \left(e^{-\lambda_1 t} - e^{-\lambda_2 t} \right),$$

and we can see immediately that it satisfies $y(0) = 0$. The LHS of equation $(*)$ is

$$\frac{dy}{dt} = \frac{\lambda_1 N}{\lambda_2 - \lambda_1} \left(-\lambda_1 e^{-\lambda_1 t} + \lambda_2 e^{-\lambda_2 t} \right).$$

The equation for the decay of X nuclei has the standard result $x(t) = N e^{-\lambda_1 t}$, and so the RHS of equation $(*)$ is

$$\lambda_1 x - \lambda_2 y = \lambda_1 N e^{-\lambda_1 t} - \frac{\lambda_2 \lambda_1 N}{\lambda_2 - \lambda_1} \left(e^{-\lambda_1 t} - e^{-\lambda_2 t} \right)$$

$$= \frac{N}{\lambda_2 - \lambda_1} \left(\lambda_1 \lambda_2 e^{-\lambda_1 t} - \lambda_1^2 e^{-\lambda_1 t} - \lambda_1 \lambda_2 e^{-\lambda_1 t} + \lambda_1 \lambda_2 e^{-\lambda_2 t} \right)$$

$$= \frac{\lambda_1 N}{\lambda_2 - \lambda_1} \left(-\lambda_1 e^{-\lambda_1 t} + \lambda_2 e^{-\lambda_2 t} \right).$$

This is the same as the LHS of $(*)$, thus establishing that the required equation is satisfied.

For the second part of the question we first note that half-lives are inversely related to decay constants.

(a) If the half-life of X is very much smaller than that of Y, then $\lambda_1 \gg \lambda_2$, and

$$y(t) \approx \frac{\lambda_1 N}{-\lambda_1} \left(0 - e^{-\lambda_2 t} \right) = Ne^{-\lambda_2 t}.$$

Physically this corresponds to the rapid decay of all X-nuclei to give N of type Y, which then decay as if they were of a single initial nuclear type.

(b) If the half-life of Y is very much smaller than that of X, then $\lambda_1 \ll \lambda_2$, and

$$y(t) \approx \frac{\lambda_1 N}{\lambda_2} \left(e^{-\lambda_1 t} - 0 \right) = \frac{\lambda_1 N}{\lambda_2} e^{-\lambda_1 t}.$$

In this situation the number of nuclei are in their equilibrium ratio (the ratio of their half-lives) with very few Y nuclei and many of type X, but with an overall slowly diminishing supply of nuclei.

(c) With equal half-lives (both equal to λ, say) a more 'delicate' approach is needed, as directly setting $\lambda_1 = \lambda_2$ leads to an expression for $y(t)$ of the indeterminate form $0/0$. To evaluate the expression for $y(t)$ we set $\lambda_1 = \lambda$ and $\lambda_2 = \lambda + \delta$, and then let $\delta \to 0$.[33]

$$
\begin{aligned}
y(t) &= \frac{\lambda_1 N}{\lambda_2 - \lambda_1} \left(e^{-\lambda_1 t} - e^{-\lambda_2 t} \right) \\
&= \frac{\lambda N}{\delta} \left(e^{-\lambda t} - e^{-\lambda t} e^{-\delta t} \right) \\
&= N\lambda e^{-\lambda t} \lim_{\delta \to 0} \frac{1 - e^{-\delta t}}{\delta} \\
&= N\lambda e^{-\lambda t} \lim_{\delta \to 0} \frac{1 - (1 - \delta t + \cdots)}{\delta} \\
&= N\lambda t e^{-\lambda t}.
\end{aligned}
$$

This is a function that initially increases linearly, but later falls to zero; the number of Y-nuclei, starting from zero, peaks at Ne^{-1} at time $t = \lambda^{-1}$, but falls monotonically thereafter.

---

[33] As it must be to be a true limit, the final expression is independent of the sign of δ. Physically, λ_1 and λ_2 have been treated on an equal footing.

S 213   For the meson to remain in the circular orbit, the Coulomb attraction between it and the nucleus must provide the necessary centripetal force, i.e.

$$\frac{Zee}{4\pi\epsilon_0 r^2} = \frac{mv^2}{r}.$$

The given constraint on the angular momentum[34] allows us to express v in terms of the other variables:

$$v = \frac{h}{2\pi mr} \quad \Rightarrow \quad Ze^2 = 4\pi\epsilon_0 mr \frac{h^2}{4\pi^2 m^2 r^2} = \frac{\epsilon_0 h^2}{\pi mr}.$$

From the given expression for R and the above equation, we see that as Z increases, R increases but r decreases. When they are approximately equal the meson will interact with the nucleus (the short-range strong nuclear force) and a π-mesic atom will not be possible. The limit will be an *upper* limit on Z.

The limiting value will satisfy

$$R_0 Z^{1/3} = r = \frac{\epsilon_0 h^2}{\pi m Z e^2},$$

leading to

$$Z^{4/3} = \frac{\epsilon_0 h^2}{\pi me^2 R_0} = \frac{8.85 \times 10^{-12} \times (6.63 \times 10^{-34})^2}{\pi \times 2.4 \times 10^{-28} \times (1.60 \times 10^{-19})^2 \times 1.7 \times 10^{-15}} = 118.6.$$

Taking the $\frac{3}{4}$th root of this result, gives an estimate for the upper limit on Z of 36.

S 214   (i) Clearly the masses of such small oil drops ($\sim 10^{-15}$ kg) cannot be measured directly, but the observations of terminal speeds of gravitational fall offer an indirect measure. The two forces that are equal when terminal speed v is attained, the drop's weight and the viscous air-drag, depend on the radius r of the drop (assumed spherical because of the surface tension of the oil) in different ways, and so an expression for r can found. From Stokes' law, the viscous drag is $F = 6\pi\eta rv$, where η is the viscosity of air, and so

$$6\pi\eta rv = \frac{4\pi}{3}r^3(\rho_{oil} - \rho_{air})g.$$

The mass can then be found from

$$r^2 = \frac{9\eta v}{2g(\rho_{oil} - \rho_{air})} \quad \text{and} \quad m = \frac{4\pi}{3}r^3\rho_{oil}.$$

---

[34] Which has its origin in quantum physics.

With $\rho_{\text{oil}} \approx 950$ kg m^{-3}, $\eta = 1.6 \times 10^{-5}$ kg m^{-1} s^{-1}, ρ_{air} negligible, and oil-drop masses $\sim 10^{-15}$ kg, we find that $r \approx 8 \times 10^{-7}$ m and v is of the order of 10^{-4} m s^{-1}.

(ii) If n_1 is the number of electron charges initially on a drop, then, for it to be stationary, $n_1 e E_1 = mg$ and so $n_1 = cm/V_1$ for some constant c. With only electrons present, the six n_1-values should be in whole number ratios. That they are is demonstrated in the fourth column of the augmented table below, the ratios being $2 : 1 : 3 : 1 : 2 : 2$.

Drop	$m/10^{-15}$ kg	V_1/V	$n_1 = cm/V_1$	V_2/V	$n_2 = cm/V_2$	$n_2 - n_1$	q/e
a	3	459	$6.53c$	690	$4.35c$	$-2.18c$	-0.667
b	1	306	$3.27c$	230	$4.35c$	$1.08c$	0.330
c	3	307	$9.77c$	230	$13.04c$	$3.27c$	1.000
d	2	612	$3.27c$	1836	$1.09c$	$-2.18c$	-0.667
e	2	305	$6.56c$	204	$9.80c$	$3.24c$	0.991
f	4	613	$6.53c$	525	$7.62c$	$1.09c$	0.333

The sixth column gives the new number of electron charges after the 'quark-material' has been introduced, and the difference (potentially due to quarks) is recorded in the seventh. From the values obtained in column 4, one electron charge corresponds to $3.27c$, and, on this basis the final column gives the number of additional charges (electron charges) acquired.

No conclusion can be drawn from cases c and e, as the observed effect could be explained by the acquisition of an additional electron. However, drops a and d do strongly suggest quark charges of $-\frac{2}{3}e$, i.e. $+\frac{2}{3}$ of an elementary charge. Similarly, cases b and f suggest a quark charge of $-\frac{1}{3}$ of an elementary charge.

These two possible quark charges are in agreement with current theories, with the first corresponding to an 'up quark' and the second to a 'down' or to a 'strange' quark. However, although many efforts have been made, no *isolated* quark, such as those implied here, has ever been positively identified – and the data quoted is a figment of the author's imagination!

S 215 (a) Because both the masses and speeds of the two particles are equal, the overall momentum of the system is initially zero – and so it must be zero finally. This implies that the two gamma rays have equal but opposite momenta, and therefore equal energies. The energy equation is therefore

$$E_e + E_p = 2E_\gamma, \quad \text{where} \quad E_e = E_p = \frac{m_e c^2}{\sqrt{1 - v^2/c^2}}.$$

With $v = 0.8c$, $E_p = E_e = m_e c^2 / 0.6$, and so

$$\frac{m_e c^2}{0.6} = hf = \frac{hc}{\lambda} \quad \Rightarrow \quad \lambda = 1.5 \times 10^{-12} \text{ m}.$$

(b) For the same reasons as in part (a), the overall momentum is zero, and so, at the threshold for producing antiprotons, all the final particles can be at rest. However, the production of an antiproton as a single additional particle is not possible; like the total lepton-antilepton number, the total baryon-antibaryon number must be conserved. So the simplest antiproton-producing reaction is

$$p + p \to p + p + p + \bar{p}.$$

In energy terms

$$\frac{2m_p c^2}{\sqrt{1 - v^2/c^2}} = 4m_p c^2.$$

It follows that, at this threshold, $v^2/c^2 = 3/4$, and that the minimum value for v is $0.866c = 2.60 \times 10^8$ m s^{-1}.

S 216 (i) We clearly have to use equations involving ^{30}P and ^{31}P, and reaction (a) is an obvious starting place. However, the γ-ray and neutron are not involved in the target reaction; but they do appear in reaction (d), which (promisingly) also involves a proton and a deuteron. If we add (in our abbreviated notation) the equations for reactions (a) and (d), we obtain

$$^{31}\text{P}(\gamma, \text{n})^{30}\text{P} + \text{p}(\text{n}, \gamma)\text{d} = {}^{31}\text{Pp}(n\gamma, \gamma\text{n})^{30}\text{Pd} = {}^{31}\text{P}(\text{p}, \text{d})^{30}\text{P} = {}^{30}\text{P}(\text{d}, \text{p})^{31}\text{P}$$
$$\text{(reversed)}.$$

The corresponding Q value equation is

$$Q(a) + Q(d) = -Q(i),$$
$$-12.313 + 2.226 = -Q(i),$$
$$Q(i) = 10.087 \text{ MeV}.$$

(ii) This reaction is not possible – whatever the energies involved – as it does not conserve charge; Si + α has 16 positive elementary charges, whilst P + n has only 15.

(iii) This question requires an extended version of the method used in (i), and the ordering of the calculation is not unique. However, all orderings should lead to the same final answer. Below is one possibility.

add (b) and (c) together $^{31}P\,^{28}Si(pd,\alpha p)^{28}Si\,^{29}Si = ^{31}P(d,\alpha)^{29}Si,$

add in (a) reversed $^{30}P\,^{31}P(dn,\alpha\gamma)^{29}Si\,^{31}P = ^{30}P(dn,\alpha\gamma)^{29}Si,$

add in (e) reversed $\alpha\,^{30}P(dn\gamma,d\alpha\gamma)d\,^{29}Si = ^{30}P(n,d)^{29}Si.$

This gives the reverse of the target reaction, and so the Q value of the latter is

$$Q(iii) = -[Q(b)+Q(c)-Q(a)-Q(e)]$$
$$= -[1.916+6.251-(-12.313)-23.847]$$
$$= 3.367\ \text{MeV}.$$

S 217 (a) We first note that a decay constant λ is related to the corresponding half life τ by

$$\lambda = \frac{\ln 2}{\tau},$$

and that, quite generally, the rate of decay of radioactive nuclei is given by

$$\frac{dN}{dt} = -\lambda N(0)e^{-\lambda t}.$$

For radium, the half life is so long – certainly much longer than the lifetime of a watch, or its wearer – that the exponential factor can be taken as unity. The relevant equation now becomes

$$10 = -\frac{dN}{dt} = \frac{\ln 2}{\tau} N(0) \quad \Rightarrow \quad N(0) = \frac{10 \times 1620 \times 3.16 \times 10^7}{\ln 2} = 7.34 \times 10^{11}.$$

The mass needed to provide this number of radium nuclei is

$$m = \frac{N(0) \times A}{N_A} = \frac{7.34 \times 10^{11} \times 226}{6.0 \times 10^{26}} = 2.8 \times 10^{-13}\ \text{kg}.$$

(b) We first establish the consequences of 'radioactive equilibrium'. If x_n is the number of nuclei of the nth isotope present in the chain, then

$$\frac{dx_n}{dt} = \lambda_{n-1}x_{n-1} - \lambda_n x_n.$$

But, in equilibrium $dx_n/dt = 0$ for all n except the first and last in the chain; the first establishes the overall scale, and the last merely reflects the number of nuclei that have passed through the chain. The implication of $dx_n/dt = 0$ is that

$$\frac{x_n}{x_{n-1}} = \frac{\lambda_{n-1}}{\lambda_n} = \frac{\tau_n}{\tau_{n-1}},$$

i.e. the number x_n of nuclei of type n that are present is proportional to the half-life τ_n of that type. In particular,

$$\frac{x(\text{Ra})}{x(\text{U})} = \frac{1620}{4.5 \times 10^9}.$$

But, in 1 kg of uranium $x(\text{U}) = N_A/238$, and so

$$x(\text{Ra}) = \frac{1620}{4.5 \times 10^9} \frac{6.0 \times 10^{26}}{238} = 9.1 \times 10^{17} \text{ radium nuclei.}$$

(c) The half-lives of all the unstable isotopes in the chain are totally negligible compared to that of $^{238}_{92}\text{U}$, and so, therefore, are their abundances.[35] To a very good approximation, the isotope situation can be treated as if only uranium and lead were present, with one turning into the other with a half-life of 4.5×10^9 y.

The *original* total abundance of $^{238}_{92}\text{U}$ expressed in terms of current abundances was $A_U + A_{Pb}$, and its current abundance is A_U. Therefore

$$\frac{A_U}{A_U + A_{Pb}} = e^{-\lambda_U T},$$

where T is the estimated age of the Earth. Thus

$$T = \frac{4.5 \times 10^9}{\ln 2} \ln\left(\frac{A_U + A_{Pb}}{A_U}\right) = 6.5 \times 10^9 \ln\left(1 + \frac{A_{Pb}}{A_U}\right) \text{ years.}$$

S 218   The reason why the current decreases at particular voltages, even though the latter has been increased, is that at these voltages the accelerated electrons have enough energy to raise a sodium electron from its normal energy level within the sodium atom to a higher one. If this happens, the incident electron has given up nearly all of its energy, and can no longer overcome the (small) decelerating field between the grid and the anode to reach the latter; it therefore no longer contributes to the measured current.

Relative to the ground state, taken as 0 eV, the electron energy levels in sodium are therefore

Level	0	1	2	3	4
E/eV	0.00	2.10	3.18	3.75	4.34

---

[35] This can be justified by assuming that $^{238}_{92}\text{U}$ was produced as a consequence of the decay of another isotope of very short half-life, and then applying the result derived in part (b).

When the sodium gas is heated, its emission spectrum is that generated by the relaxation of (thermally induced) temporary transitions to higher level states within the atom. The spectrum therefore consists of lines corresponding to (downwards) transitions between these states. If the energy difference between a pair of states is ΔE, then the corresponding wavelength λ of the light emitted is given by $\Delta E = hf = hc/\lambda$, with, in this situation, $\Delta E = e\Delta V$, i.e. a wavelength λ corresponds to a voltage difference between the two states involved of $\Delta V = (hc)/(\lambda e)$.

The following figure and table show corresponding values, and the transitions giving rise to the lines.

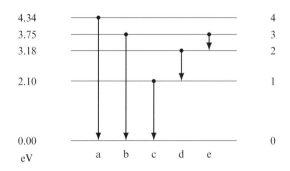

Line	a	b	c	d	e
$\lambda/10^{-7}$ m	2.853	3.303	5.893	11.39	22 07
$\Delta V/$eV	4.33	3.75	2.10	1.09	0.561
Transition	$4 \to 0$	$3 \to 0$	$1 \to 0$	$2 \to 1$	$3 \to 2$

Based on what has been examined so far, we might expect five further lines to appear in the emission spectrum, as follows:

Transition	$2 \to 0$	$3 \to 1$	$4 \to 1$	$4 \to 2$	$4 \to 3$
$\Delta V/$eV	3.18	1.65	2.24	1.16	0.591
$\lambda/10^{-7}$ m	3.892	7.500	5.525	10.67	20.97

However, only the line at 10.67×10^{-7} m occurs in practice, since quantum mechanical selection rules forbid the others.

S 219 From $E^2 = p^2c^2 + m^2c^4$, we immediately see that, since the photon has zero rest mass and energy $E = hf$, its momentum is $p = hf/c$. The figure shows the scattering process, with the initial and final momenta marked.

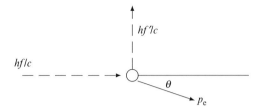

As we would for a non-relativistic situation, we apply energy and momentum conservation, but the energies must include all rest mass energies.

From momentum conservation parallel and perpendicular to the initial photon direction, we have

$$p_e \cos\theta = \frac{hf}{c} \quad \text{and} \quad p_e \sin\theta = \frac{hf'}{c} \quad \Rightarrow \quad p_e^2 c^2 = h^2(f^2 + f'^2).$$

From energy conservation

$$hf + m_e c^2 = hf' + \sqrt{p_e^2 c^2 + m_e^2 c^4}.$$

Accurate algebra is now all that is needed!

$$h(f - f') + m_e c^2 = \sqrt{h^2(f^2 + f'^2) + m_e^2 c^4},$$

$$\left[h(f - f') + m_e c^2\right]^2 = h^2(f^2 + f'^2) + m_e^2 c^4.$$

After the expansion of the LHS and some careful cancellations,

$$-2h^2 ff' + 2m_e c^2 h(f - f') = 0,$$

$$f'(hf + m_e c^2) = m_e c^2 f,$$

yielding the stated result,

$$f' = \frac{m_e c^2}{hf + m_e c^2} f.$$

S 220   (a) Although this chain of reactions is quite long, each step is well defined, the atomic mass and atomic number of each product nucleus follows directly from the previous one and the type of reaction involved.

1. $^{12}_{6}\text{C} + \text{p} \longrightarrow {}^{13}_{7}\text{N} + \gamma,$

2. $^{13}_{7}\text{N} \longrightarrow {}^{13}_{6}\text{C} + \text{e}^+ + \nu_e,$

3. $^{13}_{6}\text{C} + \text{p} \longrightarrow {}^{14}_{7}\text{N} + \gamma,$

4. $^{14}_{7}\text{N} + \text{p} \longrightarrow {}^{15}_{8}\text{O} + \gamma,$

5. $^{15}_{8}\text{O} \longrightarrow {}^{15}_{7}\text{N} + \text{e}^+ + \nu_e,$

6(a). $^{15}_{7}\text{N} + \text{p} \longrightarrow {}^{12}_{6}\text{C} + \alpha$,

6(b). $^{15}_{7}\text{N} + \text{p} \longrightarrow {}^{16}_{8}\text{O} + \gamma$,

7. $^{16}_{8}\text{O} + \text{p} \longrightarrow {}^{17}_{9}\text{F} + \gamma$,

8. $^{17}_{9}\text{F} \longrightarrow {}^{17}_{8}\text{O} + \text{e}^{+} + \nu_{\text{e}}$,

9. $^{17}_{8}\text{O} + \text{p} \longrightarrow {}^{14}_{7}\text{N} + \alpha$.

The identification parade is thus: $A = {}^{13}_{7}\text{N}$; $B = {}^{13}_{6}\text{C}$; $D = {}^{14}_{7}\text{N}$; $E = {}^{15}_{8}\text{O}$; $F = {}^{15}_{7}\text{N}$; $G = {}^{12}_{6}\text{C}$; $H = {}^{16}_{8}\text{O}$; $I = {}^{17}_{9}\text{F}$; $J = {}^{17}_{8}\text{O}$; $K = {}^{14}_{7}\text{N}$. As can be seen, D and K are the same isotope, as are G and the original $^{12}_{6}\text{C}$.

(b) As just noted, there are two 'repeat isotopes'; this offers two closed cycles, namely

(i) $4 \rightarrow 5 \rightarrow 6(\text{b}) \rightarrow 7 \rightarrow 8 \rightarrow 9 \rightarrow 4$,

(ii) $1 \rightarrow 2 \rightarrow 3 \rightarrow 4 \rightarrow 5 \rightarrow 6(\text{a}) \rightarrow 1$.

In each case, apart from the initial nucleus, the total input (on the left of the reaction equation) is 4 protons, and the total output is 1 α-particle, 2 positrons, 2 (electron) neutrinos, and energy in the form of γ-rays. The basic reaction of each cycle is thus

$$4p \longrightarrow \alpha + 2\text{e} + 2\nu_{\text{e}} + \text{energy}.$$

The isotopes of carbon, nitrogen, oxygen and fluorine are 'merely catalysts'!

(c) From the data in Q 216, we will need the Q values for $\text{p}(\text{n}, \gamma)\text{d}$ ($= 2.226$ MeV) and $\text{d}(\text{d}, \gamma)\alpha$ ($= 23.847$ MeV). Writing the equation for proton decay [36] as $(\text{p}, \text{e}^{+}\nu_{\text{e}})\text{n}$ and combining it with the first of these, gives the Q value of $\text{p}(\text{p}, \gamma)\text{de}^{+}\nu_{\text{e}}$ as $2.226 - 1.805 = 0.421$ MeV.

As two such deuterons are needed to form an α-particle, the running total for Q has to be doubled to 0.842 MeV. The final step is to combine the two deuterons to form the α-particle; this releases a further 23.847 MeV of energy in the form of γ-rays. Thus the total energy released when 4 protons are combined to make one alpha particle is $0.842 + 23.847 = 24.689$ MeV.

As 1 g of hydrogen contains N_{A} protons, the total energy released when it is completely converted to α-particles ($^{4}_{2}\text{He}$) is

$$\frac{6.0 \times 10^{23}}{4} \times 24.7 \times 1.60 \times 10^{-13} = 5.9 \times 10^{11} \text{ J}.$$

---

[36] More normally written as neutron decay, $\text{n} \longrightarrow \text{p} + \text{e}^{-} + \bar{\nu}_{\text{e}}$ with $Q = +1.805$ MeV, and actually occurring for free neutrons, albeit slowly – about 10 min.

S 221   The production and decay equations must preserve baryon and lepton numbers (B and L respectively), as well as charge Q. Since all quarks (as opposed to antiquarks) have baryon number $B = +\frac{1}{3}$, the baryons in this problem will be composed of three quarks; the mesons, having baryon number zero, will be quark-antiquark combinations.

From charge conservation, the production process for the Λ,

$$\pi^- + p \quad \rightarrow \quad \Lambda + K^0, \tag{$*$}$$

shows that it must have zero charge – and so it's quark representation will also have to be uncharged. Likewise, baryon conservation shows that the Λ is a baryon and will have to be a composite of three quarks.

We are told that K^0 is a meson with strangeness $+1$, and so it must contain an s-antiquark, i.e. $\bar{s}$. To have the correct values for B and charge Q, the accompanying quark must be the d-quark. Thus $K^0 = \bar{s}d$.

The π^- meson is non-strange, and so cannot contain s or $\bar{s}$; to give it a Q of -1 it must have structure $\bar{u}d$. With the familiar uud quark structure of the proton, the quark formulation of $(*)$ has to be

$$\bar{u}d + uud \quad \rightarrow \quad qqq + \bar{s}d.$$

From this, we see that qqq for the Λ must be uds.

Moving on to the decay processes, which do not preserve strangeness, but leave B, L and Q unaltered, we see that the W-boson from the s-quark decay must have negative charge, and therefore be a W^-. We are given the decay of a W^+, and so the decay we need, that of a W^-, can be obtained by changing every particle into its own antiparticle, i.e.

$$W^- \quad \rightarrow \quad e^- + \bar{\nu}_e.$$

Thus in total, a possible decay mode of the Λ is

$$\Lambda \quad \rightarrow \quad X + e^- + \bar{\nu}_e,$$

where X has the quark constitution udu, i.e is a proton. This mode,

$$\Lambda \quad \rightarrow \quad p + e^- + \bar{\nu}_e,$$

is observed in practice, but only at the level of less than one in a thousand of all decays.

Physical Constants

Speed of light in a vacuum	c	$3.00 \times 10^8 \text{ m s}^{-1}$
Permittivity of a vacuum	ϵ_0	$8.85 \times 10^{-12} \text{ F m}^{-1}$
Permeability of a vacuum	μ_0	$4\pi \times 10^{-7} \text{ H m}^{-1}$
Planck constant	h	$6.63 \times 10^{-34} \text{ J s}$
Elementary charge	e	$1.60 \times 10^{-19} \text{ C}$
Electron rest mass	m_e	$9.11 \times 10^{-31} \text{ kg}$
Proton rest mass	m_p	$1.673 \times 10^{-27} \text{ kg}$
Neutron rest mass	m_n	$1.675 \times 10^{-27} \text{ kg}$
Atomic mass unit	$u \text{ (AU)}$	$1.661 \times 10^{-27} \text{ kg}$
Avogadro constant	N_A	$6.02 \times 10^{23} \text{ mol}^{-1}$
Molar gas constant	R	$8.31 \text{ J mol}^{-1} \text{ K}^{-1}$
Boltzmann constant	k	$1.38 \times 10^{-23} \text{ J K}^{-1}$
Stefan constant	σ	$5.67 \times 10^{-8} \text{ W m}^{-2} \text{ K}^{-4}$
Gravitational constant	G	$6.67 \times 10^{-11} \text{ N m}^2 \text{ kg}^{-2}$
Acceleration of free fall	g	9.81 m s^{-2}
Hubble constant	H_0	$65 \text{ km s}^{-1} \text{ Mpc}^{-1}$